Challenges and Advances in Computational Chemistry and Physics

Volume 20

Series Editor
Jerzy Leszczynski
Department of Chemistry and Biochemistry
Jackson State University Chemistry, Jackson, Mississippi, USA

This book series provides reviews on the most recent developments in computational chemistry and physics. It covers both the method developments and their applications. Each volume consists of chapters devoted to the one research area. The series highlights the most notable advances in applications of the computational methods. The volumes include nanotechnology, material sciences, molecular biology, structures and bonding in molecular complexes, and atmospheric chemistry. The authors are recruited from among the most prominent researchers in their research areas. As computational chemistry and physics is one of the most rapidly advancing scientific areas such timely overviews are desired by chemists, physicists, molecular biologists and material scientists. The books are intended for graduate students and researchers.

More information about this series at http://www.springer.com/series/6918

Drahomír Hnyk • Michael L. McKee
Editors

Boron

The Fifth Element

Springer

Editors
Drahomír Hnyk
Institute of Inorganic Chemistry
of the Academy of Sciences
of the Czech Republic, v.v.i.
Husinec-Řež, Czech Republic

Michael L. McKee
Department of Chemistry and Biochemistry
Auburn University
Auburn, AL, USA

Challenges and Advances in Computational Chemistry and Physics
ISBN 978-3-319-22281-3   ISBN 978-3-319-22282-0 (eBook)
DOI 10.1007/978-3-319-22282-0

Library of Congress Control Number: 2015952454

Springer Cham Heidelberg New York Dordrecht London

Printed on acid-free paper

Springer International Publishing AG Switzerland is part of Springer Science+Business Media (www.springer.com)

Foreword

One way to bring order into the vast body of knowledge chemists keep accumulating since centuries is to group it neatly by element. Boron, "the fifth element", is one where this approach makes much sense, because its chemistry is rather unique and set apart from that of its neighbours in the periodic table. Boron chemistry is not self-contained; however, there is much potential for cross-fertilisation with other areas, and occasional "spin-offs" can have tremendous impact, as for instance with hydroboration or cross-coupling reactions in synthetic organic chemistry. It is thus useful to have the progress in the field reviewed regularly. The present monograph edited by Drahomír Hnyk and Michael McKee serves precisely this purpose, providing a snapshot of current research in the vibrant area that boron chemistry continues to be.

This chemistry is governed by the electron deficiency of boron. Diborane and its family members, the polyhedral boranes, are the epitomes of multicenter bonding. This type of bonding in turn gives rise to characteristic structural features, exemplified in the preference for clusters with shared polyhedra. In view of the rich and diverse structural chemistry that ensues, it is not surprising that structure and bonding are recurring themes throughout this book.

Another recurring theme is the concert of theory and experiment, teaming up to elucidate the details of structure, bonding and reactivity. Chemistry of boron and the boranes is an ideal playground for quantum-chemical methods. In the absence of heavy elements, a non-relativistic treatment is usually appropriate, so that "off-the-shelf", black-box methods and user-friendly software can be applied rather routinely. It also allows description and interpretation of the results in the language of molecular orbital theory. Many of the basic building blocks in boron chemistry are small enough to be treated with the most sophisticated ab initio methods, which is to say virtually exactly. This in turn allows more approximate methods, such as those mushrooming from the fertile field of density functional theory (DFT), to be reliably calibrated and to be applied to more complex systems such as large metallaboranes. If chosen properly, computational tools can provide answers at a confidence level that rivals those of established experimental techniques. The usefulness and importance of theoretical modelling tends to grow with the ever-increasing availability of computer power. In fact the largest part of this book is devoted to quantum-chemical applications and the new insights they have provided.

I have been fortunate to start my career in this field, computational boron chemistry, under the guidance of Paul Schleyer. An organic chemist by training and reputation, he did not care about the presence or absence of carbon in a compound as long as its chemistry was interesting. After very fruitful application of the emerging tools to calculate NMR parameters to carbocations in the 1980s, it was only logical for him to have the same methods applied to boron compounds. This has developed into one of the many areas in chemistry where Paul Schleyer has left a lasting mark. He had moved on since then, restlessly working on other topics, but has always kept an interest in boron chemistry. He had agreed to write the introduction to this monograph, but his sudden death in November 2014 prevented him from doing so. I am grateful to the editors for their decision to dedicate this whole book in his memory.

The present monograph is a legacy in many ways. It brings together chapters by some of the towering pioneers in the field, on whose shoulders the coming generations of boron chemists can stand, complemented by contributions from younger scientists eager to carry on the torch. As expected for a vibrant research area, the topics covered are numerous and diverse.

In Chap. 1, Alexander Boldyrev takes us into the wonderful world of boron-based chains, rings, sheets and spheres, where the continuum between localised and delocalised bonding leads to unusual and intriguing phenomena such as fluxionality reminiscent of a "molecular Wankel motor". The mature area of structure elucidation by joint gas-phase electron diffraction and quantum-chemical modelling is reviewed by Drahomír Hnyk in Chap. 2. The vast terrain of metallaborane chemistry is charted by Bruce R. King in Chap. 3 with the help of DFT. Josep Oliva goes beyond ground-state calculations in Chap. 4, exploring absorption and fluorescence properties of octadecaborane and their subtle dependence on configuration ("Dr. Jekyll and Mr. Hyde"-versions of $B_{18}H_{22}$) and on exoskeletal substituents. In Chap. 5, Michael McKee recounts his attempts to elucidate the mechanism of a classical reaction, formation of the *closo*-dodecaborane dianion, through mapping the wonderfully complex potential energy hypersurface with DFT calculations. In Chap. 6, John Kennedy embarks on a journey from the classic *nido* and *arachno* boranes via fused cluster compounds to ever more complex macropolyhedral boron species, all the way to "megaloboranes", that is, big nano-sized globules that are presented as challenging, but potentially rewarding targets for future synthesis. In Chap. 7, Pattath Pancharatna develops an understanding of the bonding in such macropolyhedral boranes based on their electronic structures, as summarised in a set of refined electron-counting rules. Chapter 8 by Narayan Hosmane illustrates how the usefulness of the "classic" hydroboration and Suzuki cross-coupling reactions can be further improved by advances in nanocatalysis. In Chap. 9, Martin Lepšík shares his insights on how seemingly weak intermolecular interactions can open up new avenues in boron chemistry, notably in relation to materials science and biomolecular or medicinal chemistry.

Through this collection of representative snapshots, the monograph conveys a good idea of the recent progress that has been made in the field of boron chemistry.

The book should be appealing and interesting for experimental and computational chemists alike. Providing highlights from the present state-of-the-art in boron chemistry, and an overview of the frontiers that are waiting to be pushed ever further, I am sure it will be a valuable source of information, but also of inspiration for further work in the years to come.

St Andrews, UK
May 2015

Michael Bühl

Preface

Professor Paul von Ragué Schleyer, who passed away on November 21, 2014, was a giant among modern scientists. He may be considered as a pioneer of computational chemistry as a whole. His imprint will be felt for generations, undoubtedly also in boron chemistry. Indeed, he won the 1996 IMEBoron Prize for Computational Boron Chemistry. Through the years, his group has been at the forefront in developing tools and applying them to the study of unusual molecules. From the first synthesis of adamantane in 1957, Paul has been on the hunt for usual molecules. His most recent quest has been for planar tetra-coordinate carbon and then later boron in a planar environment. One might argue that his extensive work on the "The Nonclassical Ion Problem" (i.e. the norbornadienyl cation) dovetailed smoothly into his studies of boranes and carbocations since they are isoelectronic. Paul obligingly agreed to write the introduction to this book. Unfortunately he passed away before he could complete the task. We think he would be very much pleased by the diversity and quality of the chapters herein. A fair number of the contributors have collaborated either directly or indirectly with his group. Therefore, we are proud to dedicate this book to his memory.

Husinec-Řež, Czech Republic
Auburn, AL, USA
May 2015

Drahomír Hnyk
Michael L. McKee

Contents

Chapter 1
Classical and Multicenter Bonding in Boron: Two Faces of Boron

Ivan A. Popov and Alexander I. Boldyrev

Abstract In this chapter we have shown that boron has two faces in chemistry: with classical and multicenter bonding. When neutral boron atoms are involved in bonding, we usually encounter domination of multicenter bonding. Such examples are planar, quasi-planar, and three dimensional pure and doped boron clusters, two-dimensional sheets as well as conventional deltahedral boranes. However, when a boron atom acquires an extra electron, it tends to form molecules similar to those of the neighboring carbon featuring classical 2c-2e σ-bonds instead of multicenter ones. Such examples are BH_4^-, analog of the CH_4 molecule; $Li_nB_nH_{2n+2}$ molecules containing $B_nH_{2n+2}^{n-}$ kernels, which are isostructural to corresponding molecules in the C_nH_{2n+2} series; $Li_6B_6H_6$, analog of benzene; linear chain of boron anions in LiB_y, analog of carbine; and 2D layer of boron in MgB_2 mimicking the graphene structure. Chemistry of boron continues to expand conquering new territories and providing us with unprecedented structures, chemical bonding, internal rotations and other unusual properties. We believe we are at the beginning of new era of boron chemistry.

1.1 Introduction

Boron and carbon are neighbors in the Periodic Table but are very different elements. Carbon is known to form strong classical two-center two-electron (2c-2e) C-C σ-bonds while π-bonding can be delocalized in aromatic organic compounds. Boron, on the other hand, is known to avoid the formation of 2c-2e B-B σ-bonds and prefers to form multicenter σ-bonds and π-bonds. It is well illustrated on the examples of two-dimensional materials: graphene [1, 2] and all-boron α-sheet [3–5]. Graphene forms a rigid network of 2c-2e C-C σ-bonds responsible for its honeycomb structure. The α-sheet of boron has a strange derivative of the honeycomb structure with some of the hexagons being empty and some being filled with an

---

I.A. Popov • A.I. Boldyrev (✉)
Department of Chemistry and Biochemistry, Utah State University, Logan, UT, 84332, USA
e-mail: vanekpopov@gmail.com; a.i.boldyrev@usu.edu

D. Hnyk, M.L. McKee (eds.), *Boron*, Challenges and Advances in Computational Chemistry and Physics 20,
DOI 10.1007/978-3-319-22282-0_1

extra boron atom. Chemical bonding analysis revealed that there are no 2c-2e B-B σ-bonds in the boron α-sheet and that the σ-framework of this material is formed by either 3c-2e or 4c-2e σ-bonds [6, 7]. The π-bonding in both materials is similar and is due to delocalized 6c-2e or 7c-2e π-bonds. Having said that, we acknowledge that boron occasionally forms σ-bonds (four 2c-2e B-H in the BH_4^- anion, for example), but that is exactly an example of "electronic transmutation" [8], where boron acquiring an extra electron, becomes "carbon". Indeed, BH_4^- is a "copycat" of CH_4 since both species have similar chemical bonding and geometric structures.

In this chapter we would like to address the importance of both multicenter and classical (2c-2e) bonding, as well as the formation of classical 2c-2e σ-bonding in boron compounds when boron atom accepts an extra electron and electronically transmutes into "carbon".

1.2 Multicenter Bonding in Boron

1.2.1 Bonding in Pure Boron Clusters

While 2c-2e classical bonds dominate organic chemistry and are also responsible for majority of bonding in inorganic chemistry, it is boron, which is responsible for the introduction of the first multicenter 3c-2e bonds on the example of B_2H_6. The structure of B_2H_6 with bridging H-atoms was proposed in 1921 by Dilthey [9]. However, it was not considered seriously until the 1940s, when infrared spectroscopy data [10–12] supported the structure. Later, electron diffraction [13] and low-temperature X-ray diffraction [14] also confirmed the bridged structure for the diborane. The chemical bonding in boranes was first considered by Pitzer, who proposed the concept of a "protonated double bond" [15]. Further, Lipscomb and collaborators [16] put forward the concept of three-center two-electron (3c-2e) bonding, which, in the case of the B_2H_6 diborane, consisted of two 3c-2e B-H-B bonds, involving the bridging H atoms. Lipscomb also explained the structure of all known boron hydrides, in which the bridging B-H-B bond appeared to be the key structural unit [14]. In the 3c-2e bonding three atoms supply three orbitals, one from each atom. These atomic orbitals interact to form one bonding and two antibonding orbitals. Thus, the two available electrons may fill the bonding orbital to form a 3c-2e bond. In the n-atomic species, there are n atomic orbitals, and only $n/3$ bonding molecular orbitals, which can be occupied by $2n/3$ electrons. Thus, the reason for certain boranes to exhibit special stability was elucidated. In principle, Lipscomb's concept of the 3c-2e bonds, along with aromaticity, is one of the ways of describing electron deficient bonding, even though aromaticity is more common in chemistry and, in a way, more clear. The work of Lipscomb on the chemical bonding of the boranes eventually led to his winning of the Nobel Prize and opened the gateway to understanding the chemistry of boron.

Boron in three-dimensional (3D) materials flourishes with a number of polymorphs [17, 18] consisting of B_{12}-icosahedral building blocks, though only four pure elemental phases have been synthesized [19–21]. However, while 3D structural motifs are prevalent in bulk boron, atomic boron clusters are found to have planar or quasi-planar structures [22], stabilized by localized 2c-2e σ bonds on the periphery and delocalized multi-center-two-electron (nc-2e) bonds in both σ and π frameworks on the internal fragments. Thus, when chemical bonding in negatively charged boron clusters [23, 24] was studied, the authors faced the necessity to go beyond the 3c-2e bonds. Let us consider chemical bonding in boron clusters using B_9^- as an example. The anionic B_9^- has the perfect planar D_{8h} ($^1A_{1g}$, $1a_{1g}^2 1e_{1u}^4 1e_{2g}^4 1e_{3u}^4 2a_{1g}^2 1b_{2g}^2 1a_{2u}^2 2e_{1u}^4 1e_{1g}^4$) wheel-shaped structure as the global minimum (Fig. 1.1), which was established in a joint photoelectron and ab initio study by Zhai et al. [25].

The perfect octagon structure of B_9 is unprecedented in chemistry and represents the first example of octacoordinated atom in a planar environment. The remarkable

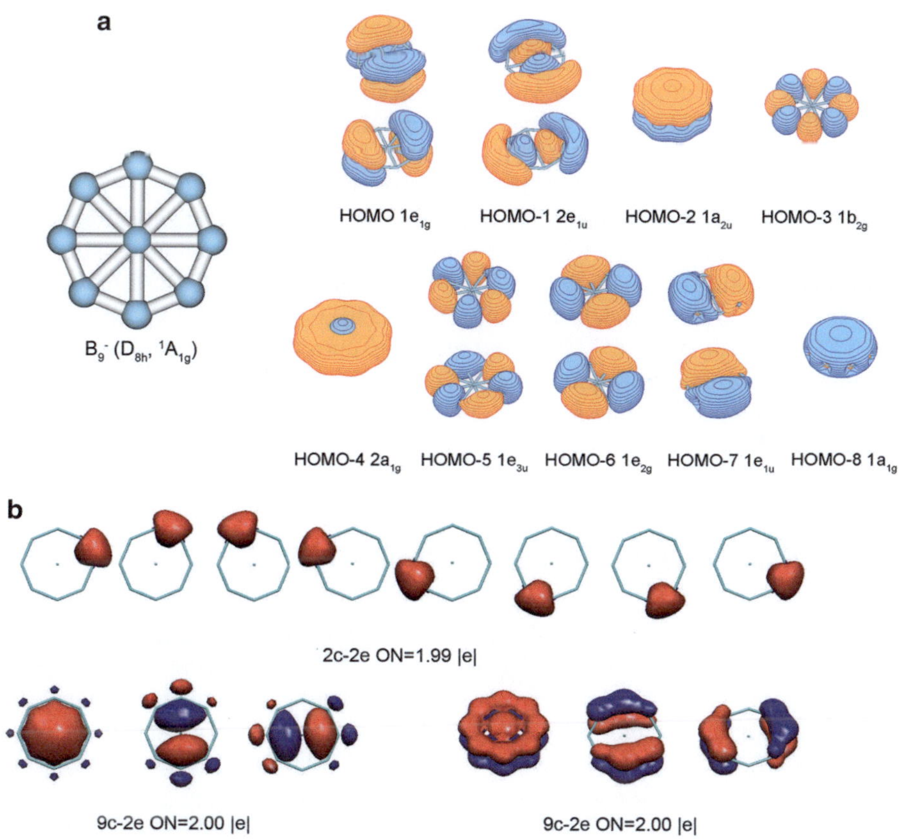

Fig. 1.1 (**a**) Global minimum structure and CMOs of B_9^- D_{8h} ($^1A_{1g}$) cluster; (**b**) results of the AdNDP localization. *Sticks* drawn between atoms represent interatomic distances <2.0 Å; they do not necessarily represent single B–B σ-bonds here and elsewhere. ON stands for occupation number (Reproduced from [26] with permission from the PCCP Owner Societies)

planar octagon structure of B_9^- can be easily rationalized on the basis of double (σ- and π-) aromaticity (Fig. 1.1). The eight MOs (Fig. 1.1a): HOMO-3 ($1b_{2g}$), HOMO-5, HOMO-5' ($1e_{3u}$), HOMO-6, HOMO-6' ($1e_{2g}$), HOMO-7, HOMO-7' ($1e_{1u}$), and HOMO-8 ($1a_{1g}$) can be localized into eight 2c-2e B-B peripheral bonds (Fig. 1.1b) using Adaptive Natural Density Partitioning (AdNDP) method [26]. In general, the AdNDP method analyzes the first-order reduced density matrix in order to obtain its local block eigenfunctions with optimal convergence properties for an electron density description. The obtained local blocks correspond to sets of *n*-atoms (*n* ranging from one to the total number of atoms in the molecule) that are tested for the presence of *n*-electron objects [*n*-center two-electron (*n*c-2e) bonds]. The AdNDP method initially searches for core electron pairs and lone pairs (1c-2e), then 2c-2e, 3c-2e, …, and finally up to *n*c-2e bonds. At every step, the density matrix is depleted of the density corresponding to the appropriate bonding elements. The user-directed form of the AdNDP analysis can be applied to specified molecular fragments and is analogous to the directed search option of the standard natural bond orbital (NBO) code [27, 28]. AdNDP accepts only those bonding elements whose occupation numbers (ONs) exceed a specified threshold value, which is usually chosen to be close to 2.0 |e|. The other valence MOs are delocalized over the octagon and they are responsible for global bonding between the central boron atom and peripheral boron atoms. The three π-MOs: HOMO, HOMO' ($1e_{1g}$) and HOMO-2 ($1a_{2u}$) are responsible for π-aromaticity and the three σ-MOs: HOMO-1, HOMO-1' ($2e_{1u}$) and HOMO-4 ($2a_{1g}$) are responsible for σ-aromaticity in B_9^-. The chemical bonding picture with double aromaticity can explain why B_9^- has a high symmetry structure with bond equalization on the periphery of the cluster, the high HOMO-LUMO gap, high first vertical electron detachment energy (VDE) for B_9^- (3.46 eV, compared to the VDE of B^- of 0.227 eV [29]), and high ring current, comparable to aromatic organic hydrocarbons [30]. This chemical bonding model was successfully applied to explain chemical bonding in many other planar and quasi-planar negative boron clusters [22–24].

1.2.2 Bonding in Doped Boron Clusters

Highly symmetric doubly aromatic boron wheels, B_8^{2-} and B_9^- [25], have inspired the discovery of a series of metal-centered monocyclic boron rings: M©B_n^- [31–34]. The electronic design principle capable of predicting which metals can replace the central B atom in either B_8^{2-} or B_9^- to render a similar doubly aromatic M©B_n^- species (n=7, 8) was proposed by Romanescu et al [31]. Based on the design principle, general geometric and electronic factors in the rational design of the novel borometallic molecular wheels were investigated [31–34]. Wang and collaborators observed and characterized the following octa- and nona-coordinated clusters: Co©B_8^- and Ru©B_9^- [31], Ru©B_9^- and Ir©B_9^- [32], Fe©B_8^- and Fe©B_9^- [33]. Tantalum and niobium were shown to possess the record-breaking coordination number in the planar metal-centered deca-coordinated Ta©B_{10}^- (Fig. 1.2) and Nb©B_{10}^- anions [34].

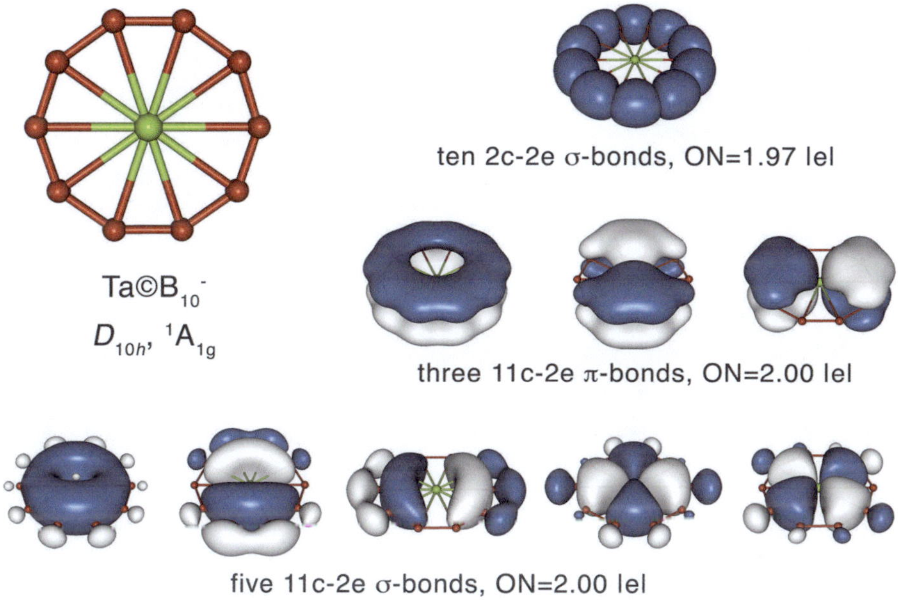

Fig. 1.2 Chemical bonding pattern of Ta©B_{10}^- revealed by the AdNDP analysis (Reproduced from [34]. Copyright 2012 Wiley)

These unprecedented results have proven that boron clusters are promising molecules for coordination chemistry as potential new ligands, as well as for material science as new building blocks. The AdNDP analysis for Ta©B_{10}^- revealed ten 2c-2e peripheral σ-bonds, five delocalized σ-bonds (satisfying the $4N+2$ rule for aromaticity with $N=2$), and three delocalized π-bonds (satisfying the $4N+2$ rule for aromaticity with $N=1$). A similar bonding pattern was found for Nb©B_{10}^-. Thus, both clusters are doubly σ- and π-aromatic. Detailed discussion of structure and chemical bonding in metal-centered monocyclic boron rings M©B_n^- was recently reviewed [35]. Theoretical study on the transition-metal stabilized exo/endo closo-borane complex $Sc[B_{24}H_{24}]^+$ is reported elsewhere [36].

Pure and metal-doped planar and quasi-planar boron clusters have a great potential to be new building blocks of solids and multi-decker sandwich complexes. In fact, two new solid-state materials: $Ti_7Rh_4Ir_2B_8$ [37] and $Nb_6Fe_{1-x}Ir_{6+z}B_8$ [38] containing planar hexagonal boron rings as building blocks have been recently synthesized by Fokwa and co-workers [37, 38]. These works show a great potential of boron chemistry extension in these two directions.

1.2.3 Boron Molecular Wankel Motors

Delocalized bonding inside boron clusters leads to the unprecedented internal rotation in planar or quasi-planar boron clusters [39–41]. This phenomenon was first discovered for the doubly concentric spider-web-like structure of B_{19}^- [42] and got

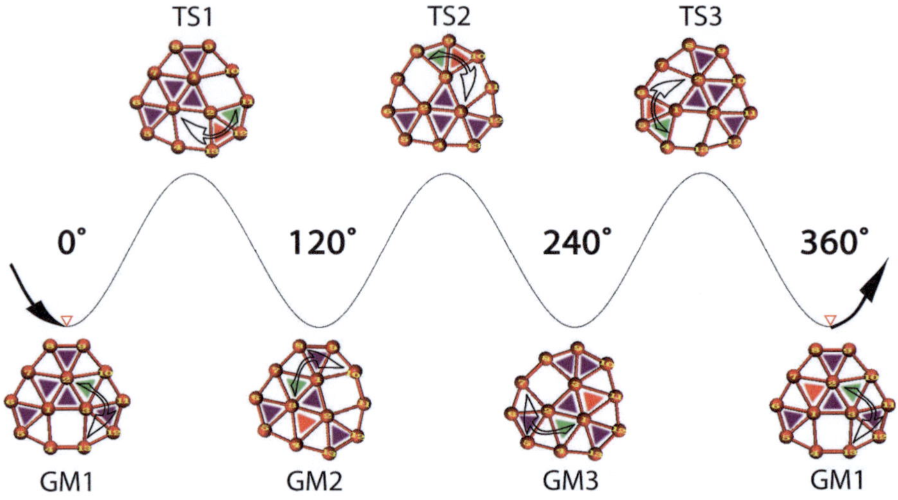

Fig. 1.3 The schematic representation of the 3c-2e σ-bonds migration during the internal rotation of B_{13}^{+} [40] (Reproduced from [40] with permission from The Royal Society of Chemistry)

the name of molecular Wankel motor [39]. The stability of B_{19}^{-} was attributed to doubly concentric π aromaticity in two concentric π systems, analogous to coronene [41]. Merino, Heine and co-workers were the first to demonstrate that B_{19}^{-} can undergo in-plane internal rotation of the inner centered pentagonal unit with respect to the peripheral boron ring [39]. B_{13}^{+} was suggested to be highly fluxional in 1998 [43] though the possibility of the internal rotation in this doubly concentric pure boron cluster was demonstrated and explained using chemical bonding analysis only recently [40]. Briefly, it was shown that the main change in chemical bonding upon rotation occurs in the delocalized σ-framework where the delocalized 3c–2e σ-bonds are symbolically presented as solid triangles (Fig. 1.3).

The electron density migrates from one 3c–2e σ-bond to other 3c–2e σ-bond (see the direction of the arrows in Fig. 1.3), while the other pairs of delocalized σ-electrons occupying 3c–2e σ-bonds stay in their places. So, the σ-electron density migration does not violate the 4n+2 rule for both concentric σ-systems. The σ-electrons number is constant over the inner triangle (two electrons) and in between the triangle and the peripheral ring (ten electrons) upon the internal rotation. The geometry of the inner triangle is rather rigid upon internal rotation. This can be explained by σ-aromaticity in this unit. The absence of localized 2c–2e σ-bonds between the inner B_3 and peripheral B_{10} moieties is the main reason why almost free internal rotation is possible. The in-plane rotation was shown to be attainable even at room temperatures due to the following factors: similarity of chemical bonds between equilibrium and transition states of the molecular motors, and prevalence of delocalized bonding inside of boron clusters [39–41]. It was shown by Alexandrova and coworkers [41] that molecular Wankel motors rotate in both directions and only the application of the circularly polarized infrared laser was shown to achieve a desirable uni-directional rotation rendering a photo-driven molecular

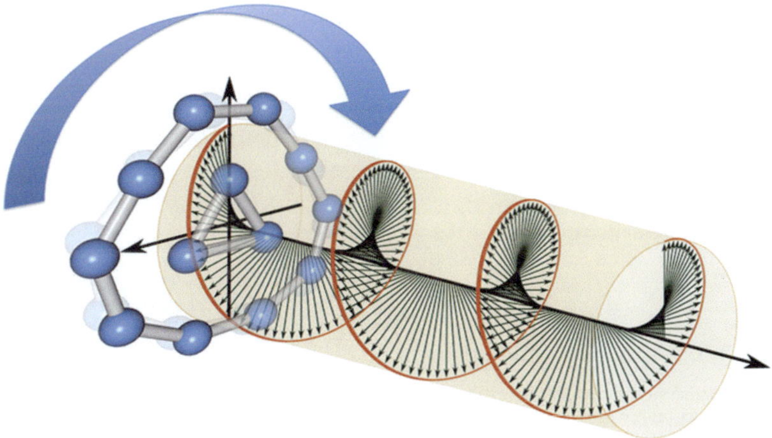

Fig. 1.4 Uni-directional rotation of a photo-driven molecular Wankel $B_{13}+$ (Reproduced from [41]. Copyright 2012 Wiley)

Wankel motor running on the electronic ground state potential energy surface with a rotational period of a few pico-seconds (Fig. 1.4).

Very recently, Merino, Heine and co-workers extended the family of the Wankel motors by a quasi-planar bowl cluster of B_{18}^{2-} [44]. Clearly, the unprecedented internal rotations are much more common in boron chemistry than we know up today. We also think that other yet unknown intra-molecular rearrangements are possible in boron clusters due to multicenter bonding in such species.

1.2.4 All-Boron Fullerenes

After the discovery of buckminsterfullerene (C_{60}) [45] researchers began their hunt for all-boron fullerene-like structures for boron clusters. Yakobson and co-workers [46] proposed that B_{80}, which is isoelectronic to C_{60} could be a candidate for the all-boron fullerene. This work sparked a new theoretical search for boron fullerenes [47–58]. The real challenge for theoreticians was to find which boron cluster has a sphere-like structure, and that is due to the need to do exhaustive machine searches for an enormous number of potential structures. Therefore, it is very difficult to predict with certainty that the computationally predicted fullerene structure can be observed in molecular beam experiments, because it must be either a global minimum structure or a low-lying isomer. So, the best way to detect all-boron fullerene cluster is through a joint computational and photoelectron spectroscopy. Indeed, recently Wang and co-workers in joint experimental and theoretical work [59] reported a ball-like structure, borospherene (Fig. 1.5) that is present in the molecular beam of B_{40}^{-} clusters.

Though, according to their results quasi-planar structure is a global minimum for B_{40}^{-}, it is the ball-like structure B_{40}^{-}, which is responsible for the low-energy part of the photoelectron spectra. Moreover, according to their theoretical calculations the

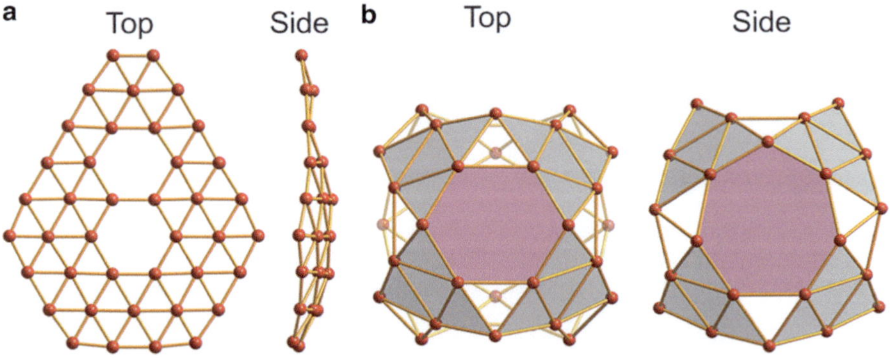

Fig. 1.5 Top and side views of the global minimum (**a**) and low-lying isomers (**b**) of B_{40}^- and B_{40} at the PBE0/6-311+G* level of theory (Reproduced from [59]. Copyright 2014, Nature Publishing Group)

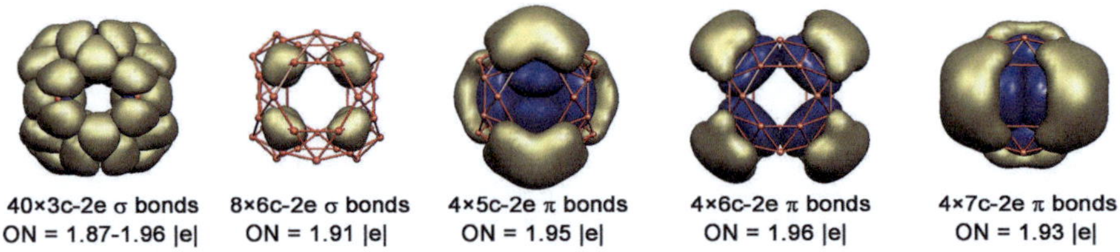

Fig. 1.6 Results of the chemical bonding analyses for the B_{40} borospherene. The analyses were done using the AdNDP method (Reproduced from [59]. Copyright 2014, Nature Publishing Group)

borospherene structure is a global minimum for the neutral B_{40} cluster. The chemical bonding analysis for the borospherene has shown a multicenter bonding character without any 1c-2e or 2c-2e bonds (Fig. 1.6).

Important difference between chemical bonding in C_{60} and B_{40} is that there are no 2c-2e neither σ- nor π-bonds in B_{40}. This first example of the all-boron fullerene is just a beginning of large all-boron fullerene chemistry, which will be very different from chemistry of carbon fullerenes. Indeed, very recently Wang and his group reported preparation of axially chiral borospherene B_{39}^- in the molecular beam (Fig. 1.7) [60], which is optically active.

1.2.5 Two-Dimensional Boron Sheet

One can construct a honeycomb crystal lattice of neutral boron sheet assuming that every boron is sp^2-hybridized and forms three 2c-2e σ-bonds. Such structure was shown to be less stable than the truly remarkable α-sheet structure (Fig. 1.8a), computationally predicted by Tang and Ismail-Beigi [3, 4] and Yang, Ding and Ni [5]. This structure is formed of two types of hexagons: empty hexagons and ones with an additional boron atom at the center.

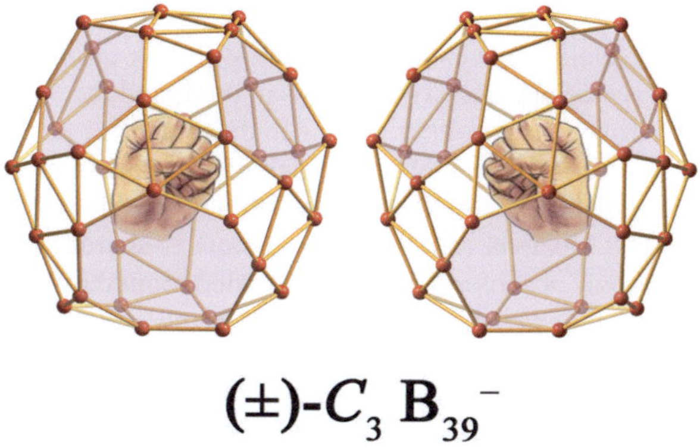

Fig. 1.7 Axially chiral structure of B_{39}^- borospherene (Reproduced from [60]. Copyright 2015, American Chemical Society)

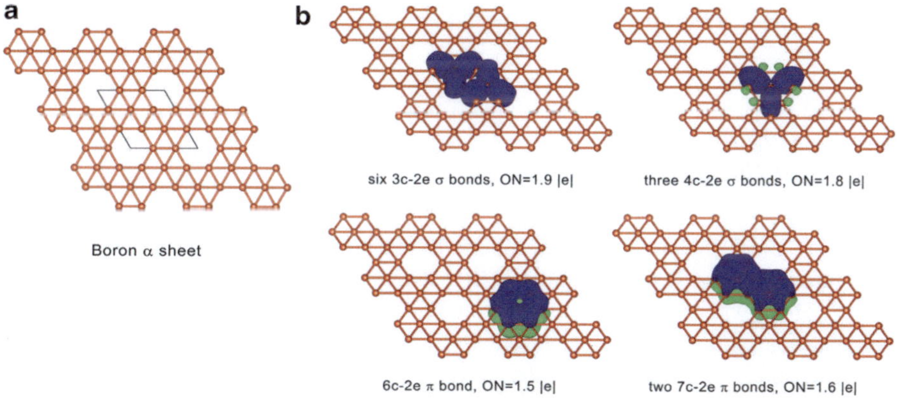

Fig. 1.8 (**a**) Structure and (**b**) SSAdNDP chemical bonding pattern of boron α-sheet. The unit cell is shown in *black* (Reproduced from [7] with permission from the PCCP Owners)

The spotting 2D-lattice with hexagon holes and filled hexagon motifs in the α-sheet was rationalized using Solid State Adaptive Natural Density Partitioning method [6, 7]. The resulting chemical bonding pattern is presented in Fig. 1.8b. There are 8 boron atoms and 24 valence electrons per unit cell, thus one can anticipate 12 two-electron bonds. Six 3c-2e σ-type bonds with occupation number (ON) of 1.9 |e| were found on every boron triangle bordering a vacant hexagon. Three 4c-2e σ bonds were revealed in the rhombi connecting two centered hexagons. Thus nine electron pairs were found via general search over three and four centers, leaving three more to be accounted for. The next smallest tuple, which maintains the symmetry of the system, is a six-center fragment over the hexagonal hole. Using a directed search a π-bond with ON = 1.5 |e| was found over this hexagonal vacancy. Similarly, two 7c-2e π-bonds were found via directed search over each centered hexagon in the unit cell with ON = 1.6 |e|. With this chemical bonding for each B_7 fragment we have six valence electrons coming from three 3c-2e σ-bonds, three electrons coming from three 4c-2e σ-bonds and two electrons coming from the 7c-2e π-bond with the total number of eleven electrons. On the other hand, if we consider a filled hexagon as a part of the lattice we can calculate the total number of valence electrons as follows: each of the six peripheral boron atoms brings half of its valence electrons (9 electrons in total) and the central atom brings all its valence electrons (3 electrons) resulting in the total of 12 electrons per filled hexagon. Thus, there is one extra electron on each filled hexagon motif not involved in the bonding presented above. As one can see from the whole lattice picture the extra electron on a filled hexagon (an electronic donor) is shared by three hexagonal holes (three electronic acceptors) evenly distributed around it, while each hole is surrounded by six filled hexagons, resulting in two 'extra' electrons per hole. Those two electrons form the 6c-2e π-bond. It is interesting to notice that, unlike graphene, which contains 2c-2e C-C σ-bonds, the all-boron graphene α-sheet possesses no localized 2c-2e B-B σ-interactions. Despite the theoretical prediction of the 2D boron sheet was made in 2008, there is no experimental confirmation of it. Thus, a new world of two-dimensional boron still awaits us ahead.

1.2.6 Competition Between 2D and 3D Structures in the B_nH_{n+2} Species

One may think that if boron was in sp^2 hybridization it could form boron chain structures with the B_nH_{n+2} formula, which would be similar to saturated hydrocarbons C_nH_{2n+2}, where carbon chain is formed by the sp^3 hybridized carbon. Theoretical calculations on the B_nH_{n+2} species (n=2–5) have shown that chain structures starting from n=3 with classical 2c-2e bonding are significantly less stable than nonclassical structures with multicenter bonding [61]. Moreover, the 3D structures are favored starting from n=4. The major reason why classical structures are significantly less stable is the need to fill up all three p-AO orbitals on boron atom and avoid sp^2 hybridization. This result is similar to what is discussed above in 2D boron sheets where honeycomb structure formed by sp^2 hybridized boron is appreciably less stable than the α-sheet, where 30 % of sigma-electrons were transferred into the π-system. Thus, the multicenter bonding in boron systems discussed above is due to the electron deficiency of boron with three electrons and four valence atomic orbitals.

1.3 Electronic Transmutation of Boron into "Carbon" Upon Accepting an Extra Electron

When every boron atom in boron compound accepts one extra electron, it starts to behave like neighboring carbon atom and this phenomenon is called the electronic transmutation [8]. Indeed, theoretical calculations showed that the salt-like $Li_nB_nH_{2n+2}$ molecules contain the $B_nH_{2n+2}^{n-}$ kernel, which is isostructural to corresponding molecules in the C_nH_{2n+2} series [8]. Various salt-like polyhedral borane and carborane complexes stabilized by Li atoms are reported elsewhere [62–64]. As shown in Fig. 1.9, the $Li_2B_2H_6$ molecule, which was found to be the global minimum structure, has the ethane-like kernel (Fig. 1.9).

Since this is an electronic transmutation, the nonclassical structure IV in which one electron migrated from B atoms to H forming Li_2H^+ fragment is only 17 kcal/mol higher than the global minimum. Furthermore, it was found that $Li_3B_3H_8$ also has a global minimum structure with the propane-like kernel $B_3H_8^{3-}$. However, it is important to note that the electronic transmutation concept should be used with caution. Using the geometry of polyacetylene chains, Popov and Boldyrev showed that the structurally similar all-boron polyene chains with the general formula $Li_{2n}B_{2n}H_{2n+2}$ (n=2–7) [65] are not global minima in the trans–cisoid-$Li_{2n}B_{2n}H_{2n+2}$, even though the effective NBO charges on Li atoms are in the range of 0.7–0.8 |e| (Li gives off about one electron to B). Another example of electronic transmutation could be a series of aromatic polycyclic species, which are considered to be analogues of aromatic polycyclic hydrocarbons [66]. Very recently, it was shown that aluminum atoms are also capable to form alkane-like species based on the concept

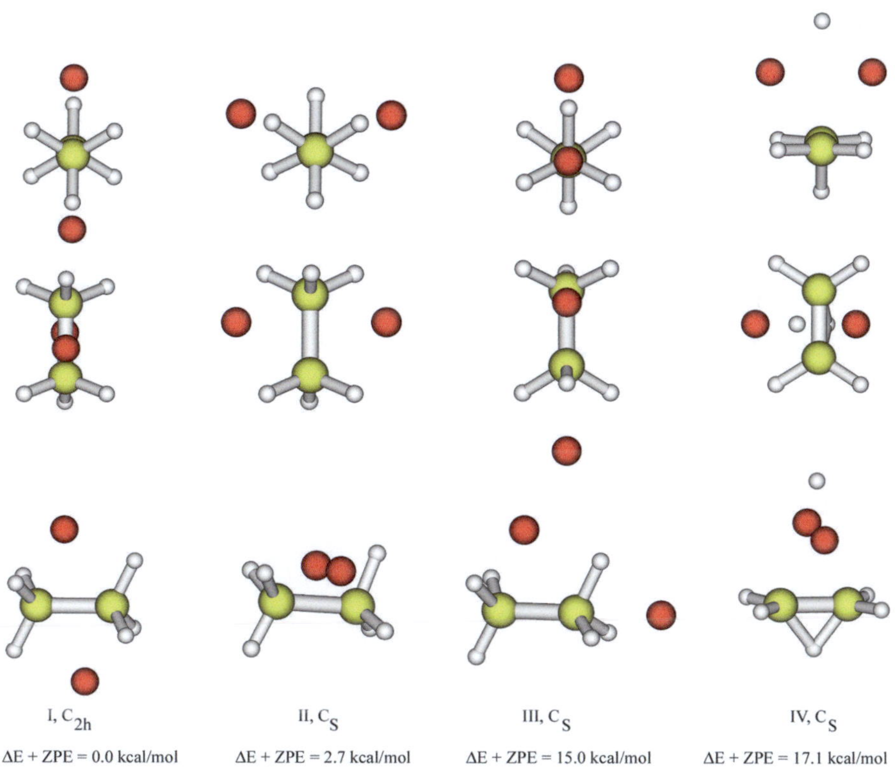

Fig. 1.9 Lowest isomers of the $Li_2B_2H_6$ molecule and their relative energies calculated at CCSD(T)/CBS//CCSD(T)/6-311++G**+ZPE (CCSD(T)/6-311++G**) (Reproduced from [8]. Copyright 2011, Elsevier B. V)

of electronic transmutation [67]. It is noteworthy that all these theoretically predicted molecules still await their experimental confirmation. However, there are some experimental examples where it was shown that boron atom accepting extra electron becomes “carbon”. The first example of such species is MgB_2 high temperature superconductor [68]. The experimentally determined MgB_2 structure is comprised of 2D layers of honeycomb structures composed of boron atoms with magnesium atoms located above and below the boron hexagons (Fig. 1.10).

In this case, Mg donates two electrons to B atoms enabling the electronic transmutation of boron atoms [7]. The 2D-lattice of boron appears exactly as the 2D-lattice of graphene. It is noteworthy that it is very different from the 2D-lattice formed by the neutral boron atoms in the α-sheet. If we assume that a complete charge transfer from Mg to B occurs, which is consistent with the stoichiometric formula of the compound, then we have the case of electronic transmutation here, too, since every boron atom acquires an extra electron and becomes a “carbon.” The σ-bonding in those 2D-sheets is found to be classical (composed out of 2c-2e B-B σ-bonds), similar to that of graphene [69]. This is a remarkable example of the electronic transmutation for the experimentally known compound. Another experimentally known example is pure-phase LiB_x samples with the approximate range $0.82 < x < 1.0$, in which Wörle and Nesper, showed the structural analogy between borynide chains in LiB_x and isoelectronic polyyne and polycumulene chains [70]. They also proposed that each boron atom in lithium boride yields one negative charge and thus becomes isoelectronic to carbon. We believe that electronic transmutation model can be used to design many new boron compounds.

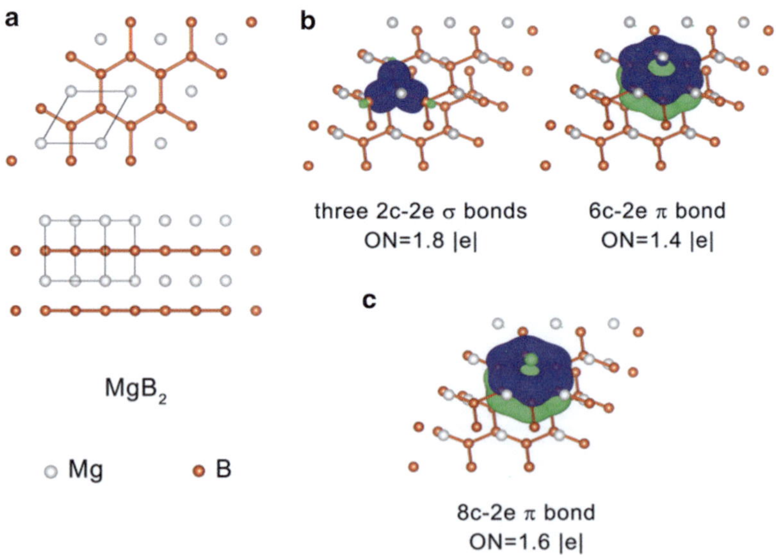

Fig. 1.10 (**a**) Structure, (**b**) SSAdNDP chemical bonding pattern and (**c**) alternative 8c-2e π bond representation of the 6c-2e π bond in magnesium diboride. The unit cell is shown in *black* (Reproduced from [7] with permission from the PCCP Owners)

1.4 Summary

In this chapter we have shown that boron continues to surprise us with unusual structures and unusual bonding because of its electron deficiency. Small and medium-sized anionic boron clusters were found to be planar or quasi-planar species though at around n = 40 the transition to 3D borospherenes occurred. In planar or quasi-planar boron clusters the multicenter bonding is dominant, though classical 2c-2e σ-bonds are responsible for bonding between peripheral boron atoms. However, in borospherenes there are no 2c-2e bonds, unlike carbon fullerenes where the σ-bonding is classical.

When boron atom accepts an extra electron it starts to behave as "carbon" (electronic transmutation) forming compounds similar to carbon, such as $Li_nB_nH_{2n+2}$ species, which are analogs of saturated hydrocarbons C_nH_{2n+2}; $Li_6B_6H_6$, analog of benzene; linear chain of boron anions in LiB_x being analogue of carbine; and 2D layer of boron in MgB_2 mimicking the graphene structure.

Chemistry of boron continues to expand conquering new territories and providing us with unprecedented structures, chemical bonding, internal rotations and other unusual properties. We believe we are at the beginning of new era of boron chemistry.

References

1. Novoselov KS, Geim AK, Morozov SV, Jiang D, Zhang Y, Dubonos SV, Grigorieva IV, Firsov AA (2004) Electric field effect in atomically thin carbon films. Science 306:666–669
2. Novoselov KS, Geim AK, Morozov SV, Jiang D, Katsnelson MI, Grigorieva IV, Dubonos SV, Firsov AA (2005) Two-dimensional gas of massless Dirac fermions in graphene. Nature 438:197–200
3. Tang H, Ismail-Beigi S (2007) Novel precursors for boron nanotubes: the competition of two-center and three-center bonding in boron sheets. Phys Rev Lett 99:115501
4. Tang H, Ismail-Beigi S (2009) Self-doping in boron sheets from first principles: a route to structural design of metal boride nanostructures. Phys Rev B 80:134113
5. Yang X, Ding Y, Ni J (2008) Ab initio prediction of stable boron sheets and boron nanotubes: structure, stability, and electronic properties. Phys Rev B 77:041402
6. Galeev TR, Chen Q, Guo J-C, Bai H, Miao C-Q, Lu H-G, Sergeeva AP, Li S-D, Boldyrev AI (2011) Deciphering the mystery of hexagon holes in an all-boron graphene α-sheet. Phys Chem Chem Phys 13:11575–11578
7. Galeev TR, Dunnington BD, Schmidt JR, Boldyrev AI (2013) Solid state adaptive natural density partitioning: a tool for deciphering multi-center bonding in periodic systems. Phys Chem Chem Phys 15:5022–5029
8. Olson JK, Boldyrev AI (2012) Electronic transmutation: boron acquiring an extra electron becomes 'carbon'. Chem Phys Lett 523:83–86
9. Dilthey WZ (1921) Personal- und Hochschulnachrichten. Angew Chem 34:596
10. Price WC (1947) The structure of diborane. J Chem Phys 15:614
11. Price WC (1948) The absorption spectrum of diborane. J Chem Phys 16:894
12. Bell RP, Longuet-Higgins HC (1945) The normal vibrations of bridged X_2Y_6 molecules. Proc R Soc (London) A 183:357–374

13. Hedberg K, Schomaker V (1951) A reinvestigation of the structures of diborane and ethane by electron diffraction. J Am Chem Soc 73:1482–1487
14. Lipscomb WN (1963) Boron hydrides. In: Benjamin WA (ed) The physical inorganic chemistry series. Verlag W. A. Benjamin Inc., New York/Amsterdam
15. Pitzer KS (1945) Electron deficient molecules. I The principles of hydroboron structures. J Am Chem Soc 67:1126–1132
16. Eberhardt WH, Crawford B, Lipscomb WN (1954) The valence structure of the boron hydrides. J Chem Phys 22:989–1001
17. Albert B, Hillebrecht H (2009) Boron: elementary challenge for experimenters and theoreticians. Angew Chem Int Ed 48:8640–8668
18. White MA, Cerqueira AB, Whitman CA, Johnson MB, Ogitsu T (2015) Determination of phase stability of elemental boron. Angew Chem Int Ed 54:3626–3629
19. Decker BF, Kasper JS (1959) The crystal structure of a simple rhombohedral form of boron. Acta Crystallogr 12:503–506
20. Hoard JL, Sullenger DB, Kennard CHL, Hughes RE (1970) The structure analysis of β-rhombohedral boron. J Solid State Chem 1:268–277
21. Oganov AR, Chen J, Gatti C, Ma YZ, Ma YM, Glass CW, Liu Z, Yu T, Kurakevych OO, Solozhenko VL (2009) Ionic high-pressure form of elemental boron. Nature 457:863–867
22. (a) Popov IA, Piazza ZA, Li W-L, Wang LS, Boldyrev AI (2013) A combined photoelectron spectroscopy and ab initio study of the quasi-planar B_{24}^- cluster. J Chem Phys 139:144307; (b) Piazza ZA, Popov IA, Li W-L, Pal R, Zeng XC, Boldyrev AI, Wang LS (2014) A photoelectron spectroscopy and ab initio study of the structures and chemical bonding of the B_{25}^- cluster. J Chem Phys 141:034303
23. Alexandrova AN, Boldyrev AI, Zhai H-J, Wang LS (2006) All-boron aromatic clusters as potential new inorganic ligands and building blocks in chemistry. Coord Chem Rev 250:2811–2866
24. Sergeeva AP, Popov IA, Piazza ZA, Li W-L, Romanescu C, Wang LS, Boldyrev AI (2014) Understanding boron through size-selected clusters: structure, chemical bonding, and fluxionality. Acc Chem Res 47:1349–1358
25. Zhai H-J, Alexandrova AN, Birch KA, Boldyrev AI, Wang LS (2003) Hepta- and octacoordinated boron in molecular wheels of eight- and nine-atom boron clusters: observation and confirmation. Angew Chem Int Ed 42:6004–6008
26. Zubarev DY, Boldyrev AI (2008) Developing paradigms of chemical bonding: adaptive natural density partitioning. Phys Chem Chem Phys 10:5207–5217
27. Foster JP, Weinhold F (1980) Natural hybrid orbitals. J Am Chem Soc 102:7211–7218
28. Reed AE, Curtiss LA, Weinhold F (1988) Intermolecular interactions from a natural bond orbital, donor-acceptor viewpoint. Chem Rev 88:899–926
29. Hotop H, Lineberger WC (1985) Binding energies in atomic negative ions. J Phys Chem Ref Data 14:731–750
30. Fowler PW, Gray BR (2007) Induced currents and electron counting in aromatic boron wheels: B_8^{2-} and B_9^-. Inorg Chem 46:2892–2897
31. Romanescu C, Galeev TR, Li WL, Boldyrev AI, Wang LS (2011) Aromatic metal-centered monocyclic boron rings: Co©B_8^- and Ru©B_9^-. Angew Chem Int Ed 50:9334–9337
32. Li WL, Romanescu C, Galeev TR, Piazza ZA, Boldyrev AI, Wang LS (2012) Transition-metal-centered nine-membered boron rings: M©B_9 and M©B_9^-. (M=Rh, Ir). J Am Chem Soc 134:165–168
33. Romanescu C, Galeev TR, Sergeeva AP, Li WL, Wang LS, Boldyrev AI (2012) Experimental and computational evidence of octa- and nona-coordinated planar iron-doped boron clusters: Fe©B_8^- and Fe©B_9^-. J Organomet Chem 721–722:148–154
34. Galeev TR, Romanescu C, Li WL, Wang LS, Boldyrev AI (2012) Observation of the highest coordination number in planar species: decacoordinated Ta©B_{10}^- and Nb©B_{10}^- anions. Angew Chem Int Ed 51:2101–2105
35. Romanescu C, Galeev TR, Li WL, Boldyrev AI, Wang LS (2013) Transition-metal-centered monocyclic boron wheel clusters (M©B_n): a new class of aromatic borometallic compounds. Acc Chem Res 46:350–358

36. Oliva JM, Vegas Á (2012) Merging boron solid state and molecular chemistry: energy landscapes in the exo/endo closo-borane complex $Sc[B_{24}H_{24}]^+$. Chem Phys Lett 533:50–55
37. Forkwa BPT, Hermus M (2012) All-boron planar B_6 ring in the solid-state phase $Ti_7Rh_4Ir_2B_8$. Angew Chem Int Ed 51:1702–1705
38. Mbarki M, Touzani RS, Fokwa BPT (2014) Unexpected synergy between magnetic iron chains and stacked B_6 rings in $Nb_6Fe_{1-x}Ir_{6+x}B_8$. Angew Chem Int Ed 53:13174–13177
39. Jimenez-Halla JOC, Islas R, Heine T, Merino G (2010) B_{19}^-: an aromatic Wankel motor. Angew Chem Int Ed 49:5668–5671
40. Martínez-Guajardo G, Sergeeva AP, Boldyrev AI, Heine T, Ugalde JM, Merino G (2011) Unravelling phenomenon of internal rotation in B_{13}^+ through chemical bonding analysis. Chem Comm 47:6242–6244
41. Zhang J, Sergeeva AP, Sparta M, Alexandrova AN (2012) B_{13}^+: a photodriven molecular Wankel engine. Angew Chem Int Ed 51:8512–8515
42. (a) Huang W, Sergeeva AP, Zhai HJ, Averkiev BB, Wang LS, Boldyrev AI (2010) A concentric planar doubly π-aromatic B_{19}^- cluster. Nature Chem 2:202–206; (b) Popov IA, Boldyrev AI (2012) Chemical bonding in coronene, isocoronene, and circumcoronene. Eur J Org Chem 2012:3485–3491
43. Gu FL, Yang X, Tang AC, Jiao H, Schleyer PR (1998) Structure and stability of B_{13}^+ clusters. J Comput Chem 19:203–214
44. Moreno D, Pan S, Zeonjuk LL, Islas R, Osorio E, Martinez-Guajardo G, Chattaraj P, Heine T, Merino G (2014) B_{18}^{2-}: a quasi-planar bowl member of the Wankel motor family. Chem Comm 50:8140–8143
45. Kroto HW, Heath JR, O'Brien SC, Curl RF, Smalley RE (1985) C_{60}: buckminsterfullerene. Nature 318:162–163
46. Szwacki NG, Sadrzadeh A, Yakobson BI (2007) B_{80} fullerene: an ab initio prediction of geometry, stability, and electronic structure. Phys Rev Lett 98:166804; (2008) erratum 100:159901
47. Yan QB, Sheng X-L, Zheng Q-R, Zhang L-Z, Su G (2008) Family of boron fullerenes: general constructing schemes, electron counting rule, and ab initio calculations. Phys Rev B 78:201401
48. Zope RR, Baruah T, Lau KC, Liu AY, Pederson MR, Dunlap BI (2009) Boron fullerenes: from B_{80} to hole doped boron sheets. Phys Rev B 79:161403
49. Sheng XL, Yan QB, Zheng QR, Su G (2009) Boron fullerenes B(32 + 8 k) with four-membered rings and B_{32} solid phases: geometrical structures and electronic properties. Phys Chem Chem Phys 11:9696–9702
50. Özdoğan C, Mukhopadhyay S, Hayami W, Güvenc ZB, Pandey R, Boustani I (2010) The unusually stable B_{100} fullerene, structural transitions in boron nanostructures, and a comparative study of α- and γ-boron and sheets. J Phys Chem C 114:4362–4375
51. Wang L, Zhao J, Li F, Chen Z (2010) Boron fullerenes with 32–56 atoms: irregular cage configurations and electronic properties. Chem Phys Lett 501:16–19
52. Muya JT, Gopakumar G, Nguyen MT, Ceulemans A (2011) The leapfrog principle for boron fullerenes: a theoretical study of structures and stability of B_{112}. Phys Chem Chem Phys 13:7524–7533
53. Zope RR, Baruah T (2011) Snub boron nanostructures: chiral fullerenes, nanotubes and planar sheet. Chem Phys Lett 501:193–196
54. Polad S, Ozay M (2013) A new hole density as a stability measure for boron fullerenes. Phys Chem Chem Phys 15:19819–19824
55. Prasad DLVK, Jemmis ED (2008) Stuffing improves the stability of fullerene-like boron clusters. Phys Rev Lett 100:165504
56. De S, Willand A, Amsler M, Pochet P, Genovese L, Goedecker S (2011) Energy landscape of fullerene materials: a comparison of boron to boron nitride and carbon. Phys Rev Lett 106:225502
57. Li F, Jin P, Jiang D, Wang L, Zhang SB, Zhao J, Chen Z (2012) B_{80} and $B_{101-103}$ clusters: remarkable stability of the core–shell structures established by validated density functional. J Chem Phys 136:074302

58. Boulanger P, Moriniere M, Genovese L, Pochet P (2013) Selecting boron fullerenes by cage-doping mechanisms. J Chem Phys 138:184302
59. Zhai H-J, Zhao Y-F, Li W-L, Chen Q, Bai H, Hu H-S, Piazza ZA, Tian W-J, Lu H-G, Wu Y-B, Mu Y-W, Wei G-F, Liu Z-P, Li J, Li S-D, Wang L-S (2014) Observation of an all-boron fullerene. Nat Chem 6:727–731
60. Chen Q, Li W-L, Zhao Y-F, Zhang S-Y, Hu H-S, Bai H, Li H-R, Tian W-J, Lu H-G, Zhai H-J, Li S-D, Li J, Wang L-S (2015) Experimental and theoretical evidence of an axially chiral borospherene. ACS Nano 9:754–760
61. Osorio E, Olson JK, Tiznado W, Boldyrev AI (2012) Analysis of why boron avoids sp^2 hybridization and classical structures in the B_nH_{n+2} series. Chem Eur J 18:9677–9681
62. Dávalos JZ, González J, Guerrero A, Hnyk D, Holub J, Oliva JM (2013) Anionic oligomerization of $Li_2[B_{12}H_{12}]$ and $Li[CB_{11}H_{12}]$: an experimental and computational study. J Phys Chem C 117:1495–1501
63. Oliva JM, Fernández-Barbero A, Serrano-Andrés L, Canle-L M, Santaballa JA, Fernández MI (2010) Energy landscapes in diexo and exo/endo isomers derived from $Li_2B_{12}H_{12}$. Chem Phys Lett 497:172–177
64. Her J-H, Yousufuddin M, Zhou W, Jalisatgi SS, Kulleck JG, Zan JA, Hwang S-J, Bowman RC Jr, Udovic TJ (2008) Crystal structure of $Li_2B_{12}H_{12}$: a possible intermediate species in the decomposition of $LiBH_4$. Inorg Chem 47:9757–9759
65. Popov IA, Boldyrev AI (2013) Computational probing of all-boron $Li_{2n}B_{2n}H_{2n+2}$ polyenes. Comp Theor Chem 1004:5–11
66. Alexandrova AN, Birch KA, Boldyrev AI (2003) Flattening the $B_6H_6^{2-}$ Octahedron. Ab initio prediction of the new family of planar all-boron aromatic molecules. J Am Chem Soc 125:10786–10787
67. Gish JT, Popov IA, Boldyrev AI (2015) Homocatenation of aluminum: alkane-like structures of $Li_2Al_2H_6$ and $Li_3Al_3H_8$. Chem Eur J. 21:5307-5310.
68. Nagamatsu J, Nakagawa N, Muranaka T, Zenitani Y, Akimitsu J (2001) Superconductivity at 39 K in magnesium diboride. Nature 410:63–64
69. Popov IA, Bozhenko KV, Boldyrev AI (2012) Is graphene aromatic? Nano Res 5:117–123
70. Wörle M, Nesper R (2000) Infinite, linear, unbranched borynide chains in LiB_x – isoelectronic to polyyne and polycumulene. Angew Chem Int Ed 39:2349–2353

Chapter 2
Molecular Structures of Free Boron Clusters

Drahomír Hnyk and Derek A. Wann

Abstract This chapter deals with gas-phase molecular structure determinations of neutral boranes and heteroboranes employing the techniques of gas-phase electron diffraction (GED) and/or modern quantum chemical calculations. Such calculations were useful for computing various observables in order to facilitate the analysis of the electron diffraction data. Additionally, microwave spectroscopy was utilized for the two thiaboranes (in conjunction with the University of Oslo). Unless otherwise stated, the samples used for the work described throughout this chapter originated from the Institute of Inorganic Chemistry of the Academy of Science of the Czech Republic, v.v.i., Řež while the GED studies were performed mainly in the School of Chemistry at the University of Edinburgh.

The structurally characterized boron clusters belong to the range of structural motifs, from *closo* to *nido*, which obey the so-called Wade's rules. Examples of boranes that do not obey Wade's rules were also studied, as were selected macropolyhedral clusters and metallaboranes. Finally, in order to gain an insight into electron density distribution, analyses of the experimental dipole moments were carried out for a few examples.

Whereas the earlier GED studies of boranes and carbaboranes ignored the calculated vibrational effects because of a lack of force fields for these clusters, the current electron diffraction investigations of boranes and various types of heteroboranes used calculated force fields to good effect. They revealed an interesting feature: the amplitudes of vibration for bonded and non-bonded cage distances are very similar, which is at odds with various empirical rules suggesting that amplitudes of vibration should be roughly proportional to the corresponding internuclear distances.

D. Hnyk (✉)
Institute of Inorganic Chemistry of the Academy of Sciences of the Czech Republic, v.v.i., Husinec-Řež 250 68, Czech Republic
e-mail: hnyk@iic.cas.cz

D.A. Wann
Department of Chemistry, University of York Heslington, York YO10 5DD, UK
e-mail: derek.wann@york.ac.uk

D. Hnyk, M.L. McKee (eds.), *Boron*, Challenges and Advances in Computational Chemistry and Physics 20,
DOI 10.1007/978-3-319-22282-0_2

Atoms colors used in the figures

hydrogen – grey
boron – green
carbon – black
nitrogen – sky blue
sulfur – yellow
fluorine/chlorine – light green
bromine – dark orange
iodine – purple
silicon – light grey
phosphorus – violet
selenium – dark yellow
nickel – orange

2.1 Introduction

Boron is one of only a few elements that is capable of forming extended binary hydrides. These so-called boranes do not occur in nature; rather they are exclusively products of man-made chemistry. Diborane (B_2H_6) can be considered as an archetype of boron hydrides; its structure was first reported in 1937 as being similar to that of ethane [1]. However, subsequent gas electron diffraction (GED) data were found to be incompatible with this ethane model, and the familiar D_{2h}-symmetric molecular geometry, involving two bridging hydrogen atoms (depicted in Scheme 2.1), was introduced [2]. The bridging hydrogen atom is an essential structural motif for boron hydrides and, together with BBB triangles, is observed for many boron hydrides and larger heteroboranes. The hydrogen bridge is an example of 3-center-2-electron bonding, the discovery of which led W. N. Lipscomb to formulate the concept of multicenter bonding, resulting in the award of the Nobel Prize in 1976.

Borane architectures are based on three-dimensional structures composed of triangular B–B–B units. The immense variety of borane structures (stable, for example, as dianions $B_nH_n^{2-}$) stems from the number of boron vertices, n (in the basic series $n=5$–12), and the electrons available. Thus, we recognize the closed-cage

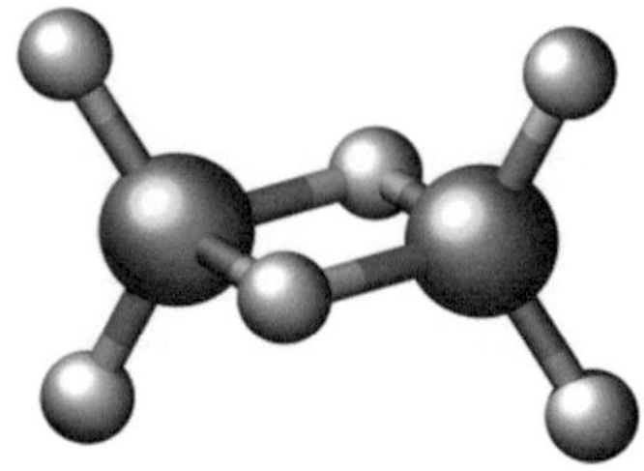

Scheme 2.1 The D_{2h}-symmetric structure of diborane, B_2H_6

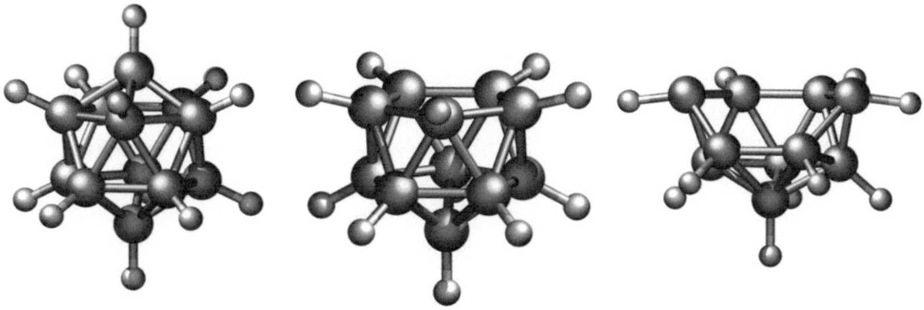

closo-$B_{12}H_{12}^{2-}$ *nido*-$B_{11}H_{11}^{4-}$ *arachno*-$B_{10}H_{10}^{6-}$

Scheme 2.2 The "*closo-nido-arachno*" relationship for deltahedral boranes

closo- species (2*n*+2 electrons) and several open-cage species, which are derived by notionally removing one (*nido*-, 2*n*+4 electrons), two (*arachno*-, 2*n*+6 electrons), or more (*hypho*-, 2*n*+8 electrons) boron vertices (as depicted in Scheme 2.2). Due to the formal electron deficiency of boron atoms, their connectivity can be as high as six. The extreme stabilities of *closo* borane cages are due to the delocalization of two surplus electrons of the dianions along the σ bonds across the entire cage.

Syntheses and studies of the molecular structures of boranes and heteroboranes at the Institute of Inorganic Chemistry of the Academy of Sciences of the Czech Republic, v.v.i. date back to the 1960s [3]. Elucidation of molecular structures can involve the determination of molecular geometries, electron distributions, and even intramolecular motions. Such structure determination can be performed in the solid state using X-ray diffraction analysis and for isolated molecules using gas electron diffraction (GED) [4] as well as, more recently, using computational methods. Efforts are made to study isolated molecular structures because single crystals of many free heteroboranes are disordered; consequently, determinations of accurate structures in the solid state are impossible. It is these isolated-molecule geometries that are the subject of this chapter, which aims to expand the body of knowledge of experimental geometries for free boranes and heteroboranes. Finally, there is an emphasis on generalizing observed structural trends.

2.2 Methodology

2.2.1 Determination of Molecular Structures

Electron diffraction is the most important technique for the determination of structures of gaseous molecules. GED is based on the scattering of a beam of high-energy electrons from a gaseous sample of randomly orientated vibrating molecules. Significant contributions to the field, specifically in relation to the study of boranes, were made in a number of laboratories across Europe. Almost every GED apparatus is unique in its design and it is common to quote the accelerating potential used, which was *ca.* 60 kV at the Hungarian Academy of Sciences in Budapest, and *ca.* 42 kV at the Universities of Oslo and Edinburgh (though Edinburgh later moved to using 40 kV electrons). Data for the vast majority of examples discussed here were collected at the School of Chemistry at the University of Edinburgh; any exceptions will be specifically mentioned in the text. Samples were typically heated to give sufficient vapor pressure, before the gas jet was intersected by the high-energy electron beam to yield a series of one-dimensional diffraction patterns. After subtracting the scattering due to individual atoms, one is left with an experimental molecular scattering pattern from which the structure can be determined. The theoretical molecular intensity $M(s)$ (see Eq. 2.1) is derived from basic scattering theory, and takes a geometrical model into account, which provides all interatomic distances, r_{ij} (Eq. 2.1). Structural analysis is performed using a least-squares refinement procedure, fitting experimental and theoretical molecular intensities; the so-called R factor that is yielded is a mathematical measure of the fit of the model to the data [5]. It should be noted that the model describing the geometry of a molecule in terms of selected refinable geometrical parameters is an essential part of the analysis of electron-diffraction data, and that writing such a model using the minimum number of parameters possible is sometimes far from routine [6].

$$M(s) = \alpha_{ij} \sum_{N}^{i=1} \sum_{\substack{N \\ i \neq j}}^{j=1} g_{ij}(s) \cdot \exp\left(-\frac{1}{2} l_{ij}^2 s^2\right) \frac{\sin\left[s\left(r_{\mathrm{a}} - \kappa_{ij} s^2\right)\right]}{s r_{ij}} \tag{2.1}$$

The theoretical molecular intensity curve is a superposition of sinusoids for each atomic pair bounded by experimental limits of the scattering variable, s [$= (4\pi/\lambda) \sin(\theta/2)$], which is often reported in Å^{-1}. The meanings of other non-refinable variables used in Eq. 2.1 are as follows: λ is the electron wavelength, θ represents the scattering angle, and g_{ij} is assumed to be known in the structure analysis and is related to scattering factors and atomic phases. α_{ij} represents the weight of each distance and is related to conformational analysis since electron diffraction represents a particularly fast timescale (*cf.* 10^{-18} s compared to, for example, NMR for which the time scale amounts to 10^{-9} to 10^{-1} s). r_{ij} is the main result from the structural analysis and represents an effective internuclear distance; in other words, it defines the molecular geometry. As well as geometric parameters, the GED refinements provide, in terms of vibrational amplitudes, l_{ij}, a good insight into relative vibrational displacements of the atomic nuclei with respect to their equilibrium positions. Initial values of vibrational amplitudes can either be calculated or estimated on the basis of data accumulated for similar compounds. Hence, this method provides valuable pieces of information about intramolecular motion. Finally, κ_{ij} is an anharmonicity constant that is significant for bonded atom pairs and far less so for non-bonded pairs.

Sine Fourier transformation of this molecular scattering pattern gives rise to a radial distribution curve consisting of a peak for each interatomic distance in the molecule (Eq. 2.2):

$$f(r) = \frac{2}{\pi} \int_0^{s_{max}} sM(s) \exp(-bs^2) \sin(sr) ds \tag{2.2}$$

where $\exp(-bs^2)$ is an artificial damping factor introduced because the range of experimental data is reduced from $s = \langle 0, \infty \rangle$ to $s = \langle s_{\min}, s_{\max} \rangle$.

Analysis of electron-diffraction data is relatively easy for small, symmetrical clusters [5], where it can provide very accurate results indeed. Conversely, larger, less symmetric molecules (such as asymmetric boranes or heteroborane clusters) may be more demanding, and such investigations often necessitate the combination of electron-diffraction data with data obtained by other methods, both experimental and theoretical, in order to obtain reliable results.

The problems associated with refining the molecular structure of a borane or heteroborane using GED data alone stem from the fact that the molecules usually contain many atom pairs separated by B–B bond lengths of around 170–190 pm. In general this can preclude the resolution of individual B–B distances with high accuracy because they are usually strongly correlated to one another. This inadequacy of GED could be overcome by supplementing the refinement with data obtained from geometry optimizations carried out at various levels of theory, and then fixing the differences between similar distances at computed values, *i.e.* as rigid constraints. This was known as the *MOCED* approach [7]. A superior approach, however, also utilizing data from theoretical geometries, has been developed to allow the refinement of all geometrical parameters [8], and is the natural extension of *MOCED*. In essence this approach known as the *SARACEN* method hinges upon (a) the use of calculated parameters as flexible restraints (rather than rigid constraints), and (b) the refinement of all geometrical parameters as a matter of principle. The restraints are entered into the GED refinements as extra observations, just as is commonly done with additional experimental data (*e.g.* rotational constants from microwave spectroscopy). More realistic estimated standard deviations are obtained as a consequence.

The architecture of a newly synthesized borane or heteroborane cluster is proposed on the basis of experimental measurements of the ^{11}B NMR spectrum (^{13}C NMR is also frequently applied in the case of carbaboranes), utilising various approaches, including decoupling and two-dimensional NMR techniques, such as COrrelated SpectroscopY. Chemical knowledge of related compounds is also considered. The chemical shifts obtained from such spectroscopic measurements are then defined relative to the usual standard of ^{11}B NMR spectroscopy, which is $BF_3 \cdot OEt_2$.

Comparison of experimental and calculated ^{11}B NMR chemical shifts may also serve as a validation of the refined geometry, as the calculated shielding tensors are quite sensitive to small changes in the geometry of a cluster, with the hydrogen positions being particularly crucial. (Cartesian coordinates serve as the input for magnetic property calculations.) There may be several models that fit the GED data almost equally well, but not all of them provide calculated values of $\delta(^{11}B)$ that are in good accordance with experimental values. A number of borane and heteroborane geometries have been refined employing this joint *ab initio*/GED method [9], the final structures having been validated by Individual Gauge Localised Orbital (IGLO) [10] or Gauge Invariant Atomic Orbital (GIAO) [11] chemical shift calculations. These efforts will be exemplified below.

The so-called *ab initio*/GIAO/NMR method, with DFT and IGLO variants, also provides the possibility of deriving internal coordinates for free boranes and heteroboranes, particularly for those that are negatively charged and/or possess no symmetry. The *ab initio*/GED method differs from this approach only in employing experimental geometries rather than theoretically derived ones. The dimensions of the proposed molecular shape are optimized by *ab initio* calculations, using Hartree-Fock theory to provide starting geometries for final computations that include the effects of electron correlation using, for example, the MP2 (Møller-Plesset second-order perturbation theory) method [12]. Density Functional Theory (DFT) methods also intrinsically involve correlation energy, but save both scratch disk space and memory as the corresponding orbitals are functions of just one variable, *i.e.* electron density, in contrast to the orbitals used for *ab initio* calculations which express the dependence on three variables, the x, y, and z coordinates for each atom of a cluster. The optimized geometry found in this way is then used as an input for the calculation of a shielding tensor, again employing IGLO- or GIAO-based methods. The so-called GIAO-MP2/II//MP2/6-31G*[1] method for clusters containing just main-group elements (the most common are C, N, S, and P, with terminal hydrogens replaced by methyl, phenyl, or *t*-Bu groups) proved to be a very successful tool. The shielding tensor is calculated using the GIAO method at the MP2 level employing a TZP Huzinaga basis set [13] (denoted as II), utilising the molecular geometry derived with a Pople-style basis set of 6-31G* [14] and with the addition of MP2-type correlation energy. Larger systems demand more CPU and memory, but the GIAO-HF/II//MP2/6-31G* method gives spectral data that are quite sufficient for the purpose of confirming the correctness of a molecular structure. The latter approach differs from the former by not including the electron correlation for the magnetic property calculations, *i.e.* the SCF level is used.

The situation is more tricky for heteroboranes that contain a metal. The choice of basis set is important both for geometry optimizations (all-electron basis set *vs.* valence basis set with relativistic pseudopotentials) and for the evaluation of the shielding tensors, for which the computational method is also crucial. The most frequent approach, justified by some examples of successful applications, relies on the GIAO-DFT/basis set//DFT/basis set scheme, where the basis set is either all-electron or valence + pseudopotentials, and the DFT method is usually represented

[1] The nomenclature used to describe these calculations gives the method and basis set for the geometry optimisation after the //, while the method and basis set used to calculate the magnetic properties are stated before it.

by the well-established functionals B3LYP [15] and BP86 [16]. The calculated ^{11}B chemical shifts (with respect to $BF_3 \cdot OEt_2$, diborane serving as a primary reference) are then compared with experimental ones. The level of agreement between computed and experimental spectra provides the basis for accepting or refusing a particular geometry, with a difference of 2–3 ppm (depending on the level of calculations) being considered acceptable. In cases where both experimental (GED) and theoretical geometries are available, ^{11}B chemical shift calculations allow the quality of the geometries to be assessed in terms of the agreement of the chemical shifts with the experimental values. Computed energies for such experimental structures may also be helpful; if one is much higher in energy (40 kJ mol^{-1} or more) than the optimized structure, then the experimental result is unlikely to be correct.

2.2.2 *Determination of Electron Distributions*

Dipole moments were measured at 25 °C in benzene (usually five solutions, weight fraction 1.8×10^{-4} to 1.1×10^{-3}) using the method published by Guggenheim and Smith [17]. Relative permittivities were measured at 6 MHz on a home-made DK meter with direct frequency reading. Refractive indices were measured on an Aerograph refractive index detector (Varian).

2.3 Structural Analyses

The success of the earlier systematic application of the *ab initio*/IGLO/NMR method and its GIAO variant for the structural studies of carbocations [18] has led to the application of the method to boranes and heteroboranes. Beaudet [19] has reviewed structures of small and medium-sized boranes and carbaboranes, determined by GED, X-ray diffraction, and microwave spectroscopy, and the validities of these structures were later checked using the *ab initio*/IGLO/NMR method [20]. Subsequently, Mastryukov reviewed gas-phase structures of parent and *exo-substituted* boranes and carboranes of larger dimensions ($n = 10, 12$) [21], though he limited the review to molecular structures that were determined by GED alone. The gas-phase structures of two heterocarboranes, *viz. closo*-1,12-$CHXB_{10}H_{10}$ (X = P, As) also appeared in the latter review. To our knowledge, there were no gas-phase structures of heteroboranes apart from the carboranes and two heterocarboranes, as confirmed by refs. 19–21. Here we aim to report molecular structures of both older and more recently prepared neutral boranes and heteroboranes determined by using GED and/or modern computational protocols. Unless otherwise stated, all the compounds presented were prepared at the Institute of Inorganic Chemistry of the ASCR, v.v.i.

2.3.1 Parent Boron Hydrides

Pentaborane(11), *arachno*-B_5H_{11} (**1a**), was prepared at the University of Leeds and was the first small borane to which the *ab initio*/IGLO/NMR method was applied (Fig. 2.1) [22]. This study revealed that the structure in which the apical bridging hydrogen is involved in a rather ordinary three-center hydrogen-bridge bond, with the molecule having C_1 symmetry, is superior to that in which this hydrogen atom bridges three boron atoms at the same time (C_s symmetry), as had been proposed in an earlier analysis of GED data [23]. There was a remarkably good fit between the calculated (DZ//MP2/6-31G*) and experimental ^{11}B values for the C_1 structure, with a maximum deviation of *ca.* 3 ppm, whereas large discrepancies, up to *ca.* 8 ppm, were found for the original GED-based C_s structure [23]. In this preliminary GED study B_5H_{11} was constrained to have overall C_s symmetry. However, when this was relaxed in the *ab initio* (MP2/6-31G*) optimization it was revealed that, for example, the B(2)–B(3) and B(4)–B(5) distances (assumed to be equal in the original GED refinement) differed considerably, at 173.7 and 181.0 pm, respectively.

Although both the GED [24] and *ab initio* geometries [25] for another small borane (also prepared at Leeds), hexaborane(12), *arachno*-B_6H_{12} (**1b**), demonstrated C_2 symmetry and the same pattern of bridging hydrogen atoms, the structural parameters that were determined differed even more noticeably than for B_5H_{11} (Fig. 2.1). For example, the assumption that the B(1)–B(6) nearest-neighbor separation is greater than B(1)–B(2) in the GED analysis was far from true in the results of the MP2/6-31G* calculations [25] [191.3 and 177.8 pm *vs.* 172.8 and 189.9 pm for the B(1)–B(6) and B(1)–B(2) separations, respectively]. It should also be noted that single-point energies calculated at the MP2/6-31G* level using the GED geometries for both molecules were much higher than those optimized at the MP2/6-31G* level [22, 25]. This was especially true for B_6H_{12} where the difference in energy was 247 kJ mol^{-1}.

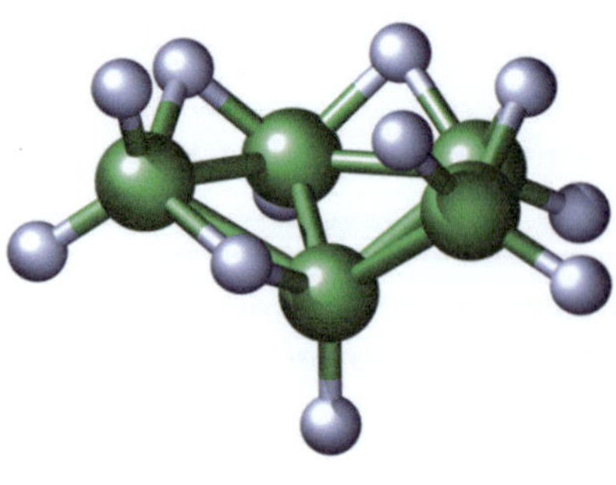

1a

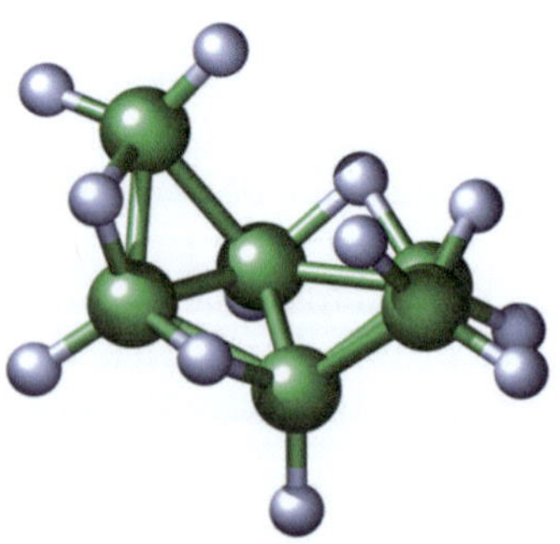

1b

Fig. 2.1 The molecular structures of *arachno* boranes B_5H_{11} (**1a**) and B_6H_{12} (**1b**)

Given the NMR and energetic evidence that the original GED structures might not be correct, the electron-diffraction data for **1a** and **1b** were reanalyzed. The new models for both B_5H_{11} and B_6H_{12} considered the theoretical geometries, but inspection of the resulting parameters for the original GED geometries revealed that some vibrational amplitudes might not be correct. Such amplitudes might be expected to have similar values for all the nearest-neighbor separations, and similarly for all the next-nearest (and even next-next-nearest) neighbor separations. In the least-squares analyses these vibrational terms were refined in groups, with little variation between members of any one group, while C_1 and C_2 symmetries were chosen for **1a** and **1b**, respectively, with differences between related bond lengths fixed at values obtained in the MP2/6-31G* calculations. These refinements yielded new optimum geometries with improved *R* factors for both molecules [26], and both energetic and NMR criteria indicated that the new structures were much more satisfactory. For hexaborane(12) the excess energy of the experimental structure dropped from 247 to 47 kJ mol^{-1}, and the maximum deviations between the DZ//new-GED calculated and experimental ^{11}B chemical shifts were reduced to around 3 ppm. The agreement for **1a** was actually better than the agreement observed for the computed (DZ//MP2/6-31G*) and the experimental values. The refined vibrational amplitudes in **1b** were also much more realistic; for example, those for B(1)–B(2) and B(1)···B(5) now refined to 7.2(2) and 7.9(4) pm, respectively.

In addition to B_5H_{11}, there is also pentaborane(9), B_5H_9, investigated by GED alone in the same study as B_5H_{11} [23]. The hydrogen atom bonded to the apical boron atom can be replaced by BX_2 groups (X = F, Cl). The compounds 1-$(F_2B)B_5H_8$ (**2a**) and 1-$(Cl_2B)B_5H_8$ (**2b**) were prepared at Oxford and investigated in the gas phase using the *ab initio*/GED/NMR method (Fig. 2.2) [27]. In both systems the dihalogenoboryl group was found to be essentially free to rotate about the adjacent B–B bond. The dihalogenoboryl groups cause slight expansion of the B_5 cages with respect to B_5H_9.

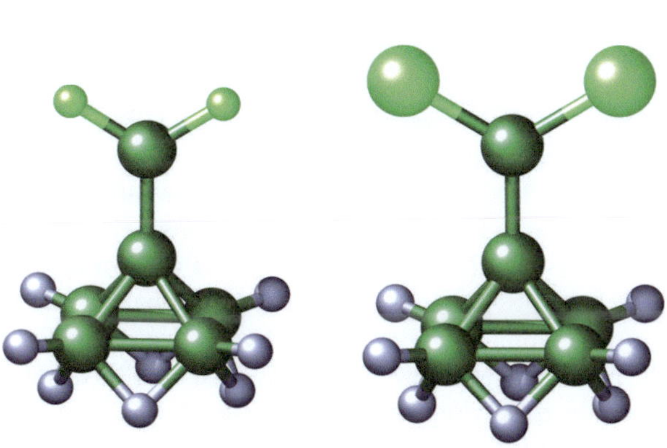

2a **2b**

Fig. 2.2 The molecular structures of *nido*-B_5H_9 derivatives 1-$(F_2B)B_5H_8$ (**2a**) and 1-$(Cl_2B)B_5H_8$ (**2b**)

2.3.2 Closo *Heteroboranes*

2.3.2.1 Icosahedral Dodecaborane(12) Derivatives

The idea that amplitudes of vibration of both closely-spaced atomic pairs and those more widely separated might have similar values in *arachno* systems has been prompted by the determination of the molecular structure of a member of another family of boron clusters known as *closo* systems, *i.e.* 1-thia-*closo*-dodecaborane(11), *closo*-1-$SB_{11}H_{11}$ (**3a**) [28] for which the electron-diffraction data were recorded in Budapest. A model assuming C_{5v} symmetry led to a distortion from a regular icosahedral structure, consisting mainly of a substantial expansion of the pentagonal belt adjacent to sulfur (Fig. 2.3). The B–B distances in this pentagon refined to 190.5(4) pm, with the other B–B distances all falling in the narrow range from 177.7 to 178.3 pm. The S–B bond is the longest in the molecule at 201.0(5) pm. Amplitudes of vibration are consistent with those found for **1** and **2**, *e.g.* 5.1(4) and 6.8(3) pm for B(2)–B(3) and B(2)···B(9), respectively. The latter value is smaller than that for S–B(2), which yielded a value of 7.1(4) pm, even though the two atoms are on opposite (rather than adjacent sides) of the molecule. This strongly supports the idea that a *closo* structure is particularly rigid. Even the HF/3-21G* and HF/6-31G* parameters [29] agree quite well with the experimental findings [*e.g.* B(2)–B(3) at HF/6-31G* is, at 190.4 pm, just 0.1 pm from the experimental value]. This observation is also reflected in the very good agreement between the DZ//GED and DZ// HF/(both basis sets) ^{11}B chemical shifts; both computed sets of shifts also compared well with the corresponding experimental values.

Geometry optimizations for **3a** have been performed at higher levels of theory in another context, where DFT calculations at the B3LYP/cc-pVTZ level were undertaken to see the effect of this computational protocol on the molecular geometry. The results of the calculations yielded S–B and B(2)–B(3) distances of 202.0 and 189.0 pm. This work also reported the results of an investigation of the structure of **3a** by microwave spectroscopy [30], but positions of hydrogen atoms were experimentally located only by GED investigation [28]. This microwave study did give a precise substitution structure for the non-hydrogen atoms, yielding an S–B bond length of 201.3(2) pm and a B(2)–B(3) bond length of 188.9(1) pm. The geometry of **3a** (calculated at the MP2/6-31G* level) is also known, along with geometries for some 12-X derivatives [31] [X=F (**3b**), Cl (**3c**), Br (**3d**), and I (**3e**), respectively, Fig. 2.3]. For the structures with very heavy atoms, instead of the 6-31G* basis set used for X=H (**3a**), F (**3b**) and Cl (**3c**), quasi-relativistic energy-consistent pseudopotentials [32] with DZP valence basis sets were employed for X=Br (**3d**) and I (**3e**).

The S–B and B(2)–B(3) separations in **3a** at the MP2/6-31G* level converged to 200.0 and 187.6 pm, respectively. It is apparent that the nearest-neighbor BB separations computed at the HF level were overestimated in relation to those derived at the correlated MP2 level of theory (and DFT). As noted earlier, errors of 5–10 % are possible, depending on the theory used [20]. Halogen substitution does not have any significant influence on the overall geometry of the icosahedral cage. A change in the chemical shift of B(12), the so-called antipodal chemical shift [33], is

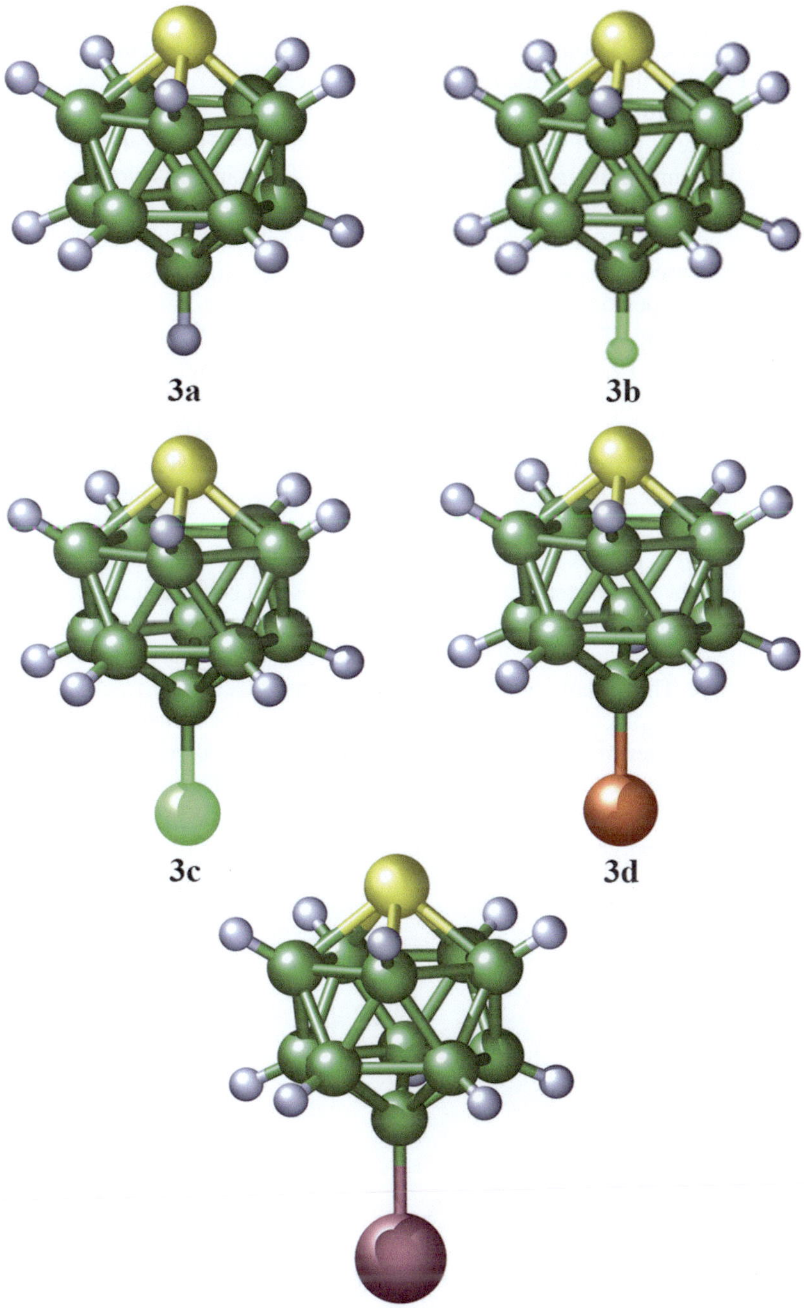

Fig. 2.3 The molecular structure of *closo*-1-$SB_{11}H_{11}$ (**3a**) and its 12-X halogen derivatives [X = F (**3b**), Cl (**3c**), Br (**3d**), and I (**3e**)]

reproduced quite well at the GIAO/II//MP2/6-31G* level for **3a** and **3c** and also, when spin-orbit corrections [34] are included, for **3d** and **3e**. Dipole moments for **3a**, **3c**, **3d** and **3e** that were measured and published in Ref. [30] showed without doubt that the sulfur atom is positively charged.

The structure of the selenium analogue of **3a**, *closo*-1-$SeB_{11}H_{11}$, **4** (Fig. 2.4a), has also been determined using the *SARACEN* method [35], and this work provided an unambiguously determined Se–B bond length without using any restraint. This finding can be used to test the quality of various computational protocols. For **4**, as was also the case for 1-$SB_{11}H_{11}$, *ab initio* geometries slightly underestimate the expansion of the pentagonal belt adjacent to the chalcogen, the center of positive charge of the cluster. For example, MP2/962(d) gives 190.9 pm for this B–B distance, compared to 192.2(2) pm as determined by GED alone.

There are other main-group elements that can replace $(BH)^{2-}$ vertices of the *symmetric* (*closo*-$B_{12}H_{12})^{2-}$. Just as S is isoelectronic with $(BH)^{2-}$ so, for example, is $(CH)^{-}$, which plays the same role. Replacement of one $(BH)^{2-}$ vertex in *closo*-$B_{12}H_{12}{}^{2-}$ will thus lead to (*closo*-1-$CB_{11}H_{12})^{-}$. The MP2/6-31G* calculated structure has been reported as well as solid-state structures with various cations [36]. In contrast, if two $(BH)^{2-}$ groups in the parent dianion are replaced by two $(CH)^{-}$ moieties, three isomeric twelve-vertex neutral dicarbaboranes can be obtained, *i.e. closo*-1,2-$C_2B_{10}H_{12}$ or *ortho*-carbaborane (**5a**, C_{2v} symmetry), *closo*-1,7-$C_2B_{10}H_{12}$ or *meta*-carbaborane (**5b**, C_{2v} symmetry), and *closo*-1,12-$C_2B_{10}H_{12}$ or *para*-carbaborane (**5c**, D_{5d} symmetry). They were ideal targets for gas-phase electron diffraction [37].

Structures of molecules derived from all three parent icosahedral carbaboranes by substitution of terminal hydrogen atoms on carbon have also been determined by the combined use of GED and *ab initio* calculations. These include 1-Ph-1,2-dicarba-*closo*-dodecaborane(12) [1-Ph-1,2-*closo*-$C_2B_{10}H_{11}$ (**5a1**, prepared at the University of Edinburgh)] [38], 1,2-dicarba-*closo*-$C_2B_{10}H_{10}$-9,12-dithiol [9,12-$(SH)_2$-

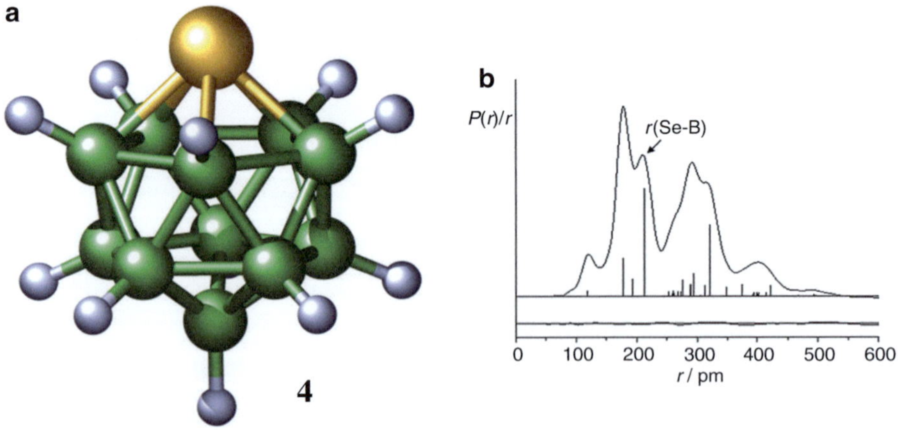

Fig. 2.4 (**a**) The molecular structure of *closo*-1-$SeB_{11}H_{11}$ (**4**). (**b**) The radial distribution curve from the GED refinement of **4**

closo-1,2-$C_2B_{10}H_{10}$(**5a2**)][39], 1,7-dichloro-1,7-dicarba-*closo*-$C_2B_{10}H_{10}$[1,7-Cl_2-1,7-*closo*-$C_2B_{10}H_{10}$ (**5b1**)] [40], 1,12-dicarba-*closo*-$C_2B_{10}H_{10}$-1,12-dithiol [1,12-$(SH)_2$-*closo*-1,12-$C_2B_{10}H_{10}$ (**5c1**)] [41], 1-ethynyl-1,12-dicarba-*closo*-$C_2B_{10}H_{11}$ [1-HC≡C-*closo*-1,12-$C_2B_{10}H_{11}$ (**5c2**)] [42], and 1-(1′-trimethyl-silyl)ethynyl-1,12-dicarba-*closo*-$C_2B_{10}H_{11}$ [1-Me_3SiC≡C-*closo*-1,12-$C_2B_{10}H_{11}$ (**5c3**)] [42], all of which are shown in Fig. 2.5. The effect of the phenyl substituent on the C–C bond length in **5a1** is marginal with a value of 162.7(8) pm compared to 162.4(8) pm for its parent **5a**. This parameter in **5a1** can probably be considered to be determined accurately, because the *SARACEN* method was used, in this case for the first time in the determination of the molecular structure of a borane or heteroborane. This observation strongly indicates that there is practically no conjugation between the two-dimensional and three-dimensional moieties comprising **5a1**. Indeed, a dipole moment study of **5a** indicated that it behaves as a slight electron acceptor [43]. Using a vector solution of a triangle within this study unambiguously revealed that the midpoint of the CC vector is the center of positive charge, with a dipole moment of 4.50 D. The molecule has overall C_1 symmetry, in which the C_6 hexagon eclipses the C(1)–B(4) bond. IGLO/II′//GED and IGLO/II′//HF/6-31G* calculations (II′ is the same TZP basis set as II but DZ is employed for hydrogens) support this finding. Similarly, the presence of two chlorine atoms does not cause any significant change to C–B and B–B nearest-neighbor separations in **5b1** with respect to **5b**, with C_{5v} symmetry for the CB_5 pentagonal pyramids in **5b1** having been assumed. Again, IGLO/II′//GED and IGLO/II′//MP2/6-31G* values agree well with one another, as well as with the experimentally determined $\delta(^{11}B)$ values. Thus the evidence strongly supports the accuracy of the experimentally and theoretically determined geometries.

Because the electron-scattering ability of sulfur is greater than that of each of hydrogen and carbon (the corresponding radial distribution curves are indeed richer in structural information), the cage geometries of 9,12-$(SH)_2$-*closo*-1,2-$C_2B_{10}H_{10}$ (**5a2**) and 1,12-$(SH)_2$-*closo*-1,12-$C_2B_{10}H_{10}$ (**5c1**) should be determined more accurately than the geometries of **5a** and **5c**, respectively, without the need for many constraints or restraints from theoretical calculations. The molecules have overall C_1 and C_2 symmetries, respectively, but local C_{2v} and D_{5d} symmetries for the carborane core were assumed, as this was shown to be a good approximation by calculations at the MP2 levels. As for **5a1** and **5b1**, the C–B and B–B distances in **5a2** and **5c1** were found to be unaffected by the substitutions at the carbon atoms. Armed with the results for a series of *p*-disubstituted benzenes [44], the possible influences of various substituents on the body diagonal, C(1)···C(12), for **5c** were tested. The structures of a series of 1,12-X_2-*closo*-1,12-$C_2B_{10}H_{10}$ molecules [41] [X=H, Li, BeH, BH_2, Me, SiH_3, NH_2, OH and F, as well as SH (**5c1**)] were optimized. The C(1)···C(12) distance ranges from 303.1 pm for X=F, to 317.1 pm for X=Li, *i.e.* the influence of electronegativity is appreciable. It is also informative to look at the B(2)–C(1)···C(12) angles in these derivatives, where a value of twice that angle is analogous to the *ipso* angle in *para*-disubstituted benzene derivatives. The values for the derivatives studies here range from 122.0° for Li to 126.0° for F, also following the trend of electronegativity. A dipole moment study of 1,12-substituted

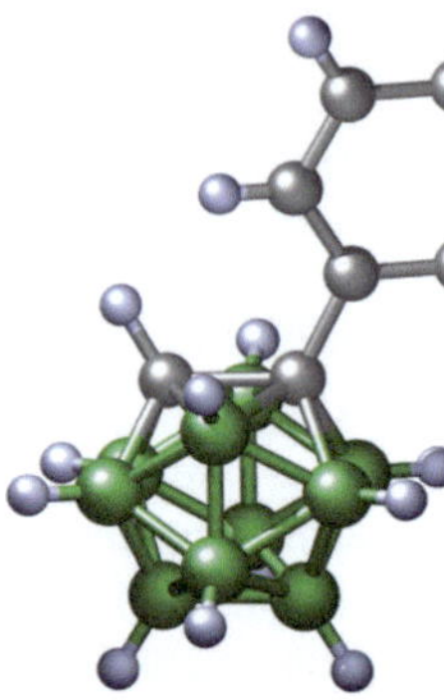

5a1

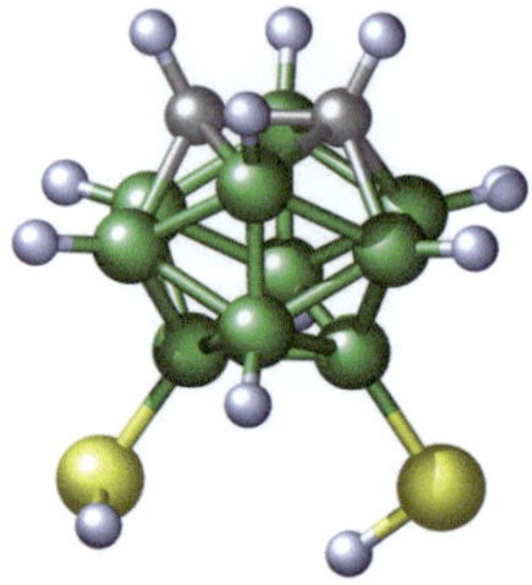

5a2

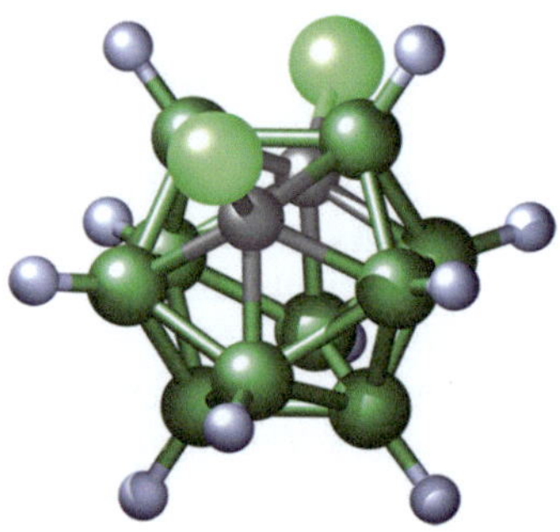

5b1

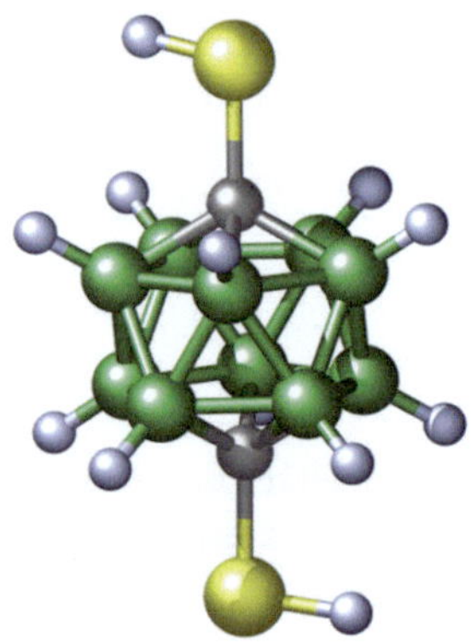

5c1

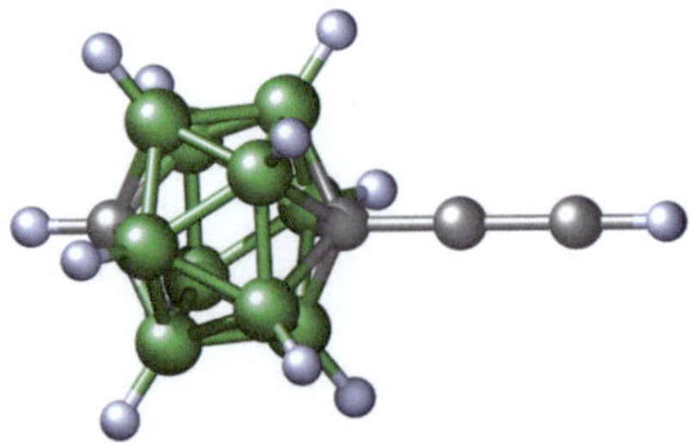

5c2

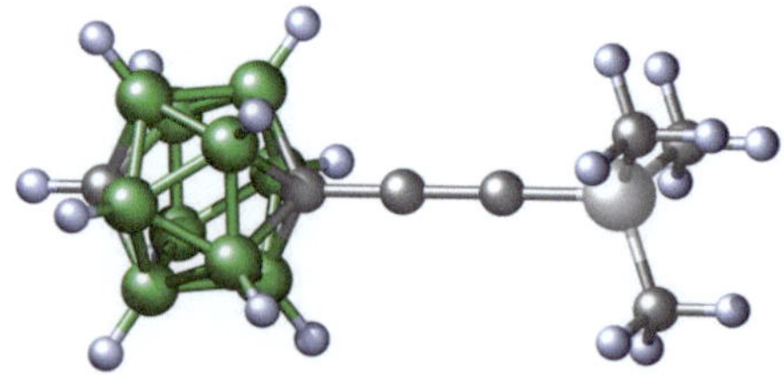

5c3

Fig. 2.5 The molecular structures of *closo*-$C_2B_{10}H_{12}$ derivatives 1-Ph-1,2-*closo*-$C_2B_{10}H_{11}$ (**5a1**), 9,12-$(SH)_2$-*closo*-1,2-$C_2B_{10}H_{10}$ (**5a2**), 1,7-Cl_2-1,7-*closo*-$C_2B_{10}H_{10}$ (**5b1**), 1,12-$(SH)_2$-*closo*-1,12-C2B$_1$0H10 (**5c1**), 1-HC≡C-*closo*-1,12-$C_2B_{10}H_{11}$ (**5c2**), and 1-Me_3SiC≡C-*closo*-1,12-$C_2B_{10}H_{11}$ (**5c3**)

derivatives of **5c** showed, as was the case for **5a**, a slight electron-acceptor ability of **5c** [45]. When a *p*-carborane cage is substituted using a *p*-substituted phenyl group (4-X-C_6H_5), the cage acts as a very weak π-acceptor toward the phenyl if X represents an uncharged group [46].

The gas-phase electron diffraction data for these two C-monoethynyl-*p*-carboranes (**5c2** and **5c3**, prepared at Durham with data collected at Edinburgh) show the same trend: there are very subtle geometrical effects of the ethynyl group on the highly symmetrical cage geometry.

Continuing along the first row of the periodic table, NH is also isoelectronic with $(BH)^{2-}$, and the structure of another compound formally related to (*closo*-$B_{12}H_{12})^{2-}$, 1-aza-*closo*-dodecaborane(12) [*closo*-1-$NB_{11}H_{12}$ (**6**), Fig. 2.6], has been determined by a combination of computational methods and GED (computers provided by the University of Aachen and data recorded at the University of Oslo) [9].

Because the nitrogen atom is so much smaller than the sulfur atom in **3a**, the BN and BB distances calculated *ab initio* lie within a relatively small range of *ca.* 10 pm, which makes the structure determination from GED data much more difficult. In fact, four models with C_{5v} symmetry fit the data almost equally well. The ^{11}B NMR chemical shifts have been used to decide which of these possibilities is the most reasonable. The final experimental geometry was selected on the basis of the best agreement between the IGLO/DZ//GED and experimental ^{11}B chemical shifts. This was the first time that this method had been employed as an additional refinement condition in conjunction with GED structure determination. The single-point energy calculations at the MP2/6-31G* level supported this observation. The elongation of the B(2)–B(3) bonds in **6** [182.5(6) pm] is less pronounced than for **3a**, and the NB_5 pentagonal pyramid is flattened as a consequence of the short B–N bond [171.6(9) pm]. All of these differences can be attributed to the smaller size of the nitrogen substituent relative to sulfur.

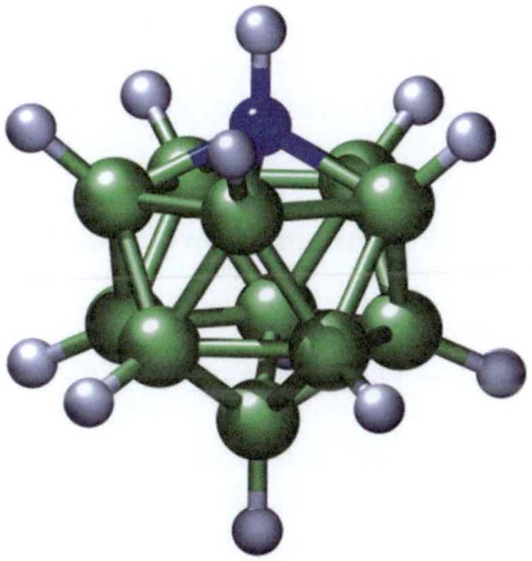

6

Fig. 2.6 The molecular structure of *closo*-1-$NB_{11}H_{12}$ (**6**)

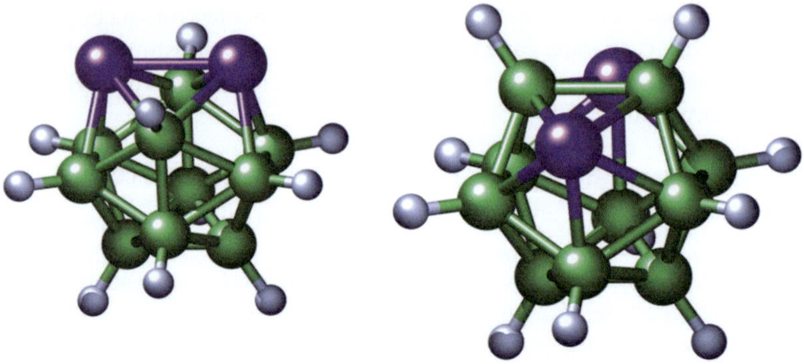

7a **7b**

Fig. 2.7 The molecular structures of *closo*-1,2-$P_2B_{10}H_{10}$ (**7a**) and *closo*-1,7-$P_2B_{10}H_{10}$ (**7b**)

It is also possible for bare phosphorus atoms to replace $(BH)^{2-}$ vertices in (*closo*-$B_{12}H_{12})^{2-}$. Substituted species that have been prepared include 1,2-diphospha-closo-dodecaborane(10) [*closo*-1,2-$P_2B_{10}H_{10}$ (**7a**)], a few of its monochloro and dichloro derivatives such as 3-Cl, 4-Cl, 3,6-Cl_2 and 3,4-Cl_2 compounds, and 1,7-diphospha-*closo*-dodecaborane(10) [*closo*-1,7-$P_2B_{10}H_{10}$ (**7b**) Fig. 2.7]. The hypothetical 1,12-diphospha-*closo*-dodecaborane(10) (*closo*-1,12-$P_2B_{10}H_{10}$) was also studied computationally.

The structures of all of these icosahedral cages have been determined by an *ab initio*/GIAO/NMR method, *viz.* GIAO-HF/II//MP2/6-31G* [47]. The presence of phosphorus at two vertices in **7a** and **7b** causes considerable distortion of the parent icosahedral skeleton (*closo*-1,12-$P_2B_{10}H_{10}$ being distorted along its body diagonal). For example, the B(3)–P(2)–B(6) and B(8)–P(7)–B(11) angles are reduced from the ideal 108.0 to 93.0 and 93.5°, reflecting the large size of the phosphorus atoms, which, like sulfur, have long bonds to their neighboring boron atoms. As in the preceding cases, the very good agreement between the computed and experimental ^{11}B NMR chemical shifts indicates that the geometries of **7a** with its monochloro and dichloro derivatives and **7b**, calculated at the MP2/6-31G* level, may be accepted as reliable representations of their solution-state structures.

2.3.2.2 Ten- and Eleven-Vertex *Closo* Structures

Just as replacement of one or two $(BH)^{2-}$ vertices of (*closo*-$B_{12}H_{12})^{2-}$ results in a number of twelve-vertex heteroboranes, the same is true for (*closo*-$B_{11}H_{11})^{2-}$, the eleven-vertex *closo* (C_{2v} symmetric) system. The corresponding heteroboranes are represented by, for example, *closo*-2,3-$C_2B_9H_{11}$ (**8**, Fig. 2.8, prepared at Durham). The structure of this C_s-symmetrical dicarbaborane has been successfully determined by using the *SARACEN* method, using restraints derived from the results of MP2/ 6-311+G* calculations [48]. It has been shown that increasing the size of the basis set from 6-31G* to 6-311G* to 6-311+G* at the MP2 level has little effect on the structural parameters, and DFT calculations behave in the same way. As pointed out earlier, this is the level of theory that is necessary for reproducing experimental data adequately.

The bicapped-square antiprismatic arrangement is known to be the basic building block for ten-vertex *closo* systems. The parent compound is represented by (*closo*-$B_{10}H_{10}$)$^{2-}$, which adopts D_{4d} symmetry. Again, by replacing $(BH)^{2-}$ vertices (although that is not as simple in the synthetic route), a number of ten-vertex *closo* species can be formed. As with **3a**, incorporation of sulfur leads to a C_{4v}-symmetric *closo*-thiaborane, 1-thia-*closo*-decaborane(9) [*closo*-1-SB_9H_9 (**9**)] as shown in Fig. 2.9.

There have been two studies dedicated to the molecular structure of **9**. First, the microwave spectrum of **9** was investigated [49] and, as was the case for **3a**, a precise substitution structure of the non-hydrogen atoms was determined. The most

Fig. 2.8 The molecular structure of *closo*-2,3-$C_2B_9H_{11}$ (**8**)

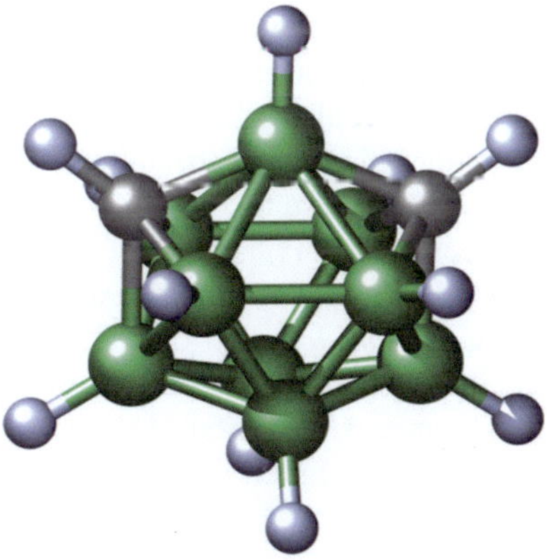

8

Fig. 2.9 The molecular structure of *closo*-1-SB_9H_9 (**9**)

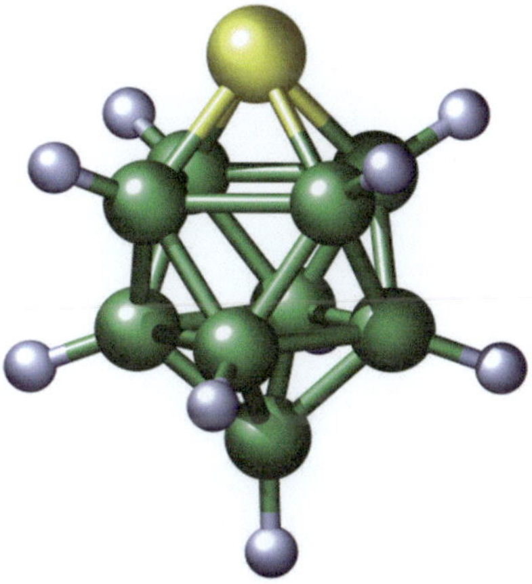

9

striking feature was a substantial expansion of the boron square adjoining sulfur, with r[B(2)–B(3)] = 193.7(1) pm, which is slightly longer than in **3a**. Supplementary high-level *ab initio* (MP2/6-311G**) and DFT calculations (B3LYP/6-311G** and B3LYP/cc-pVTZ) have confirmed this result. GED data have also been collected, and the *SARACEN* method has been applied [50], with results consistent with those obtained by theory and by rotational spectroscopy. The theoretical and GED geometries have been used for magnetic property calculations, but the MW geometry was not used, because the hydrogen-atom positions had not been determined. The three GIAO-HF and GIAO-MP2 sets of $\delta(^{11}B)$ values are in good agreement with experiment, but the latter approach is superior to the former in accounting for the chemical shift of B(10), which has the value 74.5 ppm (the difference from GIAO-MP2 is up to 1 ppm, depending on the geometry used). This is one of the most extreme ^{11}B chemical shifts to high frequency, this atom being antipodally coupled with sulfur, in a similar manner to the B(12)···S pair in **3a**, in which such a chemical shift is measured to be 18.4 ppm in $CHCl_3$. Such a difference of almost 60 ppm is accounted for by the occurrence of paramagnetic contributions to the magnetic shielding constants. These contributions arise from the coupling of suitable occupied and unoccupied molecular orbitals with large coefficients on B(10) and B(12), respectively, the latter being more pronounced in the case of ten-vertex thiaborane.

As in the case of $(B_{12}H_{12})^{2-}$, one $(BH)^{2-}$ vertex can be formally replaced by an isoelectronic $(CH)^-$ unit, resulting in (*closo*-1-$CB_9H_{10})^-$ [51], while substitution at two $(BH)^{2-}$ vertices provides three isomeric *closo* ten-vertex dicarbaboranes: *closo-1*,2-$C_2B_8H_{10}$, *closo*-1,6-$C_2B_8H_{10}$, and *closo*-1,10-$C_2B_8H_{10}$. Whereas the structure of the last molecule has been determined by GED alone [52], we have examined the first isomer, 1,2-dicarba-*closo*-decaborane(10) (**10**, Fig. 2.10) using the joint *ab initio*/IGLO/GED method [53].

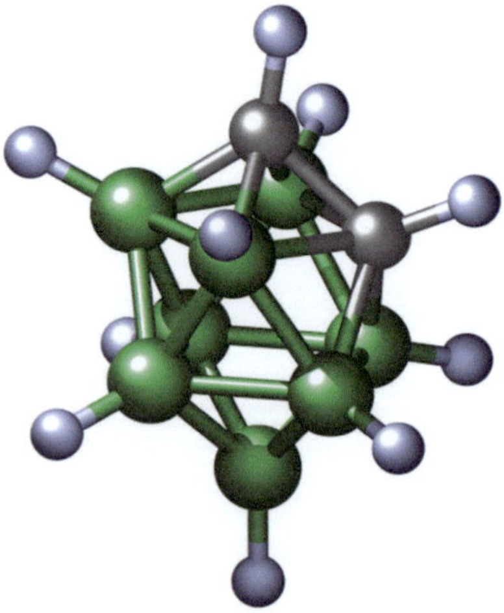

10

Fig. 2.10 The molecular structure of *closo*-1,2-$C_2B_8H_{10}$ (**10**)

In this case the electron-diffraction data were analyzed using the *MOCED* approach. This molecule has C_s symmetry only, and conversion of the molecular scattering intensity to the radial distribution curve indicated that the data were very poor in structural information. Using GED data alone would have not given a realistic chance of determining the structure, and so the differences between similar bond lengths were fixed at the values calculated at the MP2/6-31G* level. The C–C bond length refined to a typical "alkane" value of 153.8(8) pm and leads to distortion from the regular bicapped-square antiprismatic shape. The displacement of the carbon atom towards the center of the cluster results in a substantial opening of the B(3)–C(2)–B(5) bond angle to 95° from the equivalent value in the parent compound of 90°. IGLO/DZ calculations have confirmed the reliability of both the experimental and theoretical (MP2/6-31G*) geometries.

As in the twelve-vertex *closo* system, phosphorus can also occupy positions in the bicapped-square antiprismatic skeleton. Two isomeric ten-vertex *closo* phosphaboranes [2,1-*closo*-PCB_8H_9 (**11a**) and 6,1-*closo*-PCB_8H_9 (**11b**) both with C_s symmetry and shown in Fig. 2.11] have been prepared [54]. Similarly as for **10**, the phosphorus atoms are pushed away from the center of the cluster relative to the positions that they would adopt in a regular bicapped-square antiprism, and so the B(3)–P(2)–B(5) and B(7)–P(6)–B(9) angles in **11a** and **11b**, respectively, are narrowed by *ca.* 13° from the 90° expected in the regular polyhedron.

2.3.3 Nido *Heteroboranes*

By removing one BH vertex from a *closo* system, a *nido* skeleton is formally derived. As with the *closo* family, $(BH)^{2-}$ vertices can be replaced by heteroatom-based moieties such as $(CH)^-$, S, NH, and P. We have been able to elucidate the

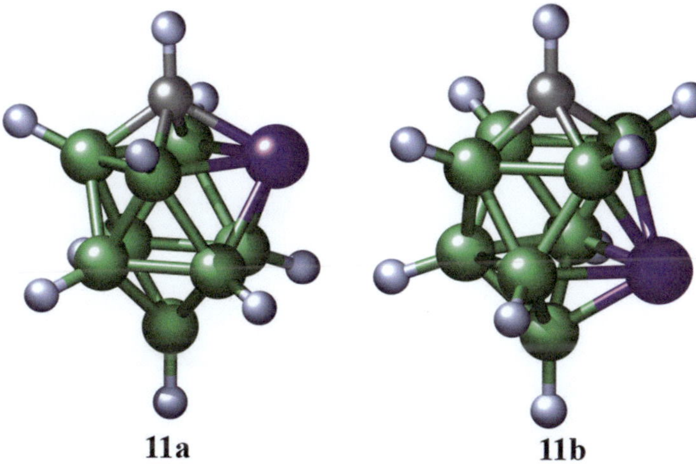

11a **11b**

Fig. 2.11 The molecular structures of the ten-vertex *closo* phosphaboranes 2,1-*closo*-PCB_8H_9 (**11a**) and 6,1-*closo*-PCB_8H_9 (**11b**)

structure of such a *nido* skeleton for the first time by using the joint *ab initio*/GED approach (data recorded in Oslo), using the *MOCED* method. The compound studied was the C_s-symmetric 7,8-dicarba-10-thia-*nido*-undecaborane(10) [7,8,10-$C_2SB_8H_{10}$ (**12**) shown in Fig. 2.12] [55].

Distortion of the C(7)C(8)B(9)S(10)B(11) open pentagonal ring is quite pronounced. For example, the B(9)–S(10)–B(11) angle is very narrow [GED: 93.1(6)°; MP2/6-31G*: 96.6°] compared to the 108.0° of a regular pentagonal ring. As a consequence, the sulfur atom lies slightly out of the plane of the open pentagonal ring that exists in the hypothetical $(nido\text{-}B_{11}H_{11})^{4-}$. IGLO/DZ calculations performed with both the GED and MP2/6-31G* geometries confirm the high quality of both structures.

The experimental structure of another eleven-vertex *nido* skeleton, that of *nido*-2,9-$C_2B_9H_{13}$ (**13**, Fig. 2.13, prepared at Durham), was refined in C_s symmetry using the *SARACEN* method. No significant deviations of nearest-neighbor BB and CB separations, or of terminal and bridging B–H bond lengths, from those calculated at the MP2/6-311+G* level of theory were observed [48].

Fig. 2.12 The molecular structure of 7,8,10-*nido*-$C_2SB_8H_{10}$ (**12**)

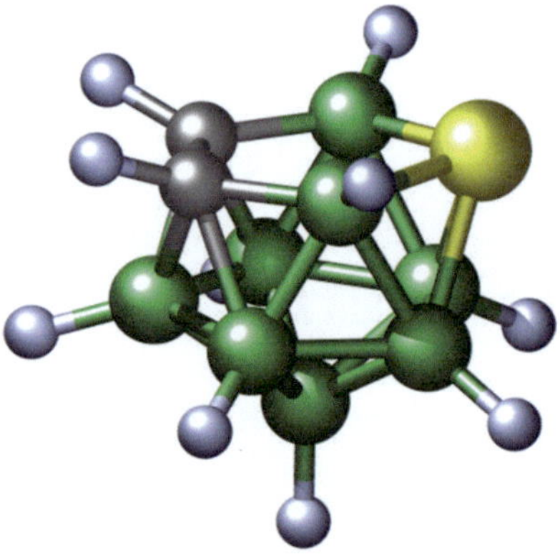

12

Fig. 2.13 The molecular structure of *nido*-$C_2B_9H_{13}$ (**13**)

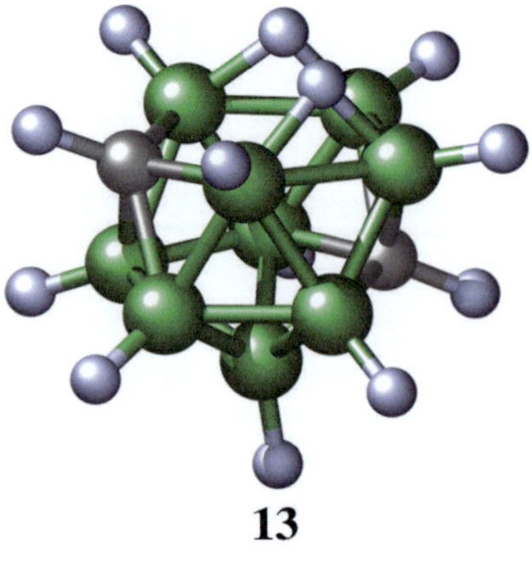

13

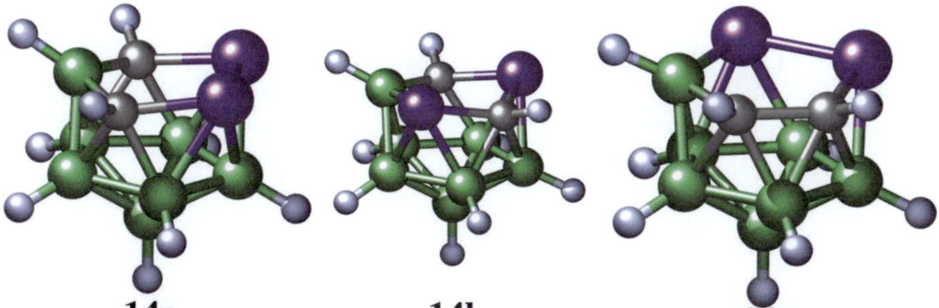

Fig. 2.14 Molecular structures of the tetrahetero analogues of 7,8,10-*nido*-$C_2SB_8H_{10}$, namely 7,8,9,11- (**14a**), 7,9,8,10- (**14b**) and 7,8,9,10-*nido*-$P_2C_2B_7H_9$ (**14c**)

There are also neutral compounds with substituents for boron at four vertices in the parent B_5 open pentagon, with two carbon and two phosphorus atoms, *i.e. nido*-$P_2C_2B_7H_9$ in the form of 7,8,9,11- (**14a**), 7,9,8,10- (**14b**) and 7,8,9,10- (**14c**) isomers (Fig. 2.14). The last possible isomer, 7,10,8,9-, has so far only been examined computationally, because its energy is the highest of the four C_2P_2 isomers, and it has not yet been accessed experimentally [56]. As expected, the presence of two phosphorus atoms leads to significant distortion of the five-atom ring, and the bond angles within it are either around 115 or 96°. Some dihedral angles within these B_5 rings are as high as 20–28°.

2.3.4 Arachno *Heteroboranes*

Arachno compounds are formally derived from *nido* ones in the same manner as *nido* structures are from *closo*, that is by removing one BH vertex from the *nido* skeleton. For example, the hypothetical *nido*-$B_{11}H_{11}^{4-}$ yields the hypothetical *arachno*-$B_{10}H_{10}^{6-}$. The latter should exhibit a hexagonal boat-like shape, which is present in *arachno*-6,9-$C_2B_8H_{14}$ (**15a**) [57] and *arachno*-6,9-CSB_8H_{12} (**15b**) [58], studied using the *SARACEN* method. Computational studies have also been performed for some analogues of **15a** and **15b**, *i.e. arachno*-6,9-$N_2B_8H_{12}$ (**15c**) [57], *arachno*-6,9-$Se_2B_8H_{10}$ (**15d**) [57], and *arachno*-6,9-CNB_8H_{13} (**15e**) [58] (all shown in Fig. 2.15). The presence of heavy atoms (*i.e.* S and Se) brings about a considerable narrowing of the B–S(Se)–B angles. On the other hand, short N–B distances are responsible for flattening **15c** and part of **15e** with respect to (*arachno*-$B_{10}H_{14}$)$^{2-}$.

Apart from ten-vertex *arachno* molecules, there are structurally characterized *arachno* systems of smaller dimensions as exemplified by *arachno*-4,6-$C_2B_7H_{13}$ (**16a**) [59], *arachno*-4,6-$S_2B_7H_9$ (**16b**) [59], and *arachno*-4,6,5-$C_2SB_6H_{10}$ (**16c**) [60]. Their structures (which have C_s symmetry) have been elucidated by the *SARACEN* method (in the case of **16a** and **16b**) and using the *ab initio*/IGLO/NMR method (for **16c**) (Fig. 2.16).

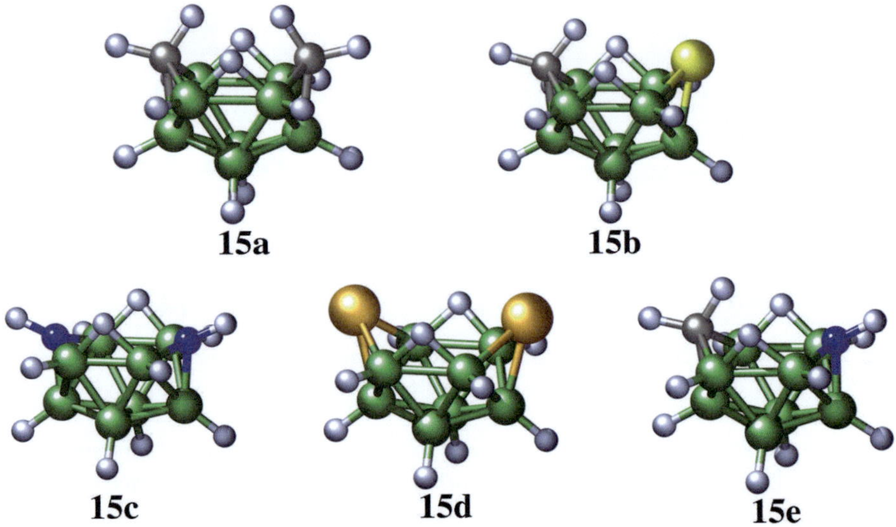

Fig. 2.15 Molecular structures of the *arachno* boat-like structures *arachno*-6,9-$C_2B_8H_{14}$ (**15a**), *arachno*-6,9-CSB_8H_{12} (**15b**), *arachno*-6,9-$N_2B_8H_{12}$ (**15c**), *arachno*-6,9-$Se_2B_8H_{10}$ (**15d**), and *arachno*-6,9-CNB_8H_{13} (**15e**)

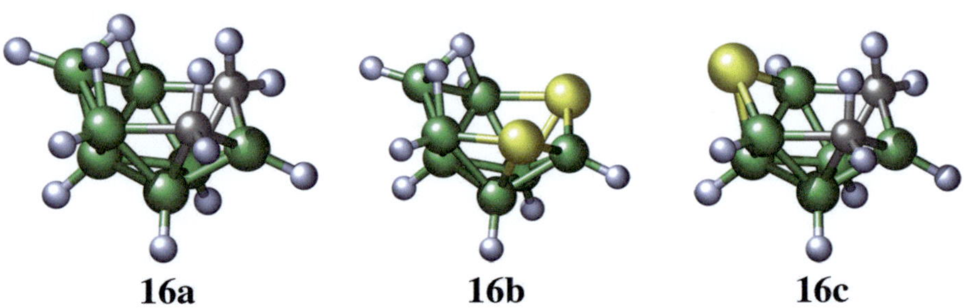

Fig. 2.16 Molecular structures of the nine-vertex *arachno* species *arachno*-4,6-$C_2B_7H_{13}$ (**16a**), *arachno*-4,6-$S_2B_7H_9$ (**16b**), and *arachno*-4,6,5-$C_2SB_6H_{10}$ (**16c**)

The derivation of the GED model for **16a** follows essentially the same route by which it was synthesized, involving the removal of three BH vertices from **5b**. In essence, the molecular shapes of **16a**-**16c** stem from the preceding family of *arachno* heteroboranes with one vertex missing.

2.3.5 *B_4 Clusters*

Tetraborane(10) or *arachno*-B_4H_{10} was reacted at the University of Leeds with ethene to produce the so-called 2,4-ethanotetraborane(10) [2,4-$CH_2CH_2B_4H_8$ (**17a**) shown in Fig. 2.17] [61], in which the "wing" boron atoms of the butterfly-shaped

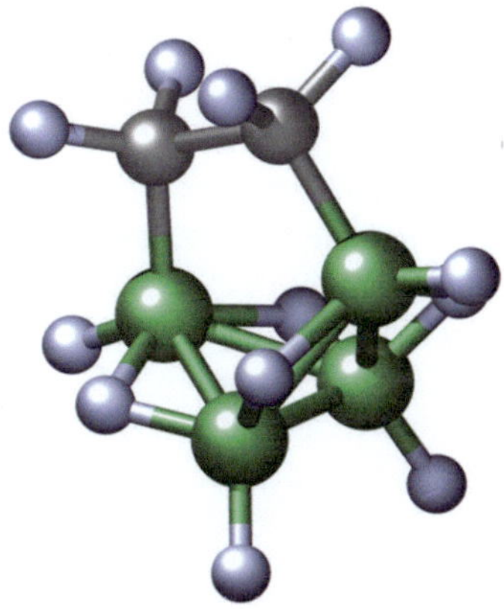

17a

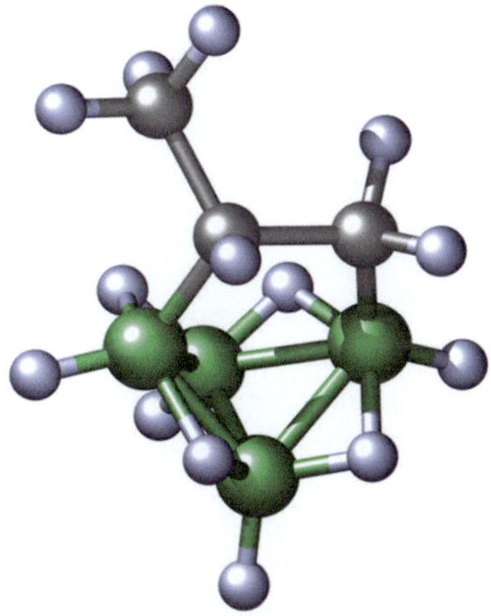

17b

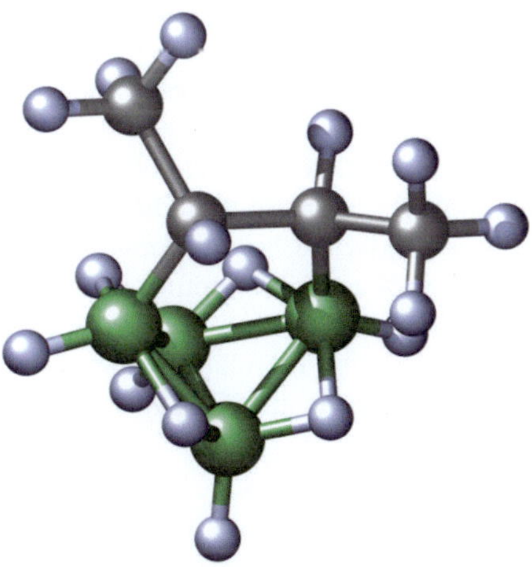

17c

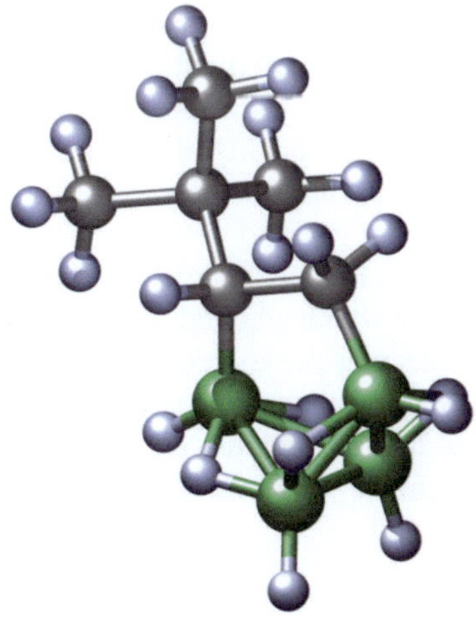

17d

Fig. 2.17 Molecular structures of ethano tetraboranes 2,4-$CH_2CH_2B_4H_8$ (**17a**), 2,4-$MeCHCH_2B_4H_8$ (**17b**), 2,4-(*trans*-MeCHCHMe)B_4H_8 (**17c**), and [2,4-(*t*-$BuCHCH_2$)B_4H_8 (**17d**)

B_4H_8 unit are attached to the C_2H_4 moiety. This structure, which has C_{2v} symmetry, has been determined by the *MOCED* approach and both GED and MP2/6-31G* geometries have been verified by IGLO ^{11}B chemical shifts. The C–C bond length is calculated at the MP2/6-31G* level to be 155.4 pm, and is 156.8 pm determined by the GED experiment, although this parameter was fixed in the final GED refinement. The value is towards the long end of the range exhibited by alkanes. Tetraborane(10) not only reacts with ethene, but also with substituted ethenes such as $MeCHCH_2$ and *trans*-MeCHCHMe, to give 2,4-$MeCHCH_2B_4H_8$ [62], (**17b**), and 2,4-(*trans*-MeCHCHMe) B_4H_8 [62], (**17c**), respectively. Both systems have been studied in the same way as **17a**, and their C_2B_4 cages are only slightly distorted from C_{2v} local symmetry, with twist angles of 0.5° and 0.8°, respectively, for the C(5)–C(6) bonds about the pseudo-C_2 axis. The *SARACEN* approach has also been applied to determine the structure of 2,4-(*t*-butylethano)tetraborane(10) [2,4-(*t*-$BuCHCH_2$) B_4H_8 (**17d**) as shown in Fig. 2.17 together with all GED structures of ethano tetraboranes; the sample was provided by the University of Leeds] [63]. The effect of the *t*-Bu group is quite marked, as the C–C bond of the C_2B_4 core is twisted by 6.6(14)°. As a consequence, the local symmetry of the C_2B_4 moiety is reduced from C_{2v} to C_2, the concomitant distortion of the B_4H_8 group from C_{2v} local symmetry being negligible.

In addition to the clusters complying with the rule that $n+1$ electron pairs are optimal for *n*-deltahedral framework bonding [64], there are boron clusters that represent exceptions to this rule; the compounds B_4X_4, X = Cl, R, are therefore of interest. Gas electron diffraction investigations (possible without using any computed data to assist the refinements) of both tetrachloro tetrabora-tetrahedrane, B_4Cl_4 (**18a**) [65], and tetra-*tert*-butyltetrabora-tetrahedrane [$B_4(^tBu)_4$, **18b**] [66] have been carried out under the conditions of T_d and T symmetry, respectively, and their structures are shown in Fig. 2.18. The high symmetries of these systems have ensured the quality of experimental geometry determined by GED alone. (The samples for these studies were provided by the University of Oxford and Aachen, respectively, and data were recorded in Edinburgh and Oslo, respectively.) In particular, torsional angles in **18b** have been determined accurately, including τ(B–B–C–C) which yielded a value of 30.3(3)°. Interestingly, the boron atoms in both molecules are extremely deshielded, with $\delta(^{11}B)$ 85.5 and 135.4 ppm for **15a** and **15b**, respectively. These values are satisfactorily reproduced computationally only by taking dynamic electron correlation into account; the calculations then yield chemical shifts of 81.4 and 139 ppm, respectively, using the GIAO-MP2/TZP′ method.

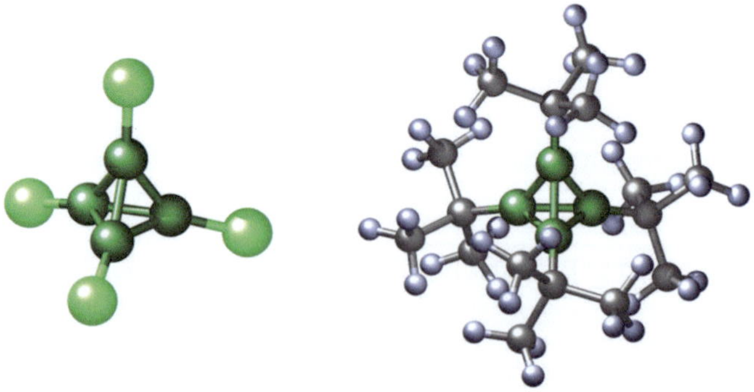

18a **18b**

Fig. 2.18 The molecular structures of tetrachloro tetrabora-tetrahedrane, B_4Cl_4 (**18a**) and tetra-tert-butyltetraboratetetrahedrane (**18b**)

2.3.6 *Shared Icosahedra*

As well as main-group elements that can be incorporated into boron frameworks, it is also possible to have metal atoms as parts of such clusters, known as metallaheteroboranes. There is a plethora of such deltahedral compounds, most prominent among them being the so-called *commo*-bis(icosahedral)metallacarbaboranes of the type $[M(1,2\text{-}C_2B_9H_{11})_2]^n$, in which two icosahedra are conjoined by a shared metal-atom vertex with the resulting charge of *n*. The most investigated and best known is the anionic $[3\text{-}Co\text{-}(1,2\text{-}C_2B_9H_{11})_2]^-$, usually referred to as the cobalt bis(dicarbollide) ion. Two dicarbollide anions are complexed with Ni^{4+} to give a neutral sandwich system. A DFT study at the BP86/AE1 level [67] (shown in Fig. 2.19) has found that the mutual rotation of the two dicarbollide moieties is facile. Note that such a molecular shape is based on sharing two icosahedra through one vertex (Ni^{4+}). Experimental ^{11}B NMR chemical shifts are reproduced to better than 3 ppm at the GIAO-B3LYP/II′ level. The same DFT approach has been employed for other heteroboranes in this study but none of them is neutral. These metallaboranes have been found to be quite promising materials in medicine and molecular electronics, with **19** appearing to act as a molecular rotor [68].

Fig. 2.19 The molecular structure of the most stable rotamer of $[3\text{-}Ni(1,2\text{-}C_2B_9H_{11})_2]$ (**19**)

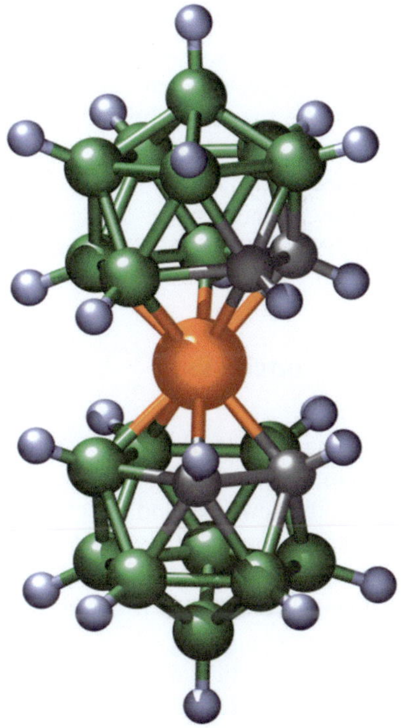

19

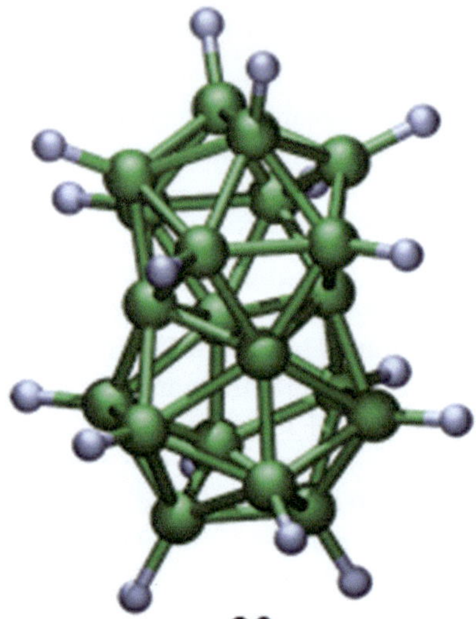

20

Fig. 2.20 Molecular structure of the macropolyhedral cluster, $B_{20}H_{16}$ (**20**)

In a similar way to two benzene rings fusing to form naphthalene, the combination of two *closo*-$B_{12}H_{12}^{2-}$ ions forms so-called macropolyhedral clusters (other parent boron hydrides can also be shared), the molecular geometries of which depend on the mode of sharing. Apart from the aforementioned one-vertex sharing, there are other modes of two-icosahedra sharing. Icosahedra can be joined through a common edge, or where three or four vertices come together. The latter is exemplified in the macropolyhedral $B_{20}H_{16}$, the first synthesized *closo* macropolyhedron [69]. Indeed, the latter has four joint vertices of two shared *closo*-$B_{12}H_{12}^{2-}$ icosahedra (see **20** as shown in Fig. 2.20). This unique cluster was obtained relatively recently in another way and was structurally characterized by applying the GIAO-B3LYP/II//MP2/6-31G* computational protocol [70].

2.4 Conclusions

This chapter has illustrated the use of structural information to better understand various series of neutral borane and heteroborane clusters, which were prepared in Řež and elsewhere, and structurally characterized in Edinburgh and/or Řež. Much of what is described involved the application of gas-phase electron diffraction and calculations of ^{11}B NMR chemical shifts (used as an additional electron-diffraction refinement condition) to yield molecular geometries of various species that were validated against experimental NMR data. Such gas-phase (and computed) structures are of particular significance since the free molecules can be considered unperturbed and the resulting macro-structure is influenced exclusively by intramolecular forces. The systems that contain a heavy element (through hydrogen or vertex substitution) are suitable targets for structural studies employing GED alone, and the corresponding results may thus verify the reliability of theoretical approaches. The gas-phase structures under scrutiny turned out to be quite rigid as demonstrated by the vibrational amplitudes obtained; this observation is particularly true for *closo* clusters, which have electron density distributions that are counterintuitive to the concept of electronegativity as also revealed by the analyses of the experimental dipole moments. (Indeed, carbon is more electronegative than boron but its charge is positive, and the same applies to sulfur.) In addition to the structural studies that have been, and will be, performed for the series of boron clusters described, there is great potential for study of the macropolyhedral clusters (see Fig. 2.17). Some of these have been known for decades, but accurate structural studies of them are entirely lacking. They are therefore a challenging target for applications both of GED and computational protocols in this demanding but important area of boron cluster chemistry.

On the basis of electron distribution in 12-Ph-*closo*-1-$SB_{11}H_{10}$, a new type of non-classical σ-hole-based non-covalent interaction, the so-called chalcogen bonding, has been very recently discovered, which offers a promising use for *closo-heteroboranes* with V and VI group elements in crystal engineering and drug design [71]. There are other areas where the use of boron clusters in medicine and materials science also looks promising. Whereas some metallaboranes appear to be potent inhibitors of various enzymes, some thiolated carbaboranes act as modifiers of layers of various metals [72]. In this context, the synthetic efforts aimed at development of further *closo-*, *nido- arachno-* and macropolyhedral clusters along with the subsequent molecular structure determinations is obvious.

Generally speaking, the joint endeavor of studying synthesis and structure in a concerted manner is worth striving for because "*there is no more basic enterprise in chemistry than the determination of the geometrical structure of a molecule. Such a determination, when it is well done, ends all speculation as to the structure and provide us with a starting point for the understanding of every physical, chemical and biological property of the molecule*", Roald Hoffmann, Nobel Prize Winner in Chemistry (1981) [73].

References

1. Bauer SH (1937) The structure of diborane. J Am Chem Soc 59:1096–1103
2. Hedberg K, Schomaker V (1951) A reinvestigation of the structures of diborane and ethane by electron diffraction. J Am Chem Soc 73:1482–1487
3. Štíbr B (1992) Carboranes other than $C_2B_{10}H_{12}$. Chem Rev 2:225–250; Plešek J (1992) Potential applications of the boron cluster compounds. Chem Rev 2:269–278; Heřmánek S (1992) ^{11}B NMR spectra of boranes, main-group heteroboranes, and substituted derivatives. Factors influencing chemical shifts of skeletal atoms. Chem Rev 2:325–362. The first IMEBORON (International Meeting on Boron Chemistry) was held in the former Czechoslovakia, Liblice Castle, 1971. This was repeated in 1987 at Bechyně Castle. The 3rd EUROBORON (European Meeting on Boron Chemistry) was held in the Czech Republic, Průhonice-by-Prague, 2004. The 15th IMEBORON was held in Prague in August 2014

4. Microwave spectroscopy is also of some importance, but only for molecules with permanent dipole moments; the large number of $^{10}B/^{11}B$ isotopomers can cause problems. For an overview of the technique see, for example, Townes CH, Schawlow AL (1975) Microwave spectroscopy. Courier Dover Publications, North Chelmsford
5. Hargittai I (1988) In: Hargittai I, Hargittai M (eds) Stereochemical applications of gas-phase electron diffraction: part A. VCH Publishers, Weinheim/New York
6. For example, the molecule (COD)Cu(hfac) is quite tricky to model: Hnyk D, Bühl M, Brain PT, Robertson HE, Rankin DWH (2002) The gas-phase structure of (1,1,1,5,5,5,-hexafluoro-2,4-pentanedionato)(η-1,5-cyclooctadiene)-copper(I), Cu(1,5-cod)(hfac), an important precursor for vapor deposition of copper. J Am Chem Soc 124:8078–8084. The same applies to, for example, CEt_4: Alder RW, Allen PR, Hnyk D, Rankin DWH, Robertson HE, Smart BA, Gillespie RJ, Bytheway I (1999) Molecular structure of 3,3-diethylpentane (tetraethylmethane) in the gas phase as determined by electron diffraction and *ab initio* calculations. J Org Chem 64:4226–4232. In contrast, just one and two parameters are needed to model CO and H_2O, respectively. Interestingly, just two parameters are also needed to model C_{60} under the symmetry of I_h, see Hedberg K, Hedberg L, Bethune DS, Brown CA, Dorn HC, Johnson RD, de Vries M (1991) Bond lengths in free molecules of buckminsterfullerene, C_{60}, from gas-phase electron diffraction. Science 254:410–412
7. Klimkowski VJ, Ewbank JD, van Alsenoy C, Schäfer L (1982) Molecular orbital constrained electron diffraction studies. 4. Conformational analysis of the methyl ester of glycine. J Am Chem Soc 104:1476–1480
8. Blake AJ, Brain PT, McNab H, Miller J, Morrison CA, Parsons S, Rankin DWH, Robertson HE, Smart BA (1996) Structure analysis restrained by *ab initio* calculations: the molecular structure of 2,5-dichloropyrimidine in gaseous and crystalline phases. J Phys Chem 100:12280–12287; Mitzel NW, Rankin DWH (2003) SARACEN-Molecular structure from theory and experiment: the best of both worlds. Dalton Trans 3650–3662
9. This approach was first introduced in the structure analysis for *closo*-1-$NB_{11}H_{12}$: Hnyk D, Bühl M, Schleyer PvR, Volden HV, Gundersen S, Müller J, Paetzold P (1993) The molecular structure of 1-aza-*closo*-dodecaborane(12). Experimental and theoretical refinement. Inorg Chem 32:2442–2445
10. Individual Gauge for Localised Orbitals: Kutzelnigg W, Fleischer U, Schindler M (1990) In: NMR basic principles and progress, vol 23. Springer, Berlin/Heidelberg, p 165
11. Gauge Including Atomic Orbitals: Ditchfield R (1974) Self-consistent perturbation theory of diamagnetism. I. A gauge-invariant LCAO method for NMR chemical shifts. Mol Phys 27:789–807
12. Binkley JS, Pople JA (1975) Møller-Plesset theory for atomic ground state energy. Int J Quantum Chem 9:229–236 and references cited therein
13. Huzinaga S (1971) Approximate atomic wave functions. University of Alberta, Edmonton
14. Hehre W, Radom L, Schleyer PR, Pople JA (1986) *Ab initio* molecular orbital theory. Wiley, New York
15. Becke AD (1993) Density-Functional Thermochemistry. III. The Role of Exact Exchange. J Chem Phys 98:5648–5652; Lee C, Yang W, Parr RG (1988) Development of the Colle-Salvetti correlation-energy formula into a functional of the electron density. Phys Rev B 37:785–789
16. Becke AD (1988) Density-functional exchange-energy approximation with correct asymptotic-behavior. Phys Rev A 38:3098–100; Perdew JP (1986) Density-functional approximation for the correlation energy of the inhomogeneous electron gas. Phys Rev B 33:8822–8824
17. Exner O (1975) Dipole moments in organic chemistry. Thieme Verlag, Stuttgart
18. See, for example, Schindler M (1987) Magnetic properties in terms of localized quantities. 5. Carbocations. J Am Chem Soc 109:1020–1033
19. Beaudet RA (1988) The molecular structures of boranes and carboranes. In: Liebmann JF, Greenberg A, Williams RE (eds) Advances in boron and the boranes. VCH Publishers, New York, p 417
20. Bühl M, Schleyer PR (1992) Application and evaluation of *ab initio* chemical shift calculations for boranes and carboranes. How reliable are "accurate" experimental structures? J Am Chem Soc 114:477–491

21. Mastryukov VS (1988) In: Hargittai I, Hargittai M (eds) Stereochemical applications of gas-phase electron diffraction Part B. VCH Publishers, Weinheim/New York
22. Schleyer PR, Bühl M, Fleischer U, Koch W (1990) Theoretical refinement of the B_5H_{11} structure. Application of IGLO chemical shift calculations. Inorg Chem 29:153–155. In fact, the C_1 structure of this pentaborane was suggested much earlier using MP2 model chemistry, see McKee ML, Lipscomb WN (1981) Probable fluxional behavior in B_5H_{11}. A theoretical study which supports C_1 symmetry. Inorg Chem 20:4442–4444
23. Greatrex R, Greenwood NN, Rankin DWH, Robertson HE (1987) The molecular structures of pentaborane(9) and pentaborane(11) in the gas phase as determined by electron diffraction and theoretical calculations. Dalton Trans 1849–1858
24. Greatrex R, Greenwood NN, Millikan MB, Rankin DWH, Robertson DWH (1988) The Molecular structure of hexaborane(12) in the gas phase as determined by electron diffraction. J Chem Soc Dalton Trans 2335–2339
25. Bühl M, Schleyer PR (1990) Application of the combined *ab initio*/IGLO/NMR method to resolve the B_6H_{12} structural question. Angew Chem Int Ed 29:886–888
26. Brain PT, Hnyk D, Rankin DWH, Bühl M, Schleyer PR (1994) The molecular structures of pentaborane(11), B_5H_{11}, and hexaborane(12), B_6H_{12}, in the gas phase as determined by electron diffraction and *ab initio* calculations. Polyhedron 13:1453–1466
27. $F_2BB_5H_8$: Brain PT, Rankin DWH, Robertson HE, Alberts IL, Hofmann M, Schleyer PvR (1994) Molecular structure of 1-(difluoroboryl)pentaborane(9), 1-$(F_2B)B_5H_8$, in the gas phase as determined by electron diffraction and supported by *ab initio* and IGLO calculations. Inorg Chem 33: 2565–2571; $Cl_2BB_5H_8$: Brain PT, Rankin DWH, Robertson HE, Alberts IL, Downs AJ, Greene TM, Hofmann M, Schleyer PvR (1995) Molecular structure of 1-(dichloroboryl) penta borane(9), 1-$(Cl_2B)B_5H_8$, in the gas phase as determined by electron diffraction and supported by theoretical calculations. J Chem Soc Dalton Trans 2193–2199
28. Hnyk D, Vajda E, Bühl M, Schleyer PR (1992) On the molecular structure of 1-thia-*closo*-dodecaborane(11) studied by electron diffraction complemented by *ab initio* calculations. Inorg Chem 31:2464–2467
29. Zahradník R, Balaji V, Michl J (1991) An SCF study of 10-vertex and 12-vertex boranes and heteroboranes. J Comput Chem 12:1147–1156
30. Møllendal H, Samdal S, Holub J, Hnyk D (2003) Structure of 1-thia-*closo*-dodecaborane(11), 1-$SB_{11}H_{11}$, as determined by microwave spectroscopy complemented by quantum chemical calculations. Inorg Chem 42:3043–3046
31. Macháček J, Plešek J, Holub J, Hnyk D, Všetečka V, Císařová I, Kaupp M, Štíbr B (2006) New route to 1-thia-*closo*-dodecaborane(11), *closo*-1-$SB_{11}H_{11}$, and its halogenation reactions. The effect of the halogen on the dipole moments and the NMR spectra and the importance of spin-orbit coupling for the ^{11}B chemical shifts. Dalton Trans 1024–1029
32. Hehre WJ, Ditchfield R, Pople JA (1972) Self-consistent molecular orbital methods. XII further extensions of Gaussian-type basis sets for use in molecular orbital studies of organic molecules. J Chem Phys 56:2257–2261
33. Heřmánek S, Hnyk D, Havlas Z (1989) Mechanism of the Antipodal effect with borane cages. J Chem Soc Chem Commun 1859–1861; Bühl M, Schleyer PvR, Havlas Z, Hnyk D, Heřmánek S (1991) On the origin of the antipodal effect in *closo*-heteroboranes. Inorg Chem 30:3107–3111
34. For BI_4^- and BI_3 the spin-orbit corrections are predominant in terms of reproducing experimental values: Hnyk D, Macháček J (2006) XIIth International Congress of Quantum Chemistry, Kyoto
35. Hnyk D, Wann DA, Holub J, Bühl M, Robertson HE, Rankin DWH (2008) The gas-phase structure of 1-selena-*closo*-dodecaborane(11), 1-$SeB_{11}H_{11}$, determined by the concerted use of electron diffraction and computational methods. Dalton Trans 96–100
36. Korbe S, Schreiber PJ, Michl J (2006) Chemistry of the carba-*closo*-dodecaborate(−) anion, $CB_{11}H_{12}^-$. Chem Rev 106:5208–5249
37. Indeed, Bohn and Bohn carried out such investigations but only the structure of **5c** was determined accurately, Bohn RK, Bohn MD (1971) Molecular structures of 1,2-, 1,7-, and 1,12-dicarba-*closo*-dodecaborane(12). Inorg Chem 10:350–355; precise structures based on *SARACEN* are in Turner AR, Robertson HE, Borisenko KB, Rankin DWH, Fox MA (2005)

gas-phase electron diffraction studies of the icosahedral carbaboranes, *ortho-*, *meta-* and *para-* $C_2B_{10}H_{12}$. Dalton Trans 1310–1318. Due to the zero dipole moment of **5c**, microwave spectroscopy was applied only to **5a** and **5b**, see: Samdal S, Møllendal H, Hnyk D, Holub J (2011) Microwave spectra and structures of 1,2-(*ortho*) and 1,7-(*meta*) carborane, $C_2B_{10}H_{12}$. J Phys Chem A 115:3380–3385

38. Brain PT, Covie J, Donohoe DJ, Hnyk D, Rankin DWH, Reed D, Reid BD, Robertson HE, Welch AJ (1996) 1-Phenyl-1,2-dicarba-*closo*-dodecaborane, 1-Ph-1,2-*closo*-$C_2B_{10}H_{11}$. Synthesis, characterization, and structure as determined in the gas phase by electron diffraction, in the crystalline phase at 199 K by X-ray diffraction, and by *ab initio* computations. Inorg Chem 35:1701–1708
39. Wann DA, Lane PD, Robertson HE, Baše T, Hnyk D (2013) The gaseous structure of *closo*-9,12-$(SH)_2$-1,2-$C_2B_{10}H_{10}$, a modifier of gold surfaces, as determined using electron diffraction and computational methods. Dalton Trans 42:12015–12019
40. Hnyk D, Brain PT, Robertson HE, Rankin DWH, Hofmann M, Schleyer PR, Bühl M (1994) The molecular structure of gaseous 1,7-dicarba-1,7-dichloro-*closo*-dodecaborane(12), 1,7-Cl_2-1,7-$C_2B_{10}H_{10}$, as studied by electron diffraction and *ab initio* calculations. J Chem Soc Dalton Trans 2885–2890
41. Hnyk D, Holub J, Hofmann M, Schleyer PvR, Robertson HE, Rankin DWH (2000) Dicarbadodecaborane(12) derivatives. 2. Synthesis and molecular structure of 1,12-dicarba-*closo*-dodecaborane(12)-1,12-dithiol, 1,12-$(SH)_2$-1,12-$C_2B_{10}H_{10}$ in the gaseous phase, determined by electron diffraction and *ab initio* calculations; geometrical consequences of three-dimensional aromaticity in carbaboranes. J Chem Soc Dalton Trans 39:4617–4622
42. Fox MA, Cameron AM, Low PL, Paterson MAJ, Batsanov AS, Goeta AE, Rankin DWH, Robertson HE, Schirlin JT (2006) Synthetic and structural studies on C-ethynyl- and C-bromo-carboranes. Dalton Trans 39:3544–3560
43. Hnyk D, Všetečka V, Drož L, Exner O (2001) Charge distribution within 1,2-dicarba-*closo*-dodecaborane: dipole moments of its phenyl derivatives. Collect Czech Chem Commun 66:1375–1379
44. Domenicano A (1988) In: Hargittai I, Hargittai M (eds) Stereochemical applications of gas-phase electron diffraction Part B. VCH Publishers, Weinheim/New York
45. Drož L, Fox MA, Hnyk D, Low PJ, MacBride JAH, Všetečka V (2009) Experimental and computed dipole moments in donor-bridge-acceptor systems with *p*-phenylene and *p*-carboranediyl bridges. Collect Czech Chem Commun 74:131–146
46. Campanelli AR, Domenicano A, Hnyk D (2015) Transmission of electronic substituent effects across the 1,12-dicarba-*closo*-dodecaborane cage: a computational study based on structural variation, atomic charges, and ^{13}C NMR chemical shifts. J Phys Chem A 119:205–214
47. Grüner B, Hnyk D, Císařová I, Plzák Z, Štíbr B (2002) Phosphaborane chemistry. Syntheses and calculated molecular structures of mono- and dichloro derivatives of 1,2-diphospha-*closo*-dodecaborane (10). J Chem Soc Dalton Trans 39:2954–2959
48. Mackie ID, Robertson HE, Rankin DWH, Fox MA, Malget JM (2004) Gas-phase electron diffraction studies on two 11-vertex dicarbaboranes, *closo*-2,3-$C_2B_9H_{11}$ and *nido*-2,9-$C_2B_9H_{13}$. Inorg Chem 43:5387–5392
49. Møllendal H, Samdal S, Holub J, Hnyk D (2002) The structure of 1-thia-*closo*-decaborane(9), 1-SB_9H_9, as determined by microwave spectroscopy and quantum chemical calculations. Inorg Chem 41:4574–4578
50. Hnyk D, Wann DA, Holub J, Samdal S, Rankin DWH (2011) Why is the antipodal effect in *closo*-1-SB_9H_9 so large? A possible explanation based on the geometry from the concerted use of gas electron diffraction and computational methods. Dalton Trans 40:5734–5737
51. Computed details of other *closo*-$CB_nH_{n+1}^-$ systems are given in Schleyer PvR, Najafian K (1998) Stability and three-dimensional aromaticity of *closo*-monocarbaborane anions, $CB_{n-1}H_n^-$ and *closo*-dicarboranes. Inorg Chem 37:3454–3470; later some were prepared and structurally characterized by the GIAO-MP2/II//MP2/6-31G* method: $CB_6H_7^-$: Štíbr B (2002) The [*closo*-2-CB_6H_7]$^-$ anion, the first representative of the seven-vertex monocarbaborane series. another stable candidate for weakly coordinating anion chemistry. Angew Chem Int Ed

41:2126–2128; $CB_8H_9^-$: Jelínek T (2001) Monocarbaborane chemistry. Preparation and characterisation of [4-CB_8H_9]$^-$, the "missing" *closo*-carbaborane anion. Chem Commun 1756–1757

52. Atavin EG, Mastryukov VS, Golubinskii AV, Vilkov LV (1980) Molecular structure of 1,10-dicarba-*closo*-decaborane(10), 1,10-$C_2B_8H_{10}$, by gas-phase electron diffraction. J Mol Struc 65:259–269
53. Hnyk D, Rankin DWH, Robertson HE, Hofmann M, Schleyer PR, Bühl M (1994) On the molecular structure of 1,2-dicarba-*closo*-decaborane(10) as studied by the concerted use of electron diffraction and *ab initio* calculations. Inorg Chem 33:4781–4786
54. Holub J, Bakardjiev M, Štíbr B, Hnyk D, Tok OL, Wrackmeyer B (2002) Two isomeric phosphacarboranes 2,1- and 6,1-PCB_8H_9, the first representatives of the ten-vertex *closo*-phosphacarborane series. Inorg Chem 41:2817–2819
55. Hnyk D, Hofmann M, Schleyer PR, Bühl M, Rankin DWH (1996) On the molecular structure of 7,8-dicarba-10-thia-*nido*-undecaborane(10), the first gaseous eleven-vertex *nido* heteroborane studied by the combined electron diffraction and *ab initio* approach: the NMR consequence of the molecular geometry. J Phys Chem 100:3435–3440
56. Holub J, Jelínek T, Hnyk D, Plzák Z, Císařová I, Bakardjiev M, Štíbr B (2001) Phosphacarborane chemistry: the 7,8,9,11-, 7,9,8,10- and 7,8,9,10-isomers of *nido*-$P_2C_2B_7H_9$-diphosphadicarbaborane analogues of 7,8,9,10-$C_4B_7H_{11}$. Chemistry 7:1546–1554
57. Hnyk D, Bühl M, Holub J, Hayes SA, Wann DA, Mackie ID, Borisenko KB, Robertson HE, Rankin DWH (2006) Molecular structures of *arachno*-decaborane derivatives 6,9-$X_2B_8H_{10}$ (X=CH_2, NH, Se) including a gas-phase electron-diffraction study of 6,9-$C_2B_8H_{14}$. Inorg Chem 45:6014–6019
58. Hnyk D, Holub J, Hayes SA, Robinson MF, Wann DA, Robertson HE, Rankin DWH (2006) Molecular structures of *arachno*-heteroboranes with decaborane frameworks: two *Cs*-symmetrical azacarba- and carbathiaboranes. Inorg Chem 45:8442–8446
59. Wann DA, Lane PD, Robertson HE, Holub J, Hnyk D (2013) Structures of, and related consequences of deprotonation on, two C_s-symmetric *arachno* nine-vertex heteroboranes, 4,6-$X_2B_7H_9$ (X=CH_2; S) studied by gas electron diffraction/quantum chemical calculations and GIAO/NMR. Inorg Chem 52:4502–4508
60. Hnyk D, Hofmann M, Schleyer PR (1999) 4,6-Dicarba-8-thia-*arachno*-nonaborane(10) revisited. Theoretical refinement of its structure. Collect Czech Chem Commun 64:993–1000
61. This compound is formally classified as a *hypho*-dicarbaborane ($4n+8=32$ valence electrons) but is usually regarded as an ethene adduct of tetraborane(8): Hnyk D, Brain PT, Rankin DWH, Robertson HE, Greatrex R, Greenwood NN, Kirk M, Bühl M, Schleyer PvR (1994) The molecular structure of dimethylentetraborane, $B_4H_8(CH_2)_2$, as determined by gas-phase electron diffraction and *ab initio* computations. Inorg Chem 33:2572–2578
62. Brain PT, Bühl M, Fox MA, Greatrex R, Leuschner E, Picton MJ, Rankin DWH, Robertson HE (1995) 2,4-Ethanotetraborane derivatives. 2. Synthesis, characterization, and gas-phase structures of 2,4-$(MeCHCH_2)B_4H_8$, 2,4-(*trans*-MeCHCHMe)B_4H_8, and 2- and 4-Pr-2,4-$(MeCHCH_2)B_4H_7$. Inorg Chem 34:2841–2849
63. Brain PT, Bühl M, Fox MA, Greatrex R, Hnyk D, Nikrahi A, Rankin DWH, Robertson HE (1998) 2,4-Ethanotetrabora derivatives. 3. [1] Determination of the molecular structure of 2,4-(*t*-butyl ethano)tetraborane(10), 2,4-$(Bu^tCHCH_2)B_4H_8$, in the gas phase by electron diffraction. J Mol Struc 445:319–334
64. Wade K (1976) Structural and bonding patterns in cluster chemistry. Adv Inorg Chem Radiochem 18:1–66
65. Brain PT, Downs AJ, Fanfarillo M, Goode M, Massey AG, Rankin DWH, Robertson HE (1989) The molecular structure of tetraboron tetrachloride in the gas phase, determined by electron diffraction. J Mol Struc 192:163–170
66. Hnyk D (1997) *T* symmetrical gaseous tetra-*tert*-butyltetrabora-tetrahedrane: an electron diffraction study. Polyhedron 16:603–606
67. Bühl M, Holub J, Hnyk D, Macháček J (2006) Computational study of structures and properties of metallaboranes. 2. Transition-metal dicarbollide complexes. Organometallics 25:2173–2181

68. Hawthorne MF, Zink JI, Skeleton JM, Bayer MJ, Liu C, Livshits E, Baer R, Neuhauser D (2004) Electrical or photocontrol of the rotary motion of a metallacarborane. Science 303:1849–1851. Similarly, protonated and deprotonated [3-Fe-(1,2-$C_2B_9H_{11})_2H$] also acts as a molecular rotor: Bühl M, Hnyk D, Macháček J (2007) Computational study of structures and properties of metallaboranes. Part 3: Protonated iron bis(dicarbollide), [3-Fe-(1,2-$C_2B_9H_{11})_2H]^-$. Inorg Chem 46:1771–1777
69. Miller NE, Muetterties EL (1963) A new boron hydride, $B_{20}H_{16}$. J Am Chem Soc 85:3506–3506
70. Hnyk D, Holub J, Jelínek T, Macháček J, Londesborough MGS (2010) Revisiting $B_{20}H_{16}$ by means of a joint computational/experimental NMR approach. Collect Czech Chem Commun 75:1115–1123
71. Fanfrlík J, Přáda A, Padělková Z, Pecina A, Macháček J, Lepšík M, Holub J, Růžička A, Hnyk D, Hobza P (2014) The dominant role of chalcogen bonding in the crystal packing of 2D/3D aromatics. Angew Chem Int Ed 53:10139–10142
72. Inhibition of HIV protease: Cígler P, Kožíšek M, Řezáčová P, Brynda J, Otwinowski Z, Pokorná J, Plešek J, Grüner B, Dolečková-Marešová L, Máša M et al (2005) From nonpeptide toward noncarbon protease inhibitors: metallacarboranes as specific and potent inhibitors of HIV protease. Proc Natl Acad Sci USA 102:15394–15399; inhibition of carbonic anhydrase: Brynda J, Mader P, Šícha V, Fábry M, Poncová K, Bakardiev M, Grüner B, Cígler P, Řezáčová P (2013) Carborane-based carbonic anhydrase inhibitors. Angew Chem Int Ed 52:13760–13763; modifications of gold layers: Baše T, Bastl Z, Plzák Z, Grygar T, Plešek J, Malina V, Šubrt J, Boháček J, Večerníková E, Kříž O (2005) Carboranethiol-modified gold surfaces. a study and comparison of modified cluster and flat surfaces. Langmuir 21:7776–7785
73. Hoffmann R, Vilkov LV, Mastryukov VS, Sadova NI (1983) Determination of the geometrical structure of free molecules. Mir Publishers, Moscow

Chapter 3
Computational Studies of Metallaboranes and Metallacarboranes

Alexandru Lupan and R. Bruce King

Abstract Computations based on quantum chemistry, particularly density functional theory methods, have provided valuable insights into the structure, bonding, thermochemistry, and chemical reactivity of diverse polyhedral metallaboranes. Examples of such computations are provided for metallaboranes having central polyhedra with five to 16 vertices.

3.1 Introduction

The capacity of boron to form self-bonded complicated molecular networks is as extensive as any element except for carbon. However, the fundamental structures of boron networks are very different from those of the carbon networks familiar in organic chemistry. Thus carbon networks are typically a combination of acyclic carbon chains and carbon rings whereas boron networks are based on three-dimensional polyhedra. In this connection the most stable polyhedral boranes and isoelectronic carboranes have structures based on the so-called most spherical deltahedra, also known as *closo* deltahedra (Fig. 3.1) [1, 2]. Such polyhedra have all triangular faces and vertices as nearly similar as possible. This means that the 6- to 12-vertex *closo* deltahedra have only degree 4 or 5 vertices except for the 11-vertex *closo* deltahedron, required by polyhedral topology [3] to have a single degree 6 vertex in addition to its degree 4 and 5 vertices (Fig. 3.1). The numbers of skeletal electrons in such structures are generally determined by the Wade-Mingos rules [4–6], which state that n-vertex *closo* deltahedral boranes are particularly stable if they contain $2n+2$ skeletal electrons. This special stability has been ascribed to

A. Lupan
Department of Chemistry, Faculty of Chemistry and Chemical Engineering, Babeş-Bolyai University, Cluj-Napoca, Romania
e-mail: alupan@chem.ubbcluj.ro

R.B. King (✉)
Department of Chemistry, University of Georgia, Athens, GA 30602, USA
e-mail: rbking@uga.edu

D. Hnyk, M.L. McKee (eds.), *Boron*, Challenges and Advances in Computational Chemistry and Physics 20,
DOI 10.1007/978-3-319-22282-0_3

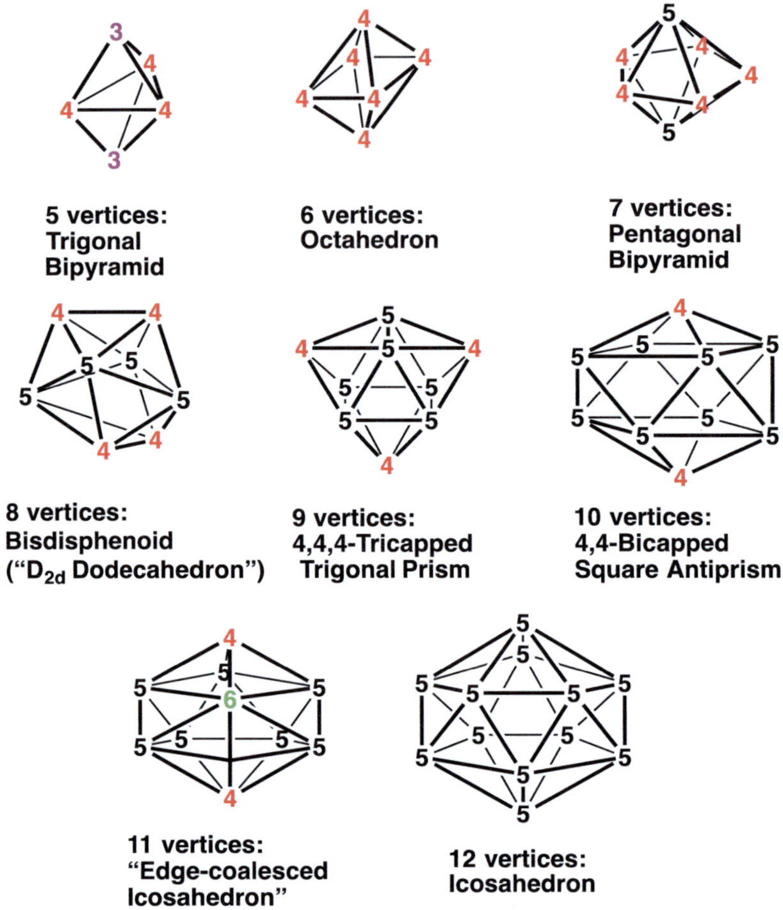

Fig. 3.1 The most spherical deltahedra with 5–12 vertices

three-dimensional aromaticity [7–9]. In accord with the Wade-Mingos rules the so-called *closo* borane anions $B_nH_n^{2-}$ ($6 \leq n \leq 12$) as well as the isoelectronic *closo* carboranes $CB_{n-1}H_n^-$ and $C_2B_{n-2}H_n$ exhibit special stability.

Hawthorne and co-workers [10] first showed that the vertices in the *closo* deltahedral boranes and related carboranes can be replaced by isolobal transition metal vertices, typically units of the type CpM or $M(CO)_3$ (Cp = η^5-cyclopentadienyl; M = transition metal), to give very stable compounds. The initial syntheses of metallaboranes typically used decaborane, $B_{10}H_{14}$ as the starting material and typically led to metallacarboranes containing a central icosahedral MC_2B_9 unit. A few years after the first metallacarborane syntheses Grimes and co-workers [11] showed that similar metallaboranes based on smaller polyhedra could be synthesized using pentaborane-9, B_5H_9 as the boron hydride starting material.

Initially it was assumed that such substitution of a light atom vertex (boron or carbon) by a transition metal vertex did not affect the underlying *closo* deltahedral geometry since the initially discovered derivatives were metallaboranes based on MC_2B_9 icosahedra, in which all vertices have the same degree, namely 5. However, as metallaborane chemistry was subsequently developed involving polyhedra other than the icosahedron, particularly by Kennedy and co-workers [12–15], a variety of deltahedral metallaborane structures were discovered based on deltahedra topologically distinct from the *closo* deltahedra. This led to the identification of a new class of less spherical deltahedra for metallaboranes and metallacarboranes called either *isocloso* [16] or *hypercloso* [17–19] deltahedra. The deviation from sphericity in such *isocloso* metallaboranes results from the preference of transition metals for degree 6 vertices, which are not present in any of the most spherical deltahedra except for the 11-vertex deltahedron (Fig. 3.1). However, these *isocloso* metallaborane deltahedra are derived from the *closo* metal-free borane deltahedra by a diamond-square-diamond process, typically generating a degree 6 vertex for the metal atom. Whereas metal-free boranes and carboranes are known for all of the *closo* deltahedra between the trigonal bipyramid and the icosahedron, namely those with 5–12 vertices, distinctive *isocloso* deltahedra are known only for a more limited vertex number range, namely 9–12 vertices (Fig. 3.2). The *isocloso* metallaborane deltahedra with n vertices are found to have $2n$ skeletal electrons rather than the $2n+2$ skeletal electrons of the *closo* borane and carborane deltahedra predicted by the Wade-Mingos rules.

The preference of metal atoms for degree 6 and sometimes even higher degree vertices leads to deltahedra for dimetallaboranes that are even less spherical than the *isocloso* metallaborane deltahedra (Fig. 3.2). A particularly noteworthy group of such species consists of the dirhenaboranes discovered by Fehlner, Ghosh, and their co-workers [20–23]. The preference of rhenium atoms for vertices of degree 6 and sometimes even degree 7 lead to highly oblate (flattened) deltahedra (Fig. 3.3). These deltahedra are characterized by the rhenium atoms at approximately antipodal (polar) degree 6 or 7 vertices at points on the deltahedral surface of lowest local curvature. Conversely, at least some of the boron atoms are located at degree 4 vertices at points on the deltahedral surface of highest local curvature. The resulting deltahedra have oblate (flattened) ellipsoidal structures rather than more nearly

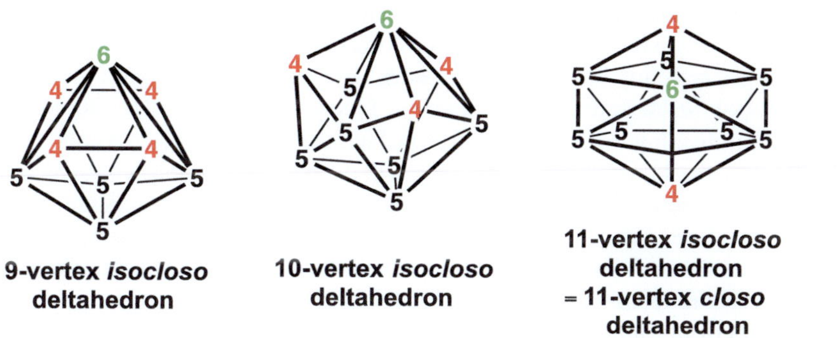

Fig. 3.2 *Isocloso* metallaborane deltahedra with 9, 10, and 11 vertices. The metal atoms in these structures are located at the unique degree 6 vertex

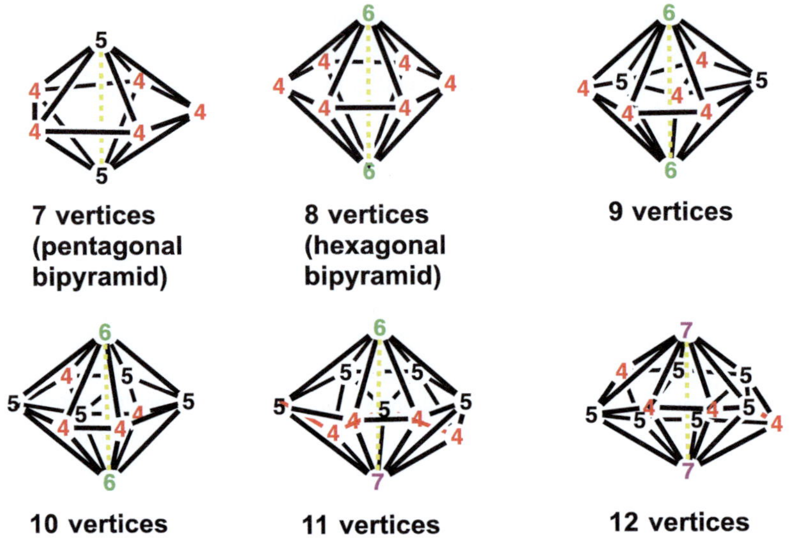

Fig. 3.3 The oblate (flattened) n-vertex deltahedra found in the dirhenaboranes $Cp^*_2Re_2B_{n-2}H_{n-2}$ ($8 \leq n \leq 12$). The rhenium atoms are located at the degree 6 (*green*) and degree 7 (*pink*) vertices

spherical structures. For this reason these polyhedra can conveniently be designated as *oblatocloso* deltahedra. The simplest example of an *oblatocloso* deltahedron is the hexagonal bipyramid (Fig. 3.3). The Wade-Mingos procedure for counting skeletal electrons [4–6] in the *oblatocloso* deltahedra arrives at $2n - 4$ skeletal electrons, which is 6 skeletal electrons short of the $2n+2$ suggested to be optimal for deltahedral structures.

Metal-free polyhedral borane chemistry includes not only the closed deltahedral borane structures $B_nH_n^{2-}$ and the isoelectronic carboranes $CB_{n-1}H_n^-$ and $C_2B_{n-2}H_n$ but also the hydrogen-rich series of neutral boranes B_nH_{n+4} ($n=5$, 6, 8, 10) and B_nH_{n+6} ($n=4$, 5, 6, 7, 8, 9) called *nido* and *arachno* boranes, respectively. The boron frameworks of the n-vertex *nido* boranes can be derived from the most spherical $(n+1)$-vertex *closo* borane deltahedra by removal of a high degree BH vertex. This leaves a "hole" in the polyhedral surface corresponding to an open face with four or more edges typically bridged by the "extra" hydrogen atoms. Figure 3.4 illustrates the *nido* polyhedra found in the structures of the known stable B_nH_{n+4} boranes ($n=5$, 6, 8, 10) indicating the deltahedra from which they are derived by vertex removal. In addition, the 11-vertex *nido* polyhedron derived from the icosahedron by removal of a vertex is depicted in Fig. 3.4. This 11-vertex *nido* polyhedron is not found in a known $B_{11}H_{15}$ borane. However, it is found in the so-called "dicarbollide" $C_2B_9H_{12}^-$ anion used for the syntheses of the first metalladicarbaboranes [10]. The n-vertex *nido* polyhedra retain the $2(n+1)+2$ Wadean skeletal electrons of the $(n+1)$-vertex deltahedra from which they are derived and thus are $2n+4$ skeletal electron systems. The process of generating an n-vertex open *nido*-like polyhedron by removal of a boron vertex from an $(n+1)$-vertex deltahedron can be applied not only to the most

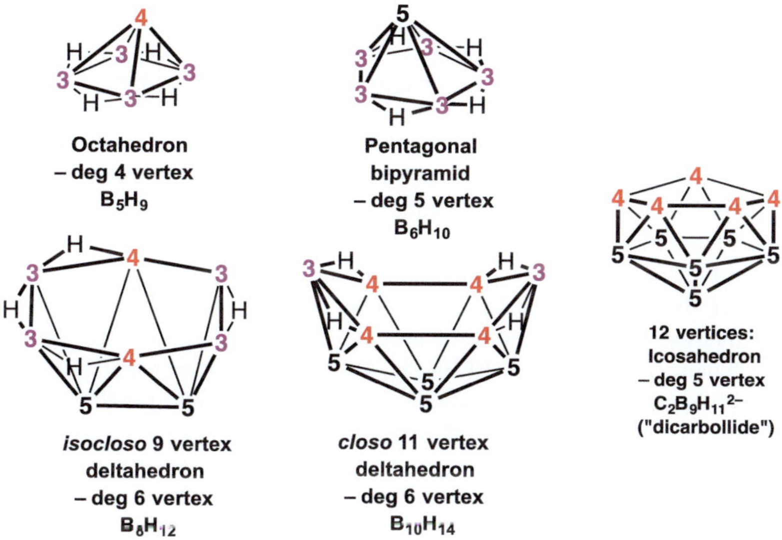

Fig. 3.4 Examples of *nido* polyhedra found in simple neutral boron hydrides. These are generated by removal of a high degree vertex from the indicated *closo* or *isocloso* deltahedron

spherical *closo* deltahedra but also to $(n+1)$-vertex *isocloso* and *oblatocloso* deltahedra (Figs. 3.2 and 3.3).

The variety of polyhedra found in metallaboranes and metallacarboranes (Figs. 3.1, 3.2, 3.3, and 3.4) combined with the variety of metal vertices and ligands attached to the metal vertices suggests that metallaboranes provide a very rich area of chemistry. The three-dimensional nature of the polyhedral borane building blocks as compared with the two-dimensional nature of the benzene and other carbon rings in organic chemistry suggests even greater complexity of polyhedral borane chemistry relative to organic chemistry. The prevalence of carbon chemistry relative to boron chemistry is a consequence of the rarity of boron relative to carbon and the difficulty of converting naturally occurring borate minerals to boron hydride derivatives. Because of the complexity of polyhedral borane chemistry and specifically metallaborane and metallacarborane chemistry, modern computational methods, particularly those using density functional theory, have been very useful both in interpreting existing experimental results and in suggesting promising new areas for future development. This area has been the subject of recent reviews [24, 25].

3.2 Computational Methods

Density functional theory (DFT) is the computational method most frequently used for modeling metallaborane clusters. Widely used functionals include the B3LYP functional composed of Becke's three parameter functional [26] with the Lee, Yang and Parr correlation functional [27] as well as the BP86 [28, 29] and PBE functionals [30]. Scalar relativistic effects have been considered by using the zero-order regular approximation (ZORA) [31] or specialized basis sets such as the SDD basis set containing the Stuttgart Dresden effective core potentials is often employed [32]. In rare cases post Hartree-Fock methods have been used such as the MP2 (Møller-Plesset) method [33]. This method is considered as the most suitable for computing first hyperpolarizabilities. The nuclear magnetic shielding tensors can be calculated using the gauge-independent atomic orbital (GIAO) approach [34]. Aromaticity can be studied by computing the nucleus-independent chemical shift (NICS) [35, 36]. Analysis of the electron density distribution function can be performed by the QTAIM method [37]. The nature of bonding can be assessed by computing the electron localization function (ELF) [38–40]. Metallaboranes and metallacarboranes have been extensively studied by theoretical methods during the last decades as indicated by the 154 articles cited in this chapter.

3.3 Summary of Some Representative Computational Studies on Metallaboranes and Metallacarboranes

3.3.1 General Comments

Modern theoretical methods have been applied so extensively to metallaboranes and metallacarboranes that a comprehensive review in a book chapter of reasonable length is not feasible. Because of the large amount of such information, only the highlights of representative studies can be mentioned in this chapter. Figures from individual papers are not repeated here; the reader is referred to the cited papers for further details. However, Figs. 3.1, 3.2, 3.3, and 3.4 in the Introduction provide views of the most prevalent polyhedra encountered in these metallaborane and metallacarborane structures.

Among the large amount of published work on computational metallaborane chemistry, papers focusing on systems with specific numbers of vertices are chosen to be cited and briefly summarized in this chapter. This allows most of the material to be organized according to the number of vertices in the central polyhedra of the metallaboranes and metallacarboranes. In this connection, most of the reported studies involve 12-vertex systems based on a central icosahedron because of the particular stability of icosahedral boranes. Furthermore, most of these studies are parts of experimental papers and focus on the interpretation of specific experimental results. In addition, to the discussion of papers focusing on systems with specific numbers of vertices in the central polyhedra, a few particularly significant general papers are discussed below.

The structures and electronic relationships of 9-, 10-, 11-, and 12-vertex *closo* and *hypercloso* (*isocloso*) metallaboranes have been explored using DFT calculations

[41]. The role of the transition metal in stabilizing the *isocloso* borane structures is explained using the concept of orbital compatibility. The *isocloso* (structures, (η^6-C_6H_6)$MB_{n-1}H_{n-1}$ (n=9–12; M = Fe, Ru, and Os) are taken as model complexes. Calculations on metal-free polyhedral boranes B_nH_n suggest that n-vertex *isocloso* structures need only n skeletal electron pairs (SEPs). However, such structures have one or more degree 6 vertices (Fig. 3.2), whereas the corresponding *closo* structures with $n+1$ SEPs have only degree 4 and degree 5 vertices (except for the required degree 6 vertex in the 11-vertex *closo* deltahedron in Fig. 3.1). This high-degree vertex of the *isocloso* structures can be effectively occupied by transition metal fragments with their highly diffuse orbitals. Calculations further show that a second- or third-row transition metal with more diffuse orbitals has a greater tendency than a first-row transition metal to exhibit *isocloso* geometry. This is consistent with the observation that there are more experimentally characterized *isocloso* structures containing second- and third-row transition metals than those containing first-row transition metals.

The size of the exopolyhedral ligands attached to the metal atom also affects the stability of *isocloso* structures. The interaction between the borane and the metal fragments in the *isocloso* geometry has been analyzed using the fragment molecular orbital approach. The interconversion of the *closo* and *isocloso* structures by the addition and removal of electrons is also discussed in terms of correlation diagrams. The *isocloso* geometry can be stabilized by substituting the degree 6 BH vertex, which caps the six-membered ring of the B_9H_9 fragment, with a cap with more diffuse orbitals. Thus a transition metal fragment with more diffuse orbitals can stabilize the *isocloso* structure, which is reflected in the experimental synthesis of many *isocloso* metallaboranes. A large number of *isocloso* metallaboranes are reported in literature with a range of metal atoms and exopolyhedral ligands.

Ruthenaborane clusters have been modeled by performing density functional theory calculations using the B3LYP functional in order to provide insights into the use of such clusters for hydrogen storage and H–H bond activation [42]. In this connection structural optimizations were performed for different ruthenaborane clusters. Transition state structures were determined for their hydrogenation addition/elimination reactions. Such calculations of the reaction pathways yielded different transition-state structures involving molecular hydrogen bonded to the cluster or formation of metal hydrides. The H–H bond of H_2 appears to be activated by the ruthenaborane clusters with calculated activation energies of 24–42 kcal/mol and Gibbs free energies of 14–27 kcal/mol for H_2 addition. The calculated Gibbs free energy range for the H_2 addition reaction ranges from 14 to 27 kcal/mol. The calculated activation energies and molecular structures of the hydrogen-rich clusters $Cp^*_2Ru_2B_{10}H_{16}$, $Cp^*_2Ru_2B_8H_{14}$, and $Cp^*_2Ru_2B_8H_{12}$ (Cp^* = η^5-C_5Me_5) with different degrees of hydrogenation are compared. The mechanisms of the H_2 addition and elimination reactions of the studied clusters suggest that they might be useful as hydrogen storage materials owing to their ability to activate the H-H bond.

3.3.2 Five-Vertex Systems

Density functional theory (DFT) calculations were used to probe the reaction of the tetragonal pyramidal diruthenaborane *nido*-1,2-$(Cp^*RuH)_2B_3H_7$ ($Cp^* = \eta^5$-C_5Me_5), with the alkyne MeC≡CMe to form *nido*-1,2-$(Cp^*Ru)_2(\mu$-H)(μ-BH_2)-4,5-Me_2-4,5-$C_2B_2H_4$ as the major product along with the minor product *nido*-$(Cp^*Ru)_2$-4,5-Me_2-4,5-$C_2B_2H_6$. The structural features of the related diruthenacarboranes, *nido*-$(Cp^*Ru)_2$-4,5-Me_2-4,5-$C_2B_2H_6$ and *closo*-1,2-$(Cp^*RuH)_2$-4,5-Me_2-4,5-$C_2B_3H_3$ were also studied [43]. The geometrical and electronic structures of the products *nido*-1,2-$(Cp^*Ru)_2(\mu$-H)(μ-BH_2)-4,5-Me_2-4,5-$E_2B_2H_4$ (E = Si (Ge, Sn) are described and compared with the ruthenacarborane analog, *nido*-1,2-$(Cp^*Ru)_2(\mu$-H)(μ-BH_2)-4,5-Me_2-4,5-$C_2B_2H_4$. The computed energetics and the geometries support the feasibility of the reaction and the stability of the products. NBO analysis was performed to delve further into the nature of the bonding in this kind of clusters.

3.3.3 Six-Vertex Systems

The reaction of CpCo(1,3-$C_3B_2Me_5H$) with CpTl affords the thallium derivative CpCo(1,3-$C_3B_2Me_5$)Tl. Both CpCo(1,3-$C_3B_2Me_5H$) and CpCo(1,3-$C_3B_2Me_5$)Tl have been structurally characterized by X-ray crystallography at 100 K [44]. According to DFT calculations, CpCo(1,3-$C_3B_2Me_5H$) exists as a mixture of two enantiomers with an enantiomerization barrier of only 0.5 kcal/mol. The transition state has C_s symmetry with an *endo*-CH hydrogen atom. The isomer with a Co–H···B bridge is less stable by 10 kcal/mol. The activation energies of formation for each of these two CpCo(1,3-$C_3B_2Me_5H$) isomers from $C_3B_2Me_5H$ and $CpCo(C_2H_4)_2$ are nearly equal.

Morokuma, Mebel, and collaborators [45] have performed an *ab initio* (RHF) and MP2 molecular orbital study of the structure, stability and rearrangements of six-vertex *nido*-metallaboranes of the type MB_5H_8 (M = Ir, Co, Fe). The optimized geometries as well as the NMR chemical shifts computed by the IGLO method agree well with the experimental data. Systematic calculations of all possible intramolecular mechanisms of B_6H_{10} reorganization were also performed. The fluxionality of B_6H_{10} arises from the bridging hydrogens. These studies also included the isomerization of the iridaborane $(IrB_5H_8)(CO)(PH_3)_2$ in which the Ir atom moves from a basal to the apical position by a mechanism similar to the apical-basal rearrangement of B_6H_{10}. The calculated barrier of 39 kcal/mol at the MP2//HF/ECP-DZ level is high enough to prevent isomerization at accessible temperatures.

The mixed "early-late" transition metal derivative $(Cp_2Zr)(Cp^*Ir)B_4H_{10}$ is the first structurally characterized dimetallaborane analogue of hexaborane(12) containing zirconium [46]. Density functional theory (DFT) calculations have been carried out on this compound as well as other hypothetical early-late transition metal combinations of hexaborane(12) analogues. The ^{11}B and ^{1}H chemical shifts of the

Cp analogue were calculated using the Gauge Including Atomic Orbital (GIAO) and the B3LYP functional and found to agree well with the experimental data. This provides a stringent test of the validity of the calculated electronic structures of the Cp model complex (maximum deviation of 3 ppm for ^{1}H NMR and 4 ppm for ^{11}B NMR at the B3LYP/def2-TZVP level). The optimized molecular structure resembles closely the experimental structure determined by X-ray crystallography. Inspection of the electron density distribution of the frontier MOs shows that the HOMO is predominantly located on iridium whereas the LUMO is located on zirconium. In addition, natural bond orbital (NBO) analysis shows a strong Ir–Zr bonding interaction with a Wiberg bond index of 0.49.

The complete series of hydrogen-rich six-vertex cyclopentadienyl dimetallaboranes $Cp_2M_2B_4H_8$ (Cp = η^5-C_5H_5; M = Ir, Ru/Os, Re, Mo/W, and Ta), including the experimentally known Ir, Ru, and Re derivatives, has been examined by DFT [47]. The nature of the central M_2B_4 polyhedra in the lowest energy $Cp_2M_2B_4H_8$ structures relates to the skeletal electron count as determined by the Wade-Mingos rules [4–6]. Thus the lowest energy $Cp_2Ir_2B_4H_8$ structures with 16 Wadean skeletal electrons have central pentagonal pyramidal Ir_2B_4 units similar to the known pentagonal pyramidal B_6H_{10}. The lowest energy $Cp_2M_2B_4H_8$ (M = Ru, Os) structures with 14 Wadean skeletal electrons have central capped tetragonal pyramidal rather than octahedral M_2B_4 units. However, isomeric $Cp_2M_2B_4H_8$ (M = Ru, Os) structures with central M_2B_4 octahedra are found at energies starting at ~15 kcal/mol (M = Ru) and ~10 kcal/mol (M = Os) above the capped tetragonal pyramid global minima. The lowest energy electron poorer $Cp_2M_2B_4H_8$ structures (M = Re, Mo, W, Ta) have central M_2B_4 bicapped tetrahedra with the metal atoms at the degree 5 vertices. Higher energy $Cp_2Re_2B_4H_8$ structures include capped tetragonal pyramidal structures with surface Re = Re double bonds and a pentagonal pyramidal structure with a surface Re ≡ Re triple bond. The lowest energy $Cp_2M_2B_4H_8$ (M = Mo, W) structures appear to have surface M = M double bonds and thus also the 12 skeletal electrons for their bicapped tetrahedral structures. However, the lowest energy likewise bicapped tetrahedral $Cp_2Ta_2B_4H_8$ structure is best interpreted in having CpTa units with 16-electron rather than 18-electron tantalum configurations and a surface Ta–Ta single bond.

3.3.4 Seven-Vertex Systems

Hosmane and collaborators [48] have synthesized the complex 1-($C_{12}H_8N_2$)Sn-2,3-$[Si(CH_3)_3]_2$-2,3-$C_2B_4H_4$. In order to understand this system MNDO-SCF molecular orbital calculations were carried out on the model compounds 1-($C_{12}H_8N_2$)Sn-2,3-$C_2B_4H_6$, 1-($C_{10}H_8N_2$)Sn-2,3-$C_2B_9H_{11}$, and $[(C_{10}H_8N_2)SnC_5(CH_3)_5]^+$ in order to determine what factors dictate the structures of these complexes. The results show that competing bonding interactions lead to a very broad energy minimum as a function of slippage and base orientation. Small energy variations, such as those produced by crystal packing forces, can produce large structural changes. Similar studies were also performed on the terpyridine complexes 1-($C_{15}H_{11}N_3$)Sn-2,3-($SiMe_3$)$_2$-2,3-$C_2B_4H_4$ and 1-($C_{15}H_{11}N_3$)Sn-2-($SiMe_3$)-3-(Me)-2,3-$C_2B_4H_4$ [49].

Satpati found the sandwich metal dimers CB_5H_6M–MCB_5H_6 (M = Si, Ge, Sn) to be minima on the potential energy surface with a characteristic M–M single bond [50]. The NBO analysis and the M–M distances (Å) indicate substantial M–M bonding. Formal generation of CB_5H_6M–MCB_5H_6 has been studied theoretically. The slip-distorted geometry was found to be preferred for MCB_5H_7 and its dehydrogenated dimer CB_5H_6M–MCB_5H_6.

Barreto et al. [51] have examined the structural distortions from *closo* geometry in various metal carborane clusters using the Fenske-Hall technique [52, 53]. They chose (2,2′-bpy)$SnB_4H_4C_2Me_2$ (bpy = bipyridine) and $(CO)_3FeB_4H_4C_2Me_2$ as model systems. An MO analysis suggested that the antibonding interactions of symmetric tin fragment orbitals with corresponding carborane orbitals lying on the B_3C_2 face are responsible for the occurrence of the slipped structure. The Fenske-Hall quantum-chemical technique was used to examine structural distortions from an idealized *closo* geometry in main-group- and transition-metal carboranes. The ΔH values were calculated for the carbons-apart to carbons-adjacent interchange reactions in these systems [54]. Both dianions were found to bond equally well with tin. However, the carbons-apart dianion was found to be considerably more stable than its carbons-adjacent isomer. The tin-bipyridine bonding arises from the interaction of the base nitrogen lone pairs with the two tangentially oriented tin p orbitals.

Ezhova et al. [55] have synthesized 1-M(THF)2-2,4-($SiMe_3$)$_2$-2,4-$C_2B_4H_5$ (M = Li, Na) and calculated their energies and geometries at the B3LYP/6-31G* level. The ^{11}B and ^{13}C NMR spectra calculated by GIAO using the 6-311G** basis set were in good agreement with experiment.

The reaction of the dichromaborane $(Cp^*Cr)_2B_4H_8$ with $BHCl_2{\cdot}SMe_2$ was found to result in cluster expansion to give $(Cp^*Cr)_2B_5H_9$ [56]. Reactions of $(Cp^*Cr)_2B_4H_8$ with binuclear metal carbonyls such as $Fe_2(CO)_9$ and $Co_2(CO)_8$ led to mixed-metal metallaboranes. MO calculations have been carried out for $(CpM)_2B_4H_8$ and $(CpM)_2B_5H_9$ (M = Cr, Mo) as well as the mixed-metal clusters $(CpCr)_2B_4H_8Fe(CO)_3$ and $(CpCr)_2B_4H_7Co(CO)_3$. The cluster $(Cp^*Cr)_2B_4H_8$ can be considered to be electronically unsaturated with too few skeletal electrons (five SEPs) to support the observed bicapped tetrahedral geometry. This has been rationalized by small but significant geometric changes causing the LUMO to move to higher energy and thus stabilize the electron-deficient species.

Full geometry optimizations were done on $Cp^*_2W_2B_5H_nCl_m$ (n=7, m=2 and n=8, m=1) using DFT [57]. Subsequent Hessians were calculated by using the B3LYP hybrid exchange functional and the LANL2DZ basis set. In order to support the assignment of isomers the ^{11}B and ^{1}H NMR chemical shifts were calculated using the GIAO method at the B3LYP level of theory and the LANL2DZ basis set. The ^{1}H NMR shifts show a systematic deviation so that a suitably chosen scaling factor would give better results.

An extended Hückel study [58] of the double-decker $CpCoC_2B_4H_6$ and triple-decker $CpCo(C_2B_3H_5)CoCp$ sandwich model complexes shows significant differences in the orbital contributions involved in the HOMO and LUMO of the former versus the latter type. The calculations afford additional insight into the electronic structures and properties of these systems as elucidated by the experimental studies. Fragment analyses were performed for two model complexes. The B3LYP functional was employed to calculate second-order nonlinear optical (NLO) responses of $(C_5H_5)Co(C_2B_4H_6)$ derivatives containing -CH=CH-C_6H_4-NO_2 and -CH=CH-C_6H_4-NH_2 substituents [59]. The geometries of isomers were optimized at the B3LYP/6-31G* level to give stable molecular configurations. Comparisons of the stabilities of molecules show that a -CH=CH–C_6H_4-NO_2 substituent improves the stability because of the electronic deficiency of the boron atom. Time-dependent DFT (TD-DFT) calculations on the electronic spectra of molecules were performed to investigate the influence of electron transfer on inner orbitals.

The iron sandwich complex $Fe(Et_6\text{-}2,4\text{-}C_2B_4H)_2$ contains two hydrido ligands, each bridging the iron and two boron atoms. The geometries of the carboranes, the boranes (all unsubstituted and permethyl-substituted), and the iron complexes (all unsubstituted) were optimized at the B3LYP/6-311+G(d,p) or B3LYP/6-31+G(d) levels [60]. The relevant NMR data (chemical shifts and coupling constants) were calculated at the same level of theory.

The linear optical properties and the static second-order nonlinear optical (NLO) properties were investigated systematically on a series of two-dimensional (2D) D-π-A-π-D/A-π-D-π-A sandwich metallacarborane-containing molecules using DFT [61]. The substituent effect was found to influence the electronic absorption and NLO responses of the 2D molecules. Time-dependent DFT calculations and the analysis of major molecular orbitals predict the sandwich metallacarboranes to act as better electron donors than electron acceptors. The observed spectral properties were rationalized by calculating the electronic absorption spectra of these systems at the TD-PBE1PBE/6-31G* and TD-PBE1PBE/6–31+G* levels with an SDD basis set for the metals. The atom population analyses of the frontier MOs were also reported. Because of the high computational cost, the electronic absorption spectra of most of the systems were calculated only at the TD-PBE1PBE/6–31G* level. The orbital transition properties associated with the bright excited states were also considered. The electron density is delocalized between the Cp and C_2B_3 rings in HOMO-2 and HOMO-1. However, the electron density is located on both the C_2B_3 and Cp rings in the LUMOs. The second-order NLO properties of 2D Λ- and W-shaped sandwich metallocarborane-containing derivatives were investigated using a variety of functionals including BP86, B3LYP, PBE1PBE, BH and HLYP, LC-BP86, CAM-B3LYP, and LC-ωPBE.

The second-order nonlinear optical (NLO) properties of the $Cp^*Co(C_2H_5)_2C_2B_4H_3$-expanded (metallo)porphyrins ($Cp^* = C_5Me_5$) were investigated using ab initio RHF and density functional theory (DFT) methods [62]. Molecules with expanded porphyrin units were found to possess remarkably large molecular hyperpolarizabilities and thus are promising second-order NLO materials. The nature of the charge transfer (CT) transition indicates that the $-Cp^*Co(C_2H_5)_2C_2B_4H_3$ unit functions as an electron donor. Furthermore, the time-dependent DFT calculation illustrates that reduced forms have significantly different charge-transfer patterns than the neutral species. This investigation provides insight into the comparison with

DFT results on estimating first hyperpolarizabilities and the NLO properties of this series of push–pull compounds. Meanwhile, the calculated values are functional-dependent: the α_{zz} values decrease with the increasing percentage of HF as well as the α_0 value in hybrid DFT functionals (B3LYP > PBE1PBE > BH and HLYP). This illustrates that the polarizability is connected to the X term of the XC functional: that of *ab initio* HF is the smallest, and the deviation between DFT and HF arises from the local nature of the DFT exchange functionals. Using the LC and CAM functionals, the α_0 values are ordered as LC-BLYP < LC-ωPBE < CAM-B3LYP. When compared with the hybrid DFT and RHF functionals, the calculated α_0 values from LC-BLYP and LC-ωPBE are lower than those from BH and HLYP but larger than those from RHF. However, the α_0 value of CAM-B3LYP lies between PBE1PBE and BH and HLYP. These numerical analyses suggest that the effect of HF exchange energy on the LC and CAM functionals is not equivalent to that of the hybrid DFTs. Importantly, the overestimation of the α_0 values using hybrid DFT methods may come from the lack of long-range exchange interaction in conventional exchange functionals.

Density functional theory (DFT) has been used to probe the bonding and electronic properties of the dimolybdaborane $(Cp^*Mo)_2B_5H_9$, as well as several other heterodimolybdaborane clusters, such as $(Cp^*Mo)_2B_5(\mu_3\text{-OEt})H_7$, $(Cp^*Mo)_2B_5(\mu_3\text{-OEt})(n\text{-BuO})H_6$, $(\eta^5\text{-}C_5H_5W)_2B_4H_4S_2$, and $(Cp^*Mo)_2B_4H_4E_2$ (E = S, Se, Te) [63]. The DFT results were also used to address some key points such as the metal–metal bond lengths, the locations and numbers of bridging and terminal hydrogen atoms, the molecular orbital analyses, and the assignments of ^{11}B and ^{1}H NMR chemical shifts.

First principles calculations on the metallacarboranes $C_2B_4H_6M$ (M = Sc, Ti, V, Cr, Mn, Fe, Co, and Ni), predict high hydrogen storage capacities since the transition metal atoms in these systems can bind up to $5H_2$-molecules [64]. The average binding energies lie within the reversible adsorption range. Among the first row transition metals, Sc and Ti are found to be the optimum for maximizing the H_2 storage capacity (8 wt%) on the metallacarborane cluster.

Density functional theory (DFT) calculations on $(Cp^*TaX)_2B_5H_{11}$, (X = Cl; Br; and I) at the BP86/TZ2P ZORA level reveal geometries in agreement with structure determinations by X-ray crystallography [65]. The predicted large gaps between the highest occupied molecular orbital (HOMO) and the lowest unoccupied molecular orbital (LUMO) are consistent with their stabilities. The B3LYP-computed ^{11}B chemical shifts are close to the experimentally measured shifts.

DFT calculations of the redox potentials have been performed on the experimentally known triple-decker sandwiches $CpCo(\mu\text{-}1,3\text{-}C_3B_2Me_5)M(\text{ring})$ [M(ring) = RuCp, RuCp*, $Co(C_4Me_4)$] with a bridging diborolyl ligand [66]. The bonding properties of the anions $[CpCo(1,3\text{-}C_3B_2R_5)]^-$ and $[C_5R_5]^-$ (R = H, Me) toward $[M(\text{ring})]^+$ cations were shown to be similar both experimentally (synthesis, electrochemistry, and X-ray diffraction) and theoretically (energy decomposition and Mulliken population analysis).

3.3.5 Eight-Vertex Systems

DFT calculations were performed to study the possibility of H_2 uptake by the electron-deficient metallaborane $Cp_2Re_2B_6H_6$ having a hexagonal bipyramidal structure (Fig. 3.3) [67]. A possible H–H insertion was indicated leading to the formation of $Cp_2Re_2B_6H_8$ isomers accompanied by B–B bond breaking in the equatorial B_6H_6 ring and breaking of Re–B bonds. A two-step pathway was calculated to be the lowest-energy route with the highest activation barrier at ca. 25 kcal/mol at the B3LYP/6-311G++(d,p) level of theory.

An unsaturated 24-valence-electron triple-decker sandwich structure is found for $(Cp^*Mo)_2\{\mu\text{-}\eta^6:\eta^6\text{-}B_3H_3TeCo_2(CO)_5\}$, in which the middle deck is composed of B, Co, and a heavy group 16 element [68]. DFT calculations using the gauge-independent atomic orbital GIAO–DFT method at the B3LYP/SDD/6-31G* level reproduced the ^1H and ^{11}B NMR and IR data satisfactorily. DFT studies also establish that the cluster $Cp^*_2Mo_2B_5H_9$ used to synthesize the triple-decker sandwich exhibits enhanced reactivity if one of the boron vertices in the open face is replaced by S, Se, or Te. The electronegativity and/or size differences of the chalcogen atoms lead to different geometries. In addition, the more spatially extended and diffuse p and d orbitals of Te favor the formation of a flatter geometry relative to S and Se.

The experimental $Cp_4M_4B_4H_4$ structures (M = Co, Ni) are both bisdisphenoids (Fig. 3.1). However, in the cobalt compound, the metal atoms are located at the four degree 5 vertices, whereas in the nickel compound the metal atoms are located at the four degree 4 vertices. DFT has shown these structures to be the lowest-energy structures by the substantial margins of around 23 and 15 kcal/mol for M = Co and Ni, respectively [69]. The unknown analogous tetraferraborane $Cp_4Fe_4B_4H_4$ was predicted to have a different type of structure, namely a tetracapped tetrahedron, with the iron atoms at the 4 degree six vertices, similar to the experimental structure of the valence isoelectronic $Cp_4Fe_4(\mu_3\text{-}CO)_4$. The experimentally unknown $Cp_4M_4B_2H_2$ and Cp_4M_4 systems have also been optimized. Addition of BH units to either $Cp_4M_4B_2H_2$ or Cp_4M_4 to give the corresponding $Cp_4M_4B_4H_4$ derivatives is predicted to be extremely exothermic.

3.3.6 Nine-Vertex Systems

The monochlorinated iridaborane 1,1,1-$(PMe_3)_2$H-*isocloso*-IrB_8H_7-8-Cl with a central IrB_8 *isocloso* deltahedron (Fig. 3.2) was found to be significantly more stable than its non-chlorinated analogue 1,1,1-$(PMe_3)_2$H-*isocloso*-IrB_8H_8 [70]. DFT calculations were carried out to help rationalize and extend the conclusions from the experimental observations. The calculations suggest that the *isocloso* cluster structure for the monochloro derivative may be both the kinetically and the thermodynamically favored structure. However, for the unsubstituted derivative a tricapped trigonal prismatic structure is thermodynamically favored. In this case the observed *isocloso* structure may well be the kinetically favored product. Definitive assignment of the ^{11}B NMR spectra is obtained using the DFT geometry calculations and GIAO NMR predictions. In order to assess the viability of iridium-containing nine-vertex cluster systems, the reported closely related nine-vertex cluster compound $(Ph_3P)_2(CO)IrCB_7H_8$ of conventional *closo* cluster geometry and conventional *closo* electron-count was optimized. Using the B3LYP level of theory, with the 6-31G* basis sets for phosphorus, carbon, boron, oxygen, and hydrogen and the LANL2DZ basis sets for iridium, the calculated interatomic distances and the GIAO-calculated boron nuclear shieldings were found to approximate very reasonably the experimentally observed values.

Extended Hückel molecular-orbital calculations have been reported for the trigonal prismatic platinaboranes and carbaplatinaboranes $(H_3P)_2PtB_8H_8^{2-}$ and $(H_3P)_2PtC_2B_6H_8$ [71]. The observed conformations of these molecules are rationalized by the nodal characteristics of the frontier orbitals of the constituent $Pt(PH_3)_2$ and carborane fragments.

3.3.7 Ten-Vertex Systems

The synergistic application of crystallography, multi-element NMR spectroscopy, and DFT calculations of structure and nuclear magnetic shielding has been used to delineate and define two basic variants in the quadrilaterally open-faced 10-vertex *isonido* IrB_9 metallaborane system [72]. One variant, exemplified by 7,7,7-$(PPh_3)_{2H\text{-}\mu\text{-}3}$,7-H-*isonido*-7-$IrB_9H_8$-9-$(PPh_3)$, has an Ir–H–B bridging hydrogen atom involving a boron atom not on the open face, and a terminal Ir-H hydrogen. The second variant, exemplified by 8-Cl-7-(PPh_3)-μ-7^P,10^C-(Ph_2P-ortho-C_6H_4)-*isonido*-7-IrB_9H_6-9-(PPh_3), has no cluster-bridging hydrogen atoms and no terminal Ir–H unit. DFT structural and nuclear magnetic shielding calculations were used to establish and confirm the precise nature of the differences. This removed previous uncertainties about the constitutions of such species and thereby defined better the characteristics of the two structural variants. DFT calculations at the B3LYP level of theory were employed, using the 6-31G* basis-sets for phosphorus, carbon, boron and hydrogen, and the LANL2DZ basis-set for iridium.

A reinvestigation of the 10-vertex cluster compounds $L_2RuC_2B_7H_9$ (L = PPh_3, PEt_3) has been carried out [73]. The central RuC_2B_7 cluster structure is shown to have *isocloso* geometry (Fig. 3.2), albeit with considerable distortion towards *isonido* geometry. Addition of excess PMe_3 leads to a $(Me_3P)_3RuC_2B_7H_9$ species of conventional 10-vertex *closo* geometry (Fig. 3.1). However, $(Me_3P)_3RuC_2B_7H_9$ undergoes cluster rearrangement, resulting in a different isomer than that originally proposed. In order to test the hypothesis that $(PPh_3)_2RuC_2B_7H_9$ might not have the presumed *hypercloso*/*isocloso* cluster geometry, DFT calculations were initially carried out on the simplified PH_3 model $(PH_3)_2RuC_2B_7H_9$ at the B3LYP level of theory using 6-31G*/LANL2DZ basis sets. GIAO calculation of the nuclear magnetic shielding of the cluster boron atoms was carried out for comparison with measured ^{11}B NMR chemical shifts. Optimizations and GIAO nuclear-shielding calculations B3LYP/6-31G*/LANL2DZ were carried out on the three 10-vertex rhodathiaboranes, 2,2,2-(H)$(PPh_3)_2$-*closo*-2,1-$RhSB_8H_8$,6,6,9-$(PPh_3)_3$-*arachno*-6,5-$RhSB_8H_9$, and 2,2,2-(Cl)(H)(PPh_3)-6-(PPh_3)-*closo*-2,1-$RhSB_8H_7$ [74].

The geometries and ^{11}B chemical shifts for *isocloso*-$(CpCo)_3B_6H_7Co(CO)_2$, *closo*-$(CpCo)_2B_2H_5Mo_2(CO)_6I$, $(CpCo)_4B_6H_6$, and $(CpRh)_4B_6H_6$ calculated by the gauge-including atomic orbital density functional theory GIAO-DFT method agree well with the experimental data [75]. The DFT frontier molecular orbitals demonstrate a substantial bonding interaction between the cobalt and boron atoms in HOMO−1 and HOMO−3. The Wiberg Bond Index of 0.46 also supports the existence of Co–B bonding interactions. Furthermore, a considerable amount of overlap between the two cobalt atoms was observed in HOMO−2.

3.3.8 11-Vertex Systems

The 11-vertex *nido* compounds 8,8-$(PPh_3)_2$-*nido*-8,7-$RhNB_9H_{11}$ and 9-(OMe)-8,8-$(PMe_2Ph)_2$-*nido*-8,7-$PtCB_9H_{10}$ have been characterized by single-crystal X-ray diffraction analysis and NMR spectroscopy [76]. Their structures are based on 11-vertex *nido* geometries (Fig. 3.4), with the metal center and the heteroatoms in adjacent positions on the pentagonal open face. The metal-to-heteroborane bonding sphere is fluxional, with a $\Delta G^{\ddagger}$ of 11.6 kcal/mol. DFT calculations on the model compounds 8,8-$(PH_3)_2$-*nido*-8,7-$RhNB_9H_{11}$ and 8,8-$(PH_3)_2$-*nido*-8,7-$RhSB_9H_{10}$ have been carried out to define the fluxional process and the intermediates involved. As part of this work, DFT calculations on the models 8,8-$(PH_3)_2$-*nido*-8,7-$RhNB_9H_{11}$ and 8,8-$(PH_3)_2$-*nido*-8,7-$RhSB_9H_{10}$ provide further insights regarding the relative stabilities of the *nido* versus the *closo* geometrical configurations in such 11-vertex systems as well as the nature of the intermediates that play a role in the fluxional process. The calculations demonstrate that agostic interactions between the *exo-polyhedral* ligands and the rhodium atom are not required to attain the observed *nido* geometrical structures. The assignments of the resonances are reasonably based on their relative intensities, and by comparison with the NMR spectrum of the previously reported rhodathiaborane, 8,8-$(PPh_3)_2$-*nido*-8,7-$RhSB_9H_{10}$. The assignments were supported by DFT calculations on the model compound 8,8-$(PH_3)_2$-*nido*-8,7-$RhNB_9H_{11}$.

The species 8,8,8-(CO)$(PMe_3)_2$-*nido*-8,7-$IrSB_9H_{10}$ has a conventional cluster structure based on classical 11-vertex *nido* geometry (Fig. 3.4) with the iridium center and the sulfur atom in adjacent positions on the pentagonal open face [77]. The species 1,1,1-(H)$(PMe_3)_2$-*isonido*-1,2-$IrSB_9H_9$ exhibits an 11-vertex *isonido* structure based on an 11-vertex *closo* deltahedron (Fig. 3.1) with the Ir(H)$(PMe_3)_2$ moiety occupying the apical position of connectivity six, but with one long nonbonding Ir···B distance generating the quadrilateral *isonido* open-face. Selected interatomic distances and angles are determined, together with DFT-calculated distances for the corresponding model compound 8,8,8-$(PH_3)_2$(CO)-*nido*-8,7-$IrSB_9H_{10}$, which represents the higher occupancy conformer. The NMR data agree with the solid-state structure together with its calculated nuclear magnetic chemical shielding values using the DFT B3LYP/6-31G*/LANL2DZ methodology. These calculations were carried out to determine if the static crystalline structure corresponded to the observed ^{11}B NMR spectrum and also to aid assignment of the NMR resonances.

The 11-vertex rhodathiaboranes 8,8-(Cl)(PPh_3)-9-(Py)-*nido*-8,7-$RhSB_9H_9$ and 8,8,8-(Cl)(CO)(PPh_3)-9-(Py)-*nido*-8,7-$RhSB_9H_8$, have recently demonstrated remarkable *nido*-to-*closo* redox flexibility and bifunctional character [78]. DFT calculations demonstrate that the *nido*-isomers are more stable than the predicted *closo* isomers. In addition, studies have been carried out on reactivity with Lewis bases. Assignments were made based on ^{1}H-{^{11}B(sel)} experiments and DFT calculations on the PH_3 model, 8,8-(Cl)(PH_3)-9-(Py)-*nido*-8,7-$RhSB_9H_9$. Such DFT calculations demonstrate that the *nido* structure is more stable than the isomeric *closo* structure, without the need to invoke intramolecular Rh···CH agostic interactions.

DFT calculations on $(PPh_3)_2$-*nido*-$RhSB_9H_{10}$ do not support the proposition that there are agostic interactions between two *ortho*-phenyl hydrogen atoms and the metal center. Thus, the claim that these interactions supply the "missing" electron pair from the formal Wadean electron-count of 13 skeletal electron pairs required for a *nido* 11-vertex cluster is not supported [79].

The $H_3N \rightarrow Rh$ interaction in 8,8,8-(NH_3)$(PPh_3)_2$-*nido*-8,7-$RhSB_9H_{10}$ induces the labilization of the PPh_3 ligands leading to the dissociation product, 8,8-(NH_3)(PPh_3)-*nido*-8,7-$RhSB_9H_{10}$, which can then react with another molecule of NH_3 to give the diammine 8,8,8-$(NH_3)_2$(PPh_3)-*nido*-8,7-$RhSB_9H_{10}$ [80]. These rhodathiaboranes exhibit reversible H_3N-promoted ligand lability, which implies weak Rh–N interactions, leading to rare examples of metal complexes that circumvent "classical" Werner chemistry. Structures were initially optimized using standard methods with the STO-3G* basis-sets for carbon, boron, phosphorus, sulfur, and hydrogen and the LANL2DZ basis-set for the rhodium atom. The final optimizations, including frequency analyses to confirm the true minima, together with GIAO nuclear-shielding calculations, were performed using the B3LYP functional, with the 6-31G* and LANL2DZ basis-sets. The GIAO nuclear shielding calculations were performed on the final optimized geometries.

Stone et al. [81] report the dianionic manganacarbaborane complex [1,1,1-$(CO)_3$-2-Ph-*closo*-1,2-$MnCB_9H_9]^{2-}$. A DFT optimization of the radical anion gave structural parameters that agree well with those obtained by X-ray diffraction. The calculated SOMO reveals an antibonding interaction between the Mn center and the carbaborane cage, consistent with the closer approach between these two units upon oxidation. Moreover, the unpaired electron was found to be distributed over both the carbaborane cage and the {$Mn(CO)_3$} moiety with Mulliken spin densities of 0.37 for Mn, 0.55 for {$PhCB_9H_9$}, and 0.08 for $(CO)_3$, in accord with the spectroscopic results.

The reduction of 1,1,1-$(CO)_3$-2-Ph-*closo*-1,2,3,4-$MnC_3B_7H_9$, to the hyperelectronic dianion has been studied by electrochemistry, spectroscopy, X-ray crystallography, and DFT calculations [82]. The monoanion was found to show ligand slippage to η^4 coordination in which relevant distances and angles are roughly intermediate. The SOMO of this monoanion is highly delocalized over the MnC_3B_7 framework. Although the tricarbadecaboranyl anion is a strongly electron-accepting ligand, the majority of the stabilization energy it imparts to nominally hyperelectronic metals originates from its flexibility for hapticity changes.

An innovative type of lithium decahydroborate ($Li@B_{10}H_{14}$) complex has been designed using quantum mechanical methods. As the lithium atom binds in a handle fashion to the terminal electrophilic boron atoms of the decaborane basket (Fig. 3.4), its NBO charge q (Li) is found to be 0.876, which is close to +1 [83]. The structure resembles a basket, and the Li atom merely forms the handle of this basket.

3.3.9 12-Vertex Systems: Metal Bis(dicarbollides)

The original metallacarboranes synthesized by Hawthorne and co-workers [10] were metal bis(dicarbollides) of the type $[M(C_2B_9H_{11})_2]^{z-}$ in which a metal vertex is shared by two MC_2B_9 icosahedra (Fig. 3.5). The cobalt bis(dicarbollide) monoanion $[Co(C_2B_9H_{11})_2]^-$ and its substitution products are particularly stable and have been studied extensively.

The $E_{1/2}$ values for the redox couple Co^{III}/Co^{II} in the $[3,3'\text{-}Co(C_2B_9H_{11})_2]^{0/-}$ system have been correlated with their HOMO and LUMO energies in order to find a theoretical basis for the redox tunability in such species [84]. The cluster total charge (CTC) was found to be a useful property to interpret the behavior of boron clusters, since it computes all charges on the cluster atoms, not only these at the periphery. In this work, the natural population analysis (NPA) method was used to compute CTC.

A series of monoansa $[\mu\text{-}1,1'\text{-}PR\text{-}3,3'\text{-}Co(1,2\text{-}C_2B_9H_{10})_2]^-$ and diansa $[8,8'\text{-}\mu\text{-}(1'',2''\text{-benzene})\text{-}\mu\text{-}1,1'\text{-}PR\text{-}3,3'\text{-}Co(1,2\text{-}C_2B_9H_9)_2]^-$ (R = Ph, tBu) cobaltabisdicarbollidephanes have been synthesized, characterized and studied by NMR, MALDI-TOF-MS, UV-visible spectroscopy, cyclic voltammetry, and DFT calculations [85].

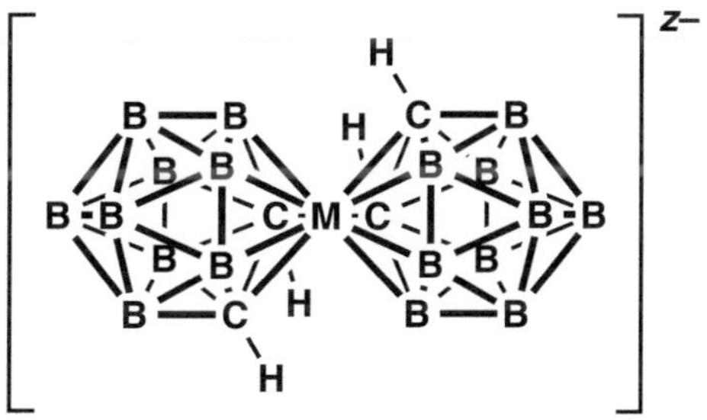

Fig. 3.5 The general structure of the metal bis(dicarbollides) $[M(C_2B_9H_{11})_2]^{z-}$. Hydrogens bonded to the boron vertices are omitted for clarity

The cluster anion 3,3′-Co(1,2-$C_2B_9H_{11})_2^-$ has been taken as a model compound to study chlorination in metallacarborane clusters [86]. In this connection, preferential chlorination sites in this cluster have been shown to have a kinetic rather than a thermodynamic origin. The substitution reaction rates have been rationalized by considering 2a-NPA on the parent compound, 3,3′-Co(1,2-$C_2B_9H_{11})_2^-$, unperturbed by chlorine substituents. The 2a-NPA value is defined as the sum of the NPA charges of the two bonded atoms. The 2a-NPA charges have been chosen because they reproduce the chlorination order of attack much more accurately than simple NPA.

The reactions of the anionic monoiodo derivatives [8-I-3,3′-Co(1,2-$C_2B_9H_{10}$)(1′,2′-$C_2B_9H_{11}$)]$^-$ and [$B_{12}H_{11}I$]$^{2-}$ with alkyl and aryl reagents has been reported [87]. Kumada carbon-carbon coupling conditions that inspired B–C bond formation have been extended to monocarborane derivatives. For these qualitative studies, and to economize on computer time, studies were conducted using density functional theory (DFT), with propylene on the monoiodo derivative of cobaltabis(dicarbollide) and a [Pd(PH_3)] catalyst. Such calculations have shown that the changes in the geometry of the cluster-Pd moiety caused by the substitution of PH_3 by PPh_3 are not significant.

The silyl-substituted cobalt bis(dicarbollide) anions [1-SiMe_2H-3,3′-Co(1,2-$C_2B_9H_{10}$)(1′,2′-$C_2B_9H_{11}$)]$^-$,[1,1′-m-SiMe_2-3,3′-Co(1,2-$C_2B_9H_{10})_2$]$^-$,[1,1′-m-SiMeH-3,3′-Co(1,2-$C_2B_9H_{10})_2$]$^-$, [1-SiMe_3-3,3′-Co(1,2-$C_2B_9H_{10}$)-(1′,2′-$C_2B_9H_{11}$)]$^-$, [1,1′-(SiMe_3)2-3,3′-Co(1,2-$C_2B_9H_{10})_2$]$^-$,[8,8′-m-(1′,2′-C_6H_4)-1,1′-m-SiMe2-3,3′-Co(1,2-$C_2B_9H_9)_2$]$^-$;[8,8′-m-(1′,2′-C_6H_4)-1,1′-m-SiMeH-3,3′-o-(1,2-$C_2B_9H_9)_2$]$^-$,and[8,8′-m-(1′,2′-C_6H_4)-1-SiMe3-3,3′-Co(1,2-$C_2B_9H_9$)(1′,2′-$C_2B_9H_{10}$)]$^-$ have been studied by DFT [88] for comparison with experimental results. In addition, DFT studies at the B3LYP/6-311G(d,p) level were used to optimize the geometries of these cobalt bis(dicarbollides) and calculate their relative energies. Racemic mixtures are found to be more stable than the corresponding *meso* isomers. Good agreement between theoretical studies and experimental results was achieved.

The recently reported crystal structure of [NMe_4][1-SiMe_2H-3,3′-Co(1,2-$C_2B_9H_{10}$)(1′,2′-$C_2B_9H_{11}$)] shows short contacts between the Si–H proton acceptor group and the C_c–H proton donor moiety in the dicarbollide ligand [89]. These short contacts were studied within the framework of the Quantum Theory of Atoms in Molecules (QTAIM) at different levels of DFT theory (B3LYP/6-311G(d,p) and BP86/TZ2P(+)) that show the existence of a bifurcated Si–H···H–C_c dihydrogen bond. This paper presents the study of an unusual function of an Si–H group as a proton acceptor in a dihydrogen bond in a role usually played by hydrides such as M–H (M = transition metal), B–H, or Al–H. Furthermore, this paper also provides a new simple method to estimate bonding energies for closed-shell intramolecular interactions using Voronoi charge population analysis and Coulomb's Law. The Quantum Theory of Atoms in Molecules of Bader (QTAIM) at different levels of theory (B3LYP/6-311G(d,p) and BP86/TZ2P(+)) has been used to study the H···H interactions found in the crystal structure of [1-SiMe_2H-3,3′-Co(1,2-$C_2B_9H_{10}$)(1′,2′-$C_2B_9H_{11}$)]$^-$. NPA and VDD are adequate basis set independent methods to compute atomic charges.

A density functional study at the BP86/AE1 level is presented for the cobalt bis(dicarbollide) ion $[3\text{-Co}(1,2\text{-}C_2B_9H_{11})_2]^-$ and selected isomers and rotamers thereof [90]. Rotation of the two dicarbollide moieties with respect to each other is facile, as judged by the small energetic separation of the three rotamers located within 2.6 kcal/mol and by the low barriers for their interconversion (at most 9.8 kcal/mol). Among the isomers containing two equivalent dicarbollide ligands but differing in carbon atom positions, the 1,7 ("carbons apart") isomer $[2\text{-Co-}(1,7\text{-}C_2B_9H_{11})_2]^-$ is the most stable. The electronic structure of $[3\text{-Co}(1,2\text{-}C_2B_9H_{11})_2]^-$ is characterized in terms of molecular orbitals, population analysis, and excitation energies from time-dependent density functional theory (TDDFT), relevant to UV/Vis spectroscopy. The salient features of the UV/Vis spectrum of $[3\text{-Co-}(1,2\text{-}C_2B_9H_{11})_2]^-$ are reproduced well by TDDFT at the B3LYP level. This allows for the theoretical interpretation of this spectrum and, presumably, the spectra of other members of the large family of metal dicarbollide complexes. Experimental ^{11}B NMR chemical shifts of $[3\text{-Co}\ (1,2\text{-}C_2B_9H_{11})_2]^-$ are reproduced to better than 5 ppm at the GIAO-B3LYP/II' level, and the computed $\delta(^{11}B)$ values are only slightly affected by rotational averaging or the presence of a polarizable continuum. The B3LYP functional was initially chosen because of its good performance for chemical shifts of transition-metal complexes, in particular for those of the metal nuclei themselves. The sensitivity of the computed NMR chemical shifts to the surrounding medium was evaluated using a popular polarizable continuum model (PCM) and employing the dielectric constant of the solvent used in the experiments, acetone. Occasionally, such significant "direct" solvation effects on chemical shifts (i.e., through the response of the electronic wave function) can be found. More often, however, such direct effects are small and can be superseded by "indirect" effects, that is, by changes in the molecular geometry upon solvation. These latter effects (which could be modeled by optimizing the geometry in the presence of a continuum) are expected to be small for $[2\text{-Co-}(1,7\text{-}C_2B_9H_{11})_2]^-$, given the slight sensitivity of the computed $\delta(^{11}B)$ values toward the details of the geometrical parameters (see above). Thus, a single-point PCM calculation was performed for the gas-phase BP86/AE1 geometry. On going from the vacuum into the continuum, no qualitative and only minor quantitative changes in the ^{11}B NMR chemical shifts (up to ca. 1.5 ppm) occurred. Likewise, counterion effects should be small, so that the values computed for the pristine ion in the gas phase (as discussed in the preceding part) should be comparable to experiments in solution.

Conformations of the cobalt biscarbollide $[3,3'\text{-Co}(1,2\text{-}C_2B_9H_{11})_2]^-$ have also been studied [91]. Three possible orientations of the carborane cages in the complexes are possible: *transoid*, *cisoid*, and *gauche*. The limiting *cisoid* and *transoid* orientations are the most commonly encountered, but it is not clear which one is the most preferred. An energy profile as a function of the rotation angle was produced by semiempirical calculations to complement the experimental NMR data. These calculations have shown the existence of two major rotamers.

The self-interaction capability of the anion $[3,3'\text{-Co}(1,2\text{-}C_2B_9H_{11})_2]^-$ through *C*luster–H···H–B (*Cc*–H···H–B) dihydrogen bonds was investigated [92]. A set of theoretical (BP86/TZ2P(+)) and empirical data aiming to establish the main rules that account for the binding mode between the negatively charged borane framework made by $[3,3'\text{-}Co(1,2\text{-}C_2B_9H_{11})_2]^-$ and the $[NMe_4]^+$ ions has been compiled. The cation-anion interaction is mainly electrostatic but the covalent contribution is also proven and quantified. The existing intermolecular H···H short contacts have been studied and are compared with available experimental data. The results show that the electronic configuration of the transition metal atom in the sandwich complex is not sufficient to define the preferred rotamer owing to the influence of the anion environment and the H···H interactions present in the solid state.

A molecular dynamics study of chlorinated cobalt bis(dicarbollide) anions $[(B_9C_2H_8Cl_3)_2Co]^-$ "CCD^-" in nitrobenzene and at the nitrobenzene–water interface has been reported by Chevrot, Wipff et al. [93] The main aim of this work is to understand the solution state of these hydrophobic species and why they act as strong synergists in assisted liquid–liquid extraction of metallic cations. Neat nitrobenzene is found to solubilize Cs^+ salts of CCD^- in the form of diluted pairs or oligomers, without displaying aggregation. In biphasic nitrobenzene–water systems, CCD^- anions mainly partition to the organic phase, thus attracting Cs^+ or even more hydrophilic counterions like Eu^{3+} into that phase. The remaining CCD^- anions adsorb at the interface, but are less surface-active than at the chloroform interface. The interfacial behaviors of $Eu(BTP)_3{}^{3+}$ in the absence and in the presence of CCD^- anions and extractant molecules are compared. In the absence of CCDs, the complex is found to be trapped at the interface, while at a sufficiently high concentration of CCDs, the complex is extracted to the nitrobenzene phase. These results are compared with those obtained with chloroform or *n*-octanol as the organic phase and discussed in the context of synergistic effect of CCDs in liquid-liquid extraction. This points to the importance of dual solvation properties of nitrobenzene or *n*-octanol to solubilize the CCD^- salts as well as the extracted complex.

Coupez and Wipff [94] have performed a molecular dynamics study (OPLS forcefield) on the effect of CCD^- [chlorinated cobalt bis(dicarbollide)] anions on the Eu^{3+} lanthanide cation extraction by a calix [4]arene-CMPO ligand focusing on the water–"oil" interface, where "oil" is modeled by chloroform. An initial important result is the high surface activity of the CCD^- anions, which contrasts with the inactivity of their corresponding neutral fictitious CCD^0 analogues. The free ligand L and its EuL^{3+} complex are also surface active. This strongly suggests that cation capture by L occurs at the interface, and that it can be promoted by CCD^- anions. In general, hard ions are instead "repelled" by the interface. However, the interfacial film formed by CCD^- anions creates a negative potential, which attracts Eu^{3+} cations and should therefore "catalyse" their capture by the ligands. The interfacial activity of the calix [4]arene ligand and its complexes is consistent with expectations based on their amphiphilic character and analogies with other extractant molecules. As far as CCD^- anions are concerned, this can be surprising, as these species are quite "spherical" and lack amphiphilic topology. Surface activity has been attributed to the fact that these hydrophobic ions still enjoy important attractive interactions with water at the interface, while avoiding payment for the cavitation energy in water.

Chevrot et al. [95] use molecular dynamics of the cobalt bis(dicarbollide) anions $[(B_9C_2H_8X_3)_2Co]^-$ (XCD^-) commonly used for liquid-liquid extractions (X = H, Me,

Cl, or Br), to show that these anions, although lacking amphiphilic topology, behave as anionic surfactants. In pure water, they display "hydrophobic attractions", leading to the formation of aggregates of different sizes and shapes depending on the counterions. When simulated at a water/"oil" interface, the different anions (HCD^-, $MeCD^-$, $ClCD^-$, and $BrCD^-$) are found to be surface active. As a result, the simulated M^+ counterions ($M^+ = Na^+$, K^+, Cs^+, H_3O^+, UO_2^{2+}, Eu^{3+}) concentrate on the aqueous side of the interface, forming a "double layer" whose characteristics are modulated by the hydrophobic character of the anion and by M^+. The highly hydrophilic Eu^{3+} or UO_2^{2+} cations that are generally "repelled" by aqueous interfaces are attracted by dicarbollides near the interface. This is crucial as far as the mechanism of assisted cation extraction to the oil phase is concerned. These cations interact with interfacial XCD^- in their fully hydrated $Eu(H_2O)_9^{3+}$ and $UO_2(H_2O)_5^{2+}$ forms, whereas the less hydrophilic monocharged cations display intimate contacts through their X substituents. The results obtained with the TIP3P and OPLS models for the solvents are confirmed with other water models (TIP5P or a polarizable 4P-Pol water) and with more polar "oil" models. The importance of interfacial phenomena is further demonstrated by simulations with a high oil-water ratio, leading to the formation of a micelle covered with CCD's. The interfacial activity of bis(dicarbollides) and related hydrophobic anions is an important feature of synergism in liquid-liquid extraction of hard cations (e.g., for nuclear waste partitioning).

HIV protease (HIV PR) is a primary target for anti-HIV drug design. In a structure-guided drug design effort, the two cobalt bis(dicarbollide) anions were connected with a linker to a substituted ammonium group to provide a set of compounds exemplified by $[H_2N\text{-}(8\text{-}(C_2H_4O)_2\text{-}1,2\text{-}C_2B_9H_{10})(1',2'\text{-}C_2B_9H_{11})\text{-}3,3'\text{-}Co)_2]$ Na [96]. The inhibition properties of these compounds with various substituents were explored. The HIV PR: inhibitor crystal structure was determined and the conformational space of the linker was explored computationally. Our results prove the capacity of linker-substituted dual-cage cobalt bis(dicarbollides) as lead compounds for design of more potent HIV PR inhibitors. A set of double (triple) cluster compounds of general formula $[^1R^2RN\text{-}(8\text{-}(C_2H_4O)_2\text{-}1,2\text{-}C_2B_9H_{10})(1',2'\text{-}C_2B_9H_{11})\text{-}3,3'\text{-}Co)_2]^{n-}$ (^1R = H, $-C_2H_5$, $-n\text{-}C_4H_9$, $-C_2H_4OH$, $-t\text{-}C_4H_9$, $-C(CH_2OH)_3$, $-C_3H_6COO^-$, $-C_2H_4SO_3^-$, $-CH_2C_6H_5$, $-SO_2C_6H_5\text{-}3\text{-}CH_3$, $-7\text{-}CB_{10}H_{11}^-$, $-1\text{-}CB_{11}H_{11}$, $-[\mu\text{-}8,8'<(1,2\text{-}C_2B_{10}H_{10})\text{-}3,3'Co]^-$, ^2R = none, H, $-C_2H_5$, $n = 1$ and 2) or $[n\text{-}C_4H_9H_2N\text{-}(10\text{-}(C_2H_4O)_2\text{-}nido\text{-}7,8\text{-}C_2B_9H_{10})_2]^-$ has been synthesized and their HIV PR inhibition properties investigated. The crystal structure of the parent compound $[H_2N\text{-}(8\text{-}(C_2H_4O)_2\text{-}1,2\text{-}C_2B_9H_{10})(1',2'\text{-}C_2B_9H_{11})\text{-}3,3'\text{-}Co)_2]Na$ in a complex with wild-type HIV-1 PR has been determined. Since the lack of a continuous electron-density map in the crystal structure suggested numerous possible linker conformations, the conformational space of the linker was explored computationally. This study demonstrates the capacity of linker-substituted dual-cage cobalt bis(dicarbollides) as lead compounds for design of more potent HIV protease inhibitors. Conformations of the cobalt bis(dicarbollide) linker within the HIV PR active site accessible by molecular-dynamics based quenching techniques (MD/Q) were computationally sampled. Quantum mechanics/molecular mechanics (QM/MM) optimizations for 20 snapshots from the MD/Q protocol resulted in a set of conformers differing substantially in energy by up to 40 kcal/mol. On the basis of a criterion of copopulation/interchangeability in the enzyme-inhibitor complex, five lowest-energy structures differing in the linker conformation with an energy span of at most 3 kcal/mol were selected. These five conformers shared the common position of the central nitrogen atom of the linker secondary amino group. Interestingly, in the crystal structure, a spherical electron density is noticeable in the vicinity of this predicted nitrogen atom position. Further refinement of the models was therefore carried out using QM/MM calculations, in which the central nitrogen atom was constrained to the position of the spherical electron density map. This yielded the final predicted linker conformations. Four resulting lowest-energy conformers fall within 3 kcal/mol and are thus predicted to be copopulated in the complex structure.

Spectral investigations of the new organic semiconductor of the formula $(TMTSF)_2[3,3'-Co(1,2-C_2B_9H_{11})_2]$, were performed using Raman and quantum chemical simulation methods [97]. The possibility of two rotamers of the Co anion has been discussed considering the DFT calculations and the experimental spectra of the reference material. The ab initio calculations of optimal geometry and normal mode frequencies of two rotamers of the cobalt bis(dicarbollide) anion (*transoid* and *cisoid* configuration) were performed. DFT calculations were carried out on an isolated molecule with the B3LYP functional and the 6–311++G(d,p) basis set. The normal mode assignment was performed based on the experimental and theoretical data.

Molecular conductors based on the 8-hydroxy cobalt bis(dicarbollide) anion, $(TMTTF)[8\text{-}HO\text{-}3,3'\text{-}Co(1,2\text{-}C_2B_9H_{10})(1',2'\text{-}C_2B_9H_{11})]$, $(BMDT\text{-}TTF)[8\text{-}HO\text{-}3,3'\text{-}Co(1,2\text{-}C_2B_9H_{10})(1',2'\text{-}C_2B_9H_{11})]$, and $(BEDT\text{-}TTF)[8\text{-}HO\text{-}3,3'\text{-}Co(1,2\text{-}C_2B_9H_{10})(1',2'\text{-}C_2B_9H_{11})]$ were synthesized, and their crystal structures and electrical conductivities were determined [98]. All of the radical-cation salts prepared were found to be semiconductors. The relative stabilities of the rotamers of the $[8\text{-}HO\text{-}3,3'\text{-}Co(1,2\text{-}C_2B_9H_{10})(1',2'\text{-}C_2B_9H_{11})]^-$ anion were estimated using DFT/BP86 quantum chemical calculations. These results demonstrate the high rotational flexibility of the dicarbollide ligands in hydroxy and alkoxy derivatives of the cobalt bis(dicarbollide) anion.

On the basis of the energies and ^{11}B NMR chemical shifts computed at the BP86/AE1(*) and GIAO-B3LYP/II' levels of density functional theory, respectively, the structure of the long-known protonated iron(II) bis(dicarbollide) can be assigned as a staggered isomer with a *cisoid* conformation of the carborane ligands. However, in the unprotonated species, these ligands adopt the usual *trans* orientation. This suggests that suitable control of protonation/deprotonation equilibria can induce rotary motion at the molecular level [99]. According to DFT computations at the BP86/AE1* level, protonation of iron(II) bis(dicarbollide) can produce one of three distinct minima, which differ in the mutual orientation of the staggered dicarbollide ligands. Assignment of the experimentally known species to one of these three structures is effected by a combination of energetic and NMR criteria. One of the three isomers with a *transoid* arrangement of the two dicarbollide ligands, can be excluded because it is quite high in energy, namely ~7 kcal/mol above the two other minima. Discrimination between the other two isomers, which are almost isoenergetic at the DFT level employed, can be achieved by the computed ^{11}B chemical shifts.

A combined first-principles theoretical as well as experimental liquid-state ^{11}B NMR study of the paramagnetic anion $[3\text{-}Fe^{III}(1,2\text{-}C_2B_9H_{11})_2]^-$ was reported [100]. This iron(III) bis(dicarbollide) is an electronically open-shell structure in which the iron center links two hemispherical boron–carbon cages. This combined theoretical and experimental analysis constitutes a firm basis for the assignment of experimental ^{11}B NMR chemical shifts in paramagnetic metallaboranes. The performance of different popular exchange–correlation DFT functionals and different basis sets was evaluated. A dynamic correction to the calculated shifts via first-principles molecular dynamics simulations was found to be important. Solvent effects on the chemical shifts were computed and found to be of minor significance. The bonding between the sandwiched metal atom and its *nido*-carbaborane ligands resembles that in ferrocene and other metallocenes, such as the paramagnetic cobaltocene. The paramagnetic ^{13}C NMR spectra of such systems have been examined computationally. In paramagnetic metallocenes, the non-contact hyperfine contributions to the pNMR shielding have been found to be significant, partly owing to delocalization of the spin density into the ligands. Thus, these contributions could also be important in metallaboranes. The calculated shifts and measured peak integrals have been used to assign the experimental shifts. Statistical error limits are shown as obtained from the dynamic correction to the results from a MD simulation. There is good overall agreement between the calculated and experimental shifts, as judged by the general sensitivity of pNMR shifts to the selection of computational methods. The results with the best functionals are of sufficient quality to settle the assignment of the measured ^{11}B NMR spectral lines.

Electron delocalization in $[3\text{-}Fe^{III}(1,2\text{-}C_2B_9H_{11})_2]^-$ is based on ligand-to-metal π bonding, with the *nido* $1,2\text{-}C_2B_9H_{11}^{2-}$ ligand containing six electrons and five molecular orbitals. The calculated spin density in $[3\text{-}Fe^{III}(1,2\text{-}C_2B_9H_{11})_2]^-$ is presented. Positive spin density is concentrated around the iron nucleus. However, significant spreading of spin density to the ligands of the molecule takes place, especially near the boron nuclei opposite to the carbon centers. The largest amount of negative spin density, on the other hand, occurs on the metal-facing side of the carbon nuclei. In order to illustrate the dependence on the structure used, the ^{11}B NMR chemical shifts as obtained from single-point calculations, i.e., without dynamic corrections, are given for different geometries. The absolute shielding values of the reference compound B_2H_6 are also indicated. The results differ by 10–30 ppm, depending on the boron site in question, between the structures optimized with hybrid (B3LYP) and GGA (BP86) functionals using the SVP basis. No clear preference between the two choices can be formed on the basis of comparison with experimental shifts. Depending on the NMR signal considered, either the BP86/SVP or the B3LYP/SVP structure provides results that are closest to the experimental data. Because of the higher computational cost of hybrid functionals and larger basis sets, the BP86 functional was pragmatically chosen for further investigations of the effect of basis set used in obtaining the structure. In the SVP, TZVP, TZVPP series of increasingly improving basis sets, only very modest changes in the calculated ^{11}B NMR chemical shifts of ~2 ppm or less are observed between the two larger basis sets. Consequently, the TZVP basis should be adequate for optimizing the structure of systems of the present kind. Introducing a scalar relativistic effective core potential (ECP) on the Fe center at the SVP basis-set level, is observed to cause a notable effect on the shifts of the boron centers close to the metal atom. Doing the same at the TZVP level produces far smaller changes. This indicates that instead of any important relativistic effects on the structure, the large effect at the SVP level is due to the quality difference of the all-electron and ECP/valence basis sets of Fe. The difference between the results obtained with the SVP(ECP) and TZVP(ECP) basis sets was minor, at most 1.2 ppm. The latter basis set was selected for the reported single-point calculations. The molecular dynamics simulation of dynamic shift corrections was carried out at the BP86/SVP(ECP) level.

The novel sandwich metallacarboranes *commo*-3,3′-Ni(8-SMe_2-1,2-$C_2B_9H_{10}$)$_2$, *commo*-[3,3′-Co(8-SMe_2-1,2-$C_2B_9H_{10}$)$_2$]$^+$, *commo*-3,3′-Ru(8-SMe_2-1,2-$C_2B_9H_{10}$)$_2$, and *commo*-3,3′-Fe(8-SMe_2-1,2-$C_2B_9H_{10}$)$_2$ have been prepared [101]. Theoretical calculations using the ZINDO/1 semiempirical method show three or four energy minima for these complexes corresponding to different rotamers. Rotamers evident from X-ray diffraction studies are in agreement with the global energy minimum calculated by the ZINDO/1 method.

Given the existence of biological motors, the interest of chemists in designing molecular motors arises from the challenge not only of making even smaller nanomachines that perform controllable motion, but also of creating systems that can be powered with light or electrical energy, rather than depending on the ATP delivery used in biological motors. The basis of the molecular device reported in this study [102] has been known since the relatively early discovery of the d^7 Ni(III) and d^6 Ni(IV) *commo*-bis-7,8-dicarbollyl metallacarboranes and their palladium analogs. These complexes are produced by the coordination of two dicarbollide ions with a Ni(II) ion, followed by subsequent oxidation. The “sandwich” species may undergo rotary motion of the ligands with respect to each other thus providing the basis for a molecular machine. This is analogous to the well-known metallocenes, but they have rotational barriers of only ~2 kcal/mol. The present metallacarboranes have barriers about three times as large, leading to temperature invariance in their nuclear magnetic resonance (NMR) spectra. Photoexcitation induces geometrical changes similar to those caused by electrochemical reduction. Reduction places an electron in the lowest unoccupied molecular orbital (LUMO) whereas photoexcitation removes an electron from the highest occupied molecular orbital (HOMO) and places it in the LUMO.

The rotary motion of metallacarboranes around a molecular axis can be controlled by simple electron transfer processes, thereby providing a basis for a structure–property relationship for nonlinear optical (NLO) switching. In this connection the nickelacarboranes $[Ni^{III/IV}(C_2B_9H_{11})_2]^{-,0}$ and their C-,B-functionalized derivatives have been investigated using DFT [103]. By calculating relative energies, the stable states were obtained before and after rotation controlled by simple electron transfer. Then, the static and frequency-dependent second-order NLO properties were calculated using several DFT functionals. According to the TDDFT results, the large NLO responses of the studied compounds are mainly caused by substituent group-to-carborane cage charge transfer (L'LCT) and substituent group-to-metal charge transfer (L'MCT) processes.

A DFT study at the BP86 level is presented for the metal bis(dicarbollides), $[3\text{-M-}(1,2\text{-}C_2B_9H_{11})_2]^n$ (M/*n* = Fe,Ru/2–, Co,Rh/1–, Ni,Pd/0), as well as selected mixed- and half-sandwich complexes $3\text{-M(L)-}(1,2\text{-}C_2B_9H_{11})$ [M(L) = $Fe(C_6H_3Me_3)$, $Ru(C_6H_6)$, $Ru(CO)_3$, $Rh(C_5Me_5)$] [104]. Available experimental ^{11}B NMR chemical shifts of these complexes with a *closo* structure of the metallacarborane moiety are reproduced reasonably well at the GIAO-B3LYP/II' level, with mean absolute deviations of ca. 3 ppm over a chemical-shift range of ca. 50 ppm.

3.3.10 Other 12-Vertex Systems

Quantum-chemical calculations of the geometry and energies of nine possible isomers of the 12-vertex cobaltacarborane $CpCoC_2B_9H_{11}$ were carried out by the DFT method PBEPBE/DGDZVP/DGA1 [105]. The thermodynamic stability of the isomers was found to increase with increasing distance between the carbon atoms in the cage but to be essentially independent of the position of the CpCo vertex. This confirms the close similarity between the isolobal CpCo and BH fragments. In accord with this similarity, a series of compounds, in which the CpCo fragments replace from one to four BH vertices, have been synthesized. Skeletal rearrangements of most metallacarboranes generally occur at lower temperatures compared to the rearrangements of carboranes. The negative charge on the carbon atoms increases with increasing distance between these atoms, and the positive charge on the cobalt atom decreases with decreasing number of the carbon atoms bound to cobalt. A decrease in the calculated dipole moment correlates well with an increase in experimental R_f.

The C-C bond lengths were found to vary continuously in a certain interval in mixed pyrrolyl/dicarbollide sandwich complexes as indicated by theoretical calculations on *closo*-$3\text{-Co}(\eta^5\text{-}NC_4H_4)(1,2\text{-}C_2B_9H_{11})$ and *closo*-$3\text{-Co}(\eta^5\text{-}NC_4H_4)\text{-}1,2\text{-}(SCH_3)_2\text{-}1,2\text{-}C_2B_9H_9$ [106]. Single-point calculations were performed at the ZINDO/1 level using crystallographic atomic coordinates. The same results were obtained at the HF/TZV level of theory. The LUMO exhibits antibonding character between the two carbon atoms of the CoC_2B_9 icosahedron. When a hydrogen atom is replaced by a S-R group, the HOMO that is formally occupied by the lone pair combines with the former LUMO of compound to form two new molecular orbitals.

Computational studies have been performed on the neutral dicarbollide complex $CpFeC_2B_9H_{11}$ and its monoanion, the charge-compensated ferradicarbollides $1\text{-Cp-4-L-}1,2,3\text{-}FeC_2B_9H_{10}$ (L = SMe_2, NMe_3, py) and their methylated analogs $1\text{-Cp-}2,3\text{-}Me_2\text{-}4\text{-}SMe_2\text{-}1,2,3\text{-}FeC_2B_9H_8$ and $1\text{-Cp*-}4\text{-}SMe_2\text{-}1,2,3\text{-}FeC_2B_9H_{10}$, the isomeric ferratricarbollides $CpFeC_3B_8H_{11}$, and the amino-substituted derivative 1-Cp-12-^{t}BuNH-1,2,4,12-$FeC_3B_8H_{10}$ [107]. Reduction of iron(II) to iron(I) becomes accessible as a consequence of the shift of the HOMO–LUMO frontier orbitals to higher energy levels. In order to arrive at a MO explanation for structural changes upon reducing $CpFeC_2B_9H_{11}$ to its monoanion, DFT calculations at the B3LYP/6-31G* level) were performed on the monoanion. The HOMO was found to be similar to the $e_{2g}/d_{x^2-y^2}$ orbital of ferrocene, with bonding for Fe···C_2B_3 and Fe···Cp interactions and antibonding for C_{Cp}–C_{Cp} and C_{Carb}–C_{Carb} interactions. Accordingly, removal of one electron from this orbital leads to the lengthening of Fe···C_2B_3 and Fe···Cp distances and shortening of C_{Cp}–C_{Cp} and C_{Carb}–C_{Carb} distances.

The existence of a dihydrogen bond (S-H···H-B) and its combination with a C-H···S hydrogen bond in an unusual cooperative effect is demonstrated using a combination of experimental and theoretical methods [108]. This cooperative effect seems to be responsible for self-assembly of mercaptan-metallacarborane complexes such as *closo*-3-Ru(η^6-C_6H_6)-8-HS-1,2-$C_2B_9H_{10}$ and *closo*-3-Co(η^5-C_5H_5)-8-HS-1,2-$C_2B_9H_{10}$ (3), which present identical supramolecular two-dimensional polymeric networks. These represent the first structural characterization and accurate DFT calculations of a new type of dihydrogen bond (S-H···H-B) and its combination with a C–H···S hydrogen bond interaction in a unprecedented cooperative effect. The existence of the S-H···H-B interactions has been further confirmed by means of Bader's atoms in molecules (AIM) theory.

The observed rearrangement sequence of (η^4-Me_4C_4)$CoPC_2B_8H_{10}$ isomers correlates well with the relative stabilities of their nonmethylated analogues as estimated by DFT calculations [109]. In order to estimate the thermodynamic stability and ^{11}B NMR spectra of the cobaltaphosphadicarbollide isomers, DFT calculations were carried out at the B3LYP/6-31G* level for their nonmethylated analogues (η^4-C_4H_4)$CoPC_2B_8H_{10}$. The stability of metallaphosphadicarbollide isomers was rationalized by the relative electronegativities of the heteroatoms. Unfortunately, the calculated ^{11}B NMR spectra for the chlorine-substituted cobaltacarborane isomers (C_4H_4)$CoPC_2B_8H_9Cl$ differ significantly from the experimental data, even for the structurally characterized isomer. The biggest deviation (5–8 ppm) was found for the signal of the boron atom connected directly to the chlorine atom. The deviations for the other boron atoms are 1–5 ppm. Increasing the theory level up to B3LYP/6-31+G* was found to provide no improvement. Similar large deviations were observed previously for calculations of 10-Cl-7,8,9-$PC_2B_8H_{10}$ at the HF/II//MP2/6-31G* level.

DFT solid-state calculations on the *closo* (18-electron) and *pseudocloso* (16-electron) Rh(III) complexes, 3-(η^5-C_5Me_5)-1,2-$(PhCH_2)_2$-*closo*-3,1,2-$RhC_2B_9H_9$, and 3-{(1–3-η^3)-C_8H_{13}}-1,2-(4′-$MeC_6H_4)_2$-*pseudocloso*-3,1,2-$RhC_2B_9H_9$ (C···C, 2.420(2) Å), show that the transfer of electron density from the carborane moiety to the rhodium center is marginally greater for the *pseudocloso* derivative. This agrees with the idea that electronics rather than sterics play a crucial role in the stabilization of 16-electron *pseudocloso*-metallacarborane species [110]. Solid-state DFT calculations were performed on the two Rh(III) complexes listed above, chosen as models of *closo* (18-electron) and *pseudocloso* (16-electron) metallacarboranes with closely related bulky cage substituents. Bader's Atoms in molecules" (AIM) method, which is based on the topological analysis of electron density distribution function ($\rho(r)$), has also been used for studies on these rhodadicarbaboranes.

The metallacarboranes 3-(η-C_5R_5)-3,1,2-$MC_2B_9H_{11}$ (M = Co, R = H; M = Co, R = Me; M = Rh, R = H; M = Rh, R = Me; M = Ir, R = H; M = Ir, R = Me) have been studied by a variety of theoretical methods [111]. Thus DFT calculations of the redox potentials and corresponding geometrical changes were performed. In addition metal–ligand bonding was analyzed by energy decomposition analysis. Computation satisfactorily predicts the potentials for the rhoda- and iridacarboranes with a maximum deviation from the experimental value of 0.23 V. However, the calculated values for the cobalt complexes agree much more poorly with the experimental values having deviations of 0.40–0.83 V. Interestingly, the behavior of the cobalt complexes proved to be significantly different from that of the rhodium and iridium analogs.

Replacing the Cp ligand in the original model complex $[CpIr^{III}PH_3CH_3]^+$ with the dicarbollide ligand η^5-7,8-$C_2B_9H_{11}{}^{2-}$ ($C_2B_9H_{11}$) produced a neutral Ir(V) intermediate, $(C_2B_9H_{11})Ir^VPH_3(CH_3)_2H$, lying 16.4 kcal/mol (ΔE, BS1) in energy above the Ir(III) complex and uncoordinated methane. This suggests that the Ir(V) complex is not viable [112].

The bonding in [(Carb)Fe(μ-η:η-*cyclo*-P_5)FeCp*]$^+$ (Carb = 9-SMe_2-7,8-$C_2B_9H_{10}$ and 1-^{t}BuNH-1,7,9-$C_3B_8H_{10}$) has been analyzed by energy decomposition analysis (BP86/TZ2P) [113].

The benzene complexes $[(\eta^5\text{-9-L-7,8-}C_2B_9R_2H_8)Fe(\eta\text{-}C_6H_6)]^+$ (L = SMe_2, R = H; L = SMe_2, R = Me; L = NMe_3, R = H) have been synthesized and converted to the tris(ligand) derivatives $[(\eta^5\text{-9-}SMe_2\text{-7,8-}C_2B_9H_{10})FeL_3]^+$ (L = ^{t}BuNC or $P(OMe)_3$) by photolysis with the ligand using visible light irradiation [114]. DFT calculations of the redox potentials and the respective geometrical changes were performed on these and related systems. The experimental and the calculated formal electrode potentials for these complexes and the related cyclopentadienyl derivatives $[CpFe(\eta\text{-}C_6H_6)]^+$, $CpFe(\eta^5\text{-}C_6H_7)$, and $FeCp_2$ have been determined. Independently from the solvation model (PCM or COSMO), computation satisfactorily predicts the potentials with a maximum deviation from the experimental value of 0.26 V for PCM and 0.22 V for COSMO. The calculations also allowed evaluation of structural changes accompanying redox processes.

DFT calculation of cluster shapes and their related boron-vertex ^{11}B nuclear shieldings for a range of substituted 12-vertex metalladicarbaborane cluster compounds 3,1,2-R_2-3-(η^6-C_6H_6)-3,1,2-$RuC_2H_9H_9$ confirms the experimentally observed C–C distances in the group of progressively more open *closo*, "*semipseudocloso*" and "*pseudocloso*" clusters. This results from the inherent propensity of the (η^6-C_6H_6)$RuC_2H_9H_9R_2$ cluster to occupy either *closo*-structured *or* so-called *pseudocloso-structured* fundamental local minima on the overall potential energy surface in the absence of steric drivers [115]. In two reported cases, the cluster flexibility is evident in solution. In these cases, GIAO calculation of the ^{11}B chemical shieldings explains the measured ^{11}B chemical shifts, which can be interpreted in terms of the dynamic interchange of the *closo* and "*pseudocloso*" species.

Structural characterization of $[HNMe_3][3,3\text{-}(CO)_2\text{-}3,3\text{-}Cl_2\text{-}closo\text{-}3,1,2\text{-}ReC_2B_9H_{11}]$ has identified it as a four-legged piano stool half-sandwich complex anion resulting from oxidation of rhenium by both the aminium and hexachloroantimonate ions and CO displacement promoted by Me_3NO in the presence of by-product chloride ions. Direct iodination of $Cs[3,3,3\text{-}(CO)_3\text{-}closo\text{-}3,1,2\text{-}ReC_2B_9H_{11}]$ has yielded $3,3,3\text{-}(CO)_3\text{-}3\text{-}I\text{-}closo\text{-}3,1,2\text{-}ReC_2B_9H_{11}$, which undergoes iodide migration in donor solvents from the metal to the boron atom in the coordinating face of the cage [116]. The ground state highest occupied and lowest unoccupied molecular orbitals were calculated at the B3LYP/6-31G*level with an initial geometry based on the crystallographic structure of the $[3,3,3\text{-}(CO)_3\text{-}closo\text{-}3,1,2\text{-}ReC_2B_9]$ core using a pseudopotential because of the presence of the heavy rhenium atom. No corrections were applied for solvation. The density functional theory calculation of the ground state highest unoccupied and lowest occupied molecular orbitals (HOMO and LUMO) reveals a wide frontier orbital energy gap of 5.27 eV. The HOMO is primarily metal-centered d in character, with significant back-bonding into the cage likely responsible for an energy stabilizing effect.

Three sets of *n*-legged piano stool shaped ($-Re(CO)_3$ or $-Re(CO)_3I$) 12-vertex rhenacarborane half-sandwich complexes ($n = 3$ or 4) have been investigated by density functional theory (DFT) [117]. Natural bond orbital (NBO) analysis shows that introducing iodine substituents changes the bond formation between rhenium and carborane cages or 3/4-legged piano stools. Furthermore, time-dependent density functional theory (TDDFT) calculations were used to account for the redox-switchable NLO properties. Theoretical investigations on geometrical structures suggest that substituting hydrogen with iodine contracts the lower part of the rhenacarborane cage. However, coordination of iodine to rhenium not only expands the upper part of the rigid rhenacarborane cage and contracts its lower part and longitudinal length, but also leads to shorter C–O bonds and longer Re-C bonds. TDDFT calculations indicate that the intramolecular interactions between iodine, rhenium, and the carborane cage play important roles in the redox processes.

The cage-alkylated metallacarborane complex $4,4,4\text{-}(Me_2N)_3\text{-}3\text{-}Me\text{-}4,1,2\text{-}closo\text{-}TaC_2B_9H_{10}$ is obtained as the only product from the reaction of $Ta(NMe_2)_5$ with *nido*-$11\text{-}Me\text{-}2,7\text{-}C_2B_9H_{12}$ [118]. The geometry of the metallaborane skeleton is supported by boron NMR shift predictions based on observed shifts of the known non-methylated analogues $3,3,3\text{-}(Me_2N)_3\text{-}closo\text{-}3,1,2\text{-}TaC_2B_9H_{10}$, and $3,3,3\text{-}(Me_2N)_3\text{-}closo\text{-}3,1,2\text{-}TaC_2B_9H_{10}$ and calculated shifts (GIAO-B3LYP/6-311G*//MP2/6-31G*) of the B-methyl carboranes, *closo*-$3\text{-}Me\text{-}1,2\text{-}C_2B_{10}H_{11}$, $4\text{-}Me\text{-}1,2\text{-}C_2B_{10}H_{11}$, and $2\text{-}Me\text{-}1,7\text{-}C_2B_{10}H_{11}$. Reliable results are obtained from *ab initio* calculations of the structures and ^{11}B NMR chemical shifts of boranes and carboranes. However, it was not possible to perform reliable calculations on metallacarboranes. Instead the calculations were performed at the highest possible level of theory for the parent *closo*-carboranes $3\text{-}Me\text{-}1,2\text{-}C_2B_{10}H_{11}$, $4\text{-}Me\text{-}1,2\text{-}C_2B_{10}H_{11}$ and $3\text{-}Me\text{-}1,7\text{-}C_2B_{10}H_{11}$, obtained by replacing the $(Me_2N)_3Ta$ vertex with a BH vertex. Then the chemical shift differences caused by methylating one boron atom, or by replacing a BH vertex with a $(Me_2N)_3Ta$ vertex were determined.

Metallaborane compounds containing two adjacent metal atoms, namely $(PMe_2Ph)_4MM'B_{10}H_{10}$ (where MM′ = Pt_2, 1; PtPd, 7; Pd_2, 8), have been synthesized [119]. Their propensities to sequester O_2, CO, and SO_2 and then to release them under pulsed and continuous irradiation are described. Only the diplatinum derivative $(PMe_2Ph)_4Pt_2B_{10}H_{10}$, is found to undergo reversible binding of O_2 to form $(PMe_2Ph)_4(O_2)Pt_2B_{10}H_{10}$. However solutions of all three complexes ({PtPt}, {PtPd}, and {PdPd}) quantitatively take up CO across their metal–metal vectors to form $(PMe_2Ph)_4(CO)MM'B_{10}H_{10}$ (MM′ = Pt_2, PtPd, and Pd_2). Crystallographically determined interatomic M–M distances and infrared CO stretching frequencies show that the CO molecule is bound progressively more weakly in the sequence {PtPt} > {PtPd} > {PdPd}. Similarly, SO_2 forms $(PMe_2Ph)_4(SO_2)Pt_2B_{10}H_{10}$, $(PMe_2Ph)_4(SO_2)PtPdB_{10}H_{10}$, and $(PMe_2Ph)_4(SO_2)Pd_2B_{10}H_{10}$ with progressively weaker binding of the SO_2 molecule. The uptake and release of gas molecules are accompanied by changes in their absorption spectra. These systems were initially assessed by DFT calculation of molecular geometry at the B3LYP/6-31G*/LANL2DZ level in order to construct a matrix of information as to how modification of the metal would be expected to weaken or strengthen the binding of the small molecules.

DFT calculations at the B3LYP/6-31G* level on the isocyanide complexes $\{(EtNC)(PMe_2Ph)_3\}(\mu\text{-}EtNC)Pt_2B_{10}H_{10}$ and $[\{(EtNC)_2(PMe_2Ph)_2\}Pd_2B_{10}H_{10}]$ agree well with the experimental values [120]. The ability to tune the $[(PR_3)_4M_2B_{10}H_{10}]$-based system to bind reversibly small molecules such as O_2, CO, and SO_2 across the metal–metal vector, as indicated above, can be extended also to alkyl isocyanides as modeled here by EtNC. However, the reaction is also accompanied by displacement of the phosphine ligand by the incoming isocyanide. This enables the assessment of the effect of the RNC-for-PR_3 substitution on the binding ability, which is to weaken the binding of the small molecules. This weakening is in accord with DFT-based calculations, although the terminal ligand displacement was not so predicted. This illustrates retrospectively a limitation of DFT predictive work when possible alternative reaction coordinates are not considered.

Crystal structure analyses of the icosahedral metallacarborane complexes, $L_2MC_nB_{11-n}$ (L = R_3P, RNC; M = Pd, Pt; n = 1, 2), have shown that the magnitude and direction of the slipped distortion depends upon the substituents on the open pentagonal face of the ligand. Extended Hückel calculations on model complexes have accounted for the observed geometries [121].

Extended-Hückel molecular-orbital calculations have also been used to study the icosahedral platinaboranes and carbaboranes $[B_{11}H_{11}Pt(PH_3)_2]^{2-}$, $[B_{10}CH_{11}Pt(PH_3)_2]^-$ and $B_9C_2H_{11}Pt(PH_3)_2$ but using the simplified PH_3 ligand rather than the alkyl and/or aryl phosphines found in the experimental systems [122]. The failure of the Wade-Mingos skeletal electron-counting rules [4–6] when applied to carbaplatinaboranes is discussed, and attributed to the unequal bonding capabilities of the platinum $5d_{xz}$ and $5d_{yz}$ orbitals in the $Pt(PH_3)_2$ fragment. The conformations of the icosahedral carbaplatinaboranes are rationalized on the basis of the symmetry characteristics of the lowest-unoccupied orbital of the carbaborane and the highest-occupied orbital of the metal-phosphine moiety.

Extended Hückel molecular orbital calculations on the *closo*-platinaboranes $[B_{11}H_{11}Pt(PH_3)_2]^{2-}$ and $[B_6H_6Pt(PH_3)_2]^{2-}$ and the *closo*-carbaplatinaboranes $B_9C_2H_{11}Pt(PH_3)_2$ and $B_4C_2H_6Pt(PH_3)_2$ have suggested that the larger "slip" distortions in the pentagonal bipyramidal derivatives can be attributed mainly to the different metal-ligand interactions that are induced by the different elevation angles of the substituents on the pentagonal faces of the ligands [123]. Differences in bond lengths associated with the two classes of platinacarbaboranes have been interpreted in terms of the different bonding capabilities of the two types of carbaborane ligands.

The geometry of the cage in 3,3-bis(triethylphosphine)-1,2-dicarba-3-platinadodecaborane(11) is that of a highly distorted icosahedron. Molecular-orbital calculations based on the extended-Hückel approximation have accounted for the slip and fold distortions observed in the related metallacarbaboranes [124].

Tricarbollide anions such as *nido*-$C_3B_8H_{11}^-$ are of interest since they have the same charge and similar bonding properties as the ubiquitous cyclopentadienide anion. The room-temperature photochemical reaction of the tricarbollide anion [*nido*-7,8,9-$C_3B_8H_{11}]^-$ with $[CpFe(C_6H_6)]^+$ proceeds without cluster rearrangement to form the 12-vertex *closo*-ferratricarbollide 1-Cp-1,2,3,4-$FeC_3B_8H_{11}$ [125]. However, heating this iron tricarbollide leads to rearrangement giving the isomeric complex 1-Cp-1,2,3,5-$FeC_3B_8H_{11}$ at 110 °C and then 1-Cp-1,2,4,10-$FeC_3B_8H_{11}$ at 165 °C. In view of the similarities of the Cp and $C_3B_8H_{11}^-$ ions, these iron complexes can be considered to be direct analogues of ferrocene. The reaction of the tricarbollide anion with $[Cp^*RuCl]_4$ is accompanied by polyhedral rearrangement giving 1-Cp*-1,2,3,5-$RuC_3B_8H_{11}$. Further isomerization of this ruthenium complex occurs slowly at room temperature and rapidly at 65 °C to give 1-Cp*-1,2,4,10-$RuC_3B_8H_{11}$. A diamond-square-diamond mechanism for the polyhedral rearrangement sequence is proposed. The relative stability of these three isomers was estimated by DFT calculations at the B3LYP/6-31G* level. The calculated structures agree well with experimental data, indicating reliability of the calculated results. The ^{11}B NMR chemical shifts of the three isomers, calculated using the GIAO method, agree well with the experimental values. This makes possible the assignment of NMR signals even in the absence of [^{11}B-^{11}B]-COSY correlations. In order to estimate the strength of the Fe-tricarbollide bond, the enthalpies of reactions of the $[CpFe]^+$ fragment with [*nido*-7,8,9-$C_3B_8H_{11}]^-$ and Cp^- anions were calculated.

DFT calculations of the hypothetical non-rearranged isomer 1-Cp-2-Bu^tNH-1,2,3,4-$FeC_3B_8H_{10}$ revealed strongly distorted geometry with the cage C–C distance (2.347 Å) being clearly non-bonding, thus explaining the mild conditions of the polyhedral rearrangement [126]. In order to provide an explanation for such a strong effect of the amino substituent, DFT calculations were carried out at the PBE/TZV level on the ferratricarbollide 1-Cp-4-Bu^tNH-1,2,3,4-$FeC_3B_8H_{10}$. The C–C bond breaking is caused by strong p donation from the nitrogen lone pair of the amino-substituent substituent. The introduction of a short bridge between carbon atoms prevents their separation in the polyhedral rearrangement.

A room-temperature reaction between the [7-^tBuNH-*nido*-7,8,9-$C_3B_8H_{10}]^-$ anion and $[Cp^*RuCl]_4$ leads to the ruthenatricarbollide 1-Cp*-12-^tBuNH-1,2,4,12-

$RuC_3B_8H_{10}$ [127]. Analogously, the room-temperature photochemical reaction of [7-^{t}BuNH-*nido*-7,8,9-$C_3B_8H_{10}]^-$ with [CpFe(C_6H_6)]PF_6 gives the previously reported iron complex 1-Cp-12-^{t}BuNH-1,2,4,12-$FeC_3B_8H_{10}$. Both reactions are associated with extensive polyhedral rearrangement, which occurs under very mild conditions and brings the carbon atoms to positions of maximum separation within the framework. Density functional theory calculations at the B3LYP/SDD level were used to estimate relative stabilities of these metallacarborane isomers.

A series of Schiff bases was prepared by the condensation of 9,9′-diaminobis-(tricarbollide)Fe(II) with appropriate 4-alkoxybenzaldehydes. The photophysical properties were investigated for the butoxy derivative and modeled using ZINDO calculations [128]. These systems provide the first liquid crystalline derivatives of a metallaborane. The extent of electronic interactions between the Schiff base substituents and the tricarbollide ligand was evaluated using electronic absorption spectroscopy augmented by ZINDO calculations. In order to assess the extent of electronic interactions between the bis(tricarbollide)Fe(II) and the Schiff base substituents the ultraviolet spectra were recorded and analyzed using quantum mechanical methods.

The 1,2-, 1,7-, and 1,12-isomers of $[CpMCB_{10}H_{11}]^{2-}$ (M = Fe, Ru, Os), $[CpMCB_{10}H_{11}]^-$ (M = Co, Rh, Ir), and $CpMCB_{10}H_{11}$ (M = Ni, Pd, Pt) have been studied computationally [129]. The 1,12- (para) isomers were found to be the lowest energy isomers. A large number of 12-vertex *closo*-ferracarborane structures from $[CpFeCB_{10}H_{11}]^{2-}$ to $CpFeC_3B_8H_{11}$, phosphaborane structures from $[CpFePB_{10}H_{10}]^{2-}$ to $CpFeP_3B_8H_8$, and the phosphacarbaborane structures $[CpFePCB_9H_{10}]^-$, $CpFeP_2CB_8H_9$, and $CpFePC_2B_8H_{10}$, were computed in order to check the additive nature of the structural increments for 12-vertex *closo*-metallaheteroboranes having CpM vertices. The results help to estimate quickly the relative thermodynamic stabilities of various 12-vertex *closo*-cyclopentadienylmetallacarboranes, where M may be Fe or any other group 8, 9, or 10 metal.

The experimental molecular structures of 1,2-*closo*-$P_2B_{10}H_{10}$ and 1,2-closo-$As_2B_{10}H_{10}$ determined by gas-phase electron diffraction have been compared with the computed structures at the MP2/6-31G** level of theory [130]. The level of agreement is good for the diarsaborane (rms misfit for As and B atoms 0.0297 Å) and very good for the diphosphaborane (rms misfit for P and B atoms 0.0082 Å). The metalladiheteroboranes 3-(η^5-C_9H_7)-3,1,2-*closo*-$CoAs_2B_9H_9$, 3-(η^6-$C_{10}H_{14}$)-3,1,2-*closo*-$RuAs_2B_9H_9$, 3-(η^5-C_5H_5)-3,1,2-*closo*-$CoP_2B_9H_9$, 3-(η^5-C_9H_7)-3,1,2-closo-$CoP_2B_9H_9$, and 3-(η^6-$C_{10}H_{14}$)-3,1,2-*closo*-$RuP_2B_9H_9$ were studied using the Gaussian program to gauge the effect of functional (BP86 and B3LYP) and the MP2 methodology and basis set (6-31G** for all atoms or 6-31G** for B and H and the Stuttgart relativistic ECP and associated basis sets, augmented with d-polarization, for As and P; BP86/SDD/6-31G**) on the structural parameters.

The molecular structure of 1-selena-*closo*-dodecaborane, 1-$SeB_{11}H_{11}$, has been determined by the concerted use of quantum chemical calculations and gas-phase electron diffraction [131]. This structure has C_{5v} symmetry and is distorted from a regular icosahedron mainly through expansion of the pentagon of boron atoms adjacent to selenium. The accuracy of the experimental structure, as well as that calculated with the MP2 method, has been gauged by comparison of the ^{11}B chemical shifts (calculated at two different gauge-including atomic orbitals (GIAO) levels) with experimental NMR values. The inclusion of electron correlation in the magnetic property calculations (GIAO-MP2) gave superior results to those carried out using GIAO-Hartree–Fock methods. The MP2/6–311+G* calculations showed that the symmetrically unique distances differed by only 0.1 pm. The good agreement between the experimental and calculated ^{11}B NMR chemical shifts, in particular at the GIAO-MP2 level, strongly suggests that the set of MP2/AE3 internal coordinates is a good representation of the molecular geometry of *closo*-1-$SeB_{11}H_{11}$. This is corroborated by a single-point energy calculation for the experimental geometry of *closo*-1-$SeB_{11}H_{11}$ at the MP2/AE3 level, which finds this structure to lie only 0.5 kcal/mol in energy above the fully optimized geometry. According to natural bond orbital analysis, the bonding between the Se and B atoms in *closo*-1-$SeB_{11}H_{11}$ has predominantly p character (86.4 %). No large spin–orbit corrections are therefore to be expected for calculations of the ^{11}B magnetic shieldings, since such effects are usually transmitted via a Fermi-contact-type relay mechanism, which is very effective when bonds with high s character are involved, as also found in *arachno*-$Se_2B_8H_{10}$. The explicit ZORA approach unambiguously confirmed this expectation, including a comparison of the non-relativistic and relativistic BP86 values. As with *arachno*-$Se_2B_8H_{10}$, scalar relativistic effects have additionally been assessed by comparing GIAO-HF/AE4 and GIAO-HF/ECP2 results, which differ at most by 0.6 ppm for any of the ^{11}B nuclei. Therefore relativistic effects can be considered as negligible.

Relativistic density functional theory (DFT) calculations were performed on model aluminaboranes such as (η^3-Cp)CH_2(η^1-NMe)]Al(Me), [{(η^5-Cp)CH_2(η^1-NMe_2)}Al(Me)]$^+$, and (η^5-C_2B_9)CH_2(η^1-NMe_2)]Al(Me) using the Amsterdam density functional (ADF) code with the Becke and Perdew functional in order to understand the bonding interactions between the aluminum ion and dicarbollyl ligand [132]. A strong π-type interaction between aluminum and the dicarbollyl unit was clearly demonstrated by the DFT calculation.

3.3.11 13-vertex Systems

The 13-vertex most spherical deltahedron is the docosahedron with 22 triangular faces, 33 edges, 2 vertices of degree 6, 10 vertices of degree 5, and 1 vertex of degree 4 (Fig. 3.6). A 13-vertex deltahedron with 12 vertices of degree 5 and 1 vertex of degree 6 is not topologically possible. Since degree 6 boron vertices are not favorable, many 13-vertex polyhedral boranes and related metallaboranes have henicosahedral structures with 1 quadrilateral face or ditrapezoidal icosahedral structures with 2 quadrilateral faces. In essentially all 13-vertex metallaboranes, the metal atoms are located at degree 6 vertices.

The 13-vertex indenyl cobaltacarboranes with 4,1,6-, 4,1,10- and 4,1,2-CoC_2B_{10} architectures have been synthesized [133]. The 4,1,6-, 4,1,8-, 4,1,10- and 4,1,12-

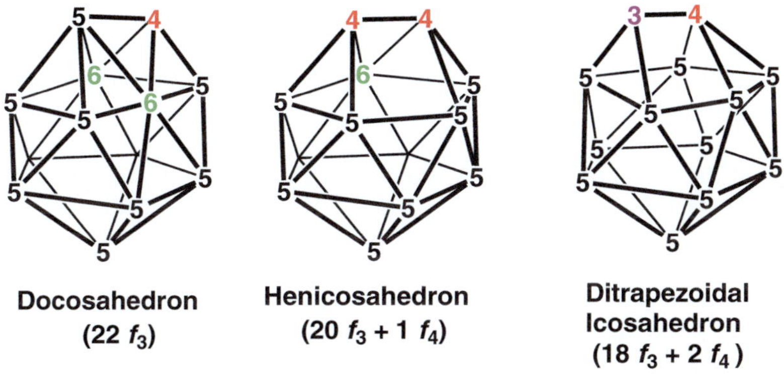

Fig. 3.6 Some 13-vertex polyhedra with mainly triangular faces

species have docosahedral cages while the 4,1,2-species is henicosahedral (Fig. 3.6). In cases where the orientation is not compromised by the presence of a non-hydrogen substituent on the face of the carborane, there is generally good agreement between the experimental orientation and that computed by DFT calculations for the related naphthalene ferracarboranes (η^6-$C_{10}H_8$)$FeC_2B_{10}H_{12}$. The presence of C-methyl substituents in the indenyl cobaltacarboranes tends to override this preference except for 1,6-Me_2-4-(η^5-C_9H_7)-4,1,6-*closo*-$CoC_2B_{10}H_{10}$ in which the indenyl ligand instead is forced to incline away from the cage methyl groups. The structural work has been supplemented with the results of DFT calculations on various isomers of the related naphthalene ferracarboranes (η-$C_{10}H_8$)$FeC_2B_{10}H_{12}$.

DFT calculations indicate that the barriers to isomerization of both 4-Cp-4,1,2-closo-$CoC_2B_{10}H_{12}$ and 4-(η^6-C_6H_6)-4,1,2-*closo*-$RuC_2B_{10}H_{12}$ to their respective 4,1,6-isomers are too high for this to be the origin of the unexpected formation of 4,1,6-MC_2B_{10} products [134]. The DFT-calculated profile for the isomerization of [7,8-*nido*-$C_2B_{10}H_{12}$]$^{2-}$ to [7,9-*nido*-$C_2B_{10}H_{12}$]$^{2-}$ suggests that the unexpected formation of 4,1,6-metallacarboranes is unlikely to result from isomerization of a reduced (*nido*) carborane.

The ruthenacarborane 4-(*p*-cymene)-4,1,6-*closo*-$RuC_2B_{10}H_{12}$ has been characterized both spectroscopically and crystallographically. Although asymmetric in the solid state with a docosahedral cage architecture with cage carbon atoms at vertices 1 and 6, this species clearly has C_s symmetry on the NMR time scale at room temperature. However, the fluxional process in operation can be arrested at low temperature, so that an activation energy of 10.3 kcal/mol is estimated [135]. A computational study of the related species 4-(η^6-C_6H_6)-4,1,6-*closo*-$RuC_2B_{10}H_{12}$ reveals that the fluxionality arises from a double diamond–square–diamond process, first suggested by Dustin, Hawthorne et al. for the analogous CpCo species [136]. These calculations yield an activation energy of 9.7 kcal/mol, in excellent agreement with that derived from experiment.

The ruthenium derivative 1-(1′,2′-*closo*-$C_2B_{10}H_{11}$)-4-{$C_{10}H_{14}$Ru(*p*-cymene)}-4,1,6-*closo*-$RuC_2B_{10}H_{11}$ consists of a 13-vertex docosahedral ruthenacarborane with a 4,1,6-RuC_2B_{10} architecture, which is the isomer obtained upon reduction and metallation of *o*-carborane [137]. The mechanism by which this compound is formed has been investigated using density functional theory calculations replacing the *p*-cymene ligand by benzene for computational expedience.

The 12-vertex naphthalene ruthenacarborane 3-(η^6-$C_{10}H_8$)-3,1,2-closo-$RuC_2B_9H_{11}$ and the analogous 13-vertex isomers 4-(η^6-$C_{10}H_8$)-4,1,6-closo-$RuC_2B_{10}H_{12}$, 4-(η^6-$C_{10}H_8$)-4,1,8-*closo*-$RuC_2B_{10}H_{12}$, 4-(η^6-$C_{10}H_8$)-4,1,10-*closo*-$RuC_2B_{10}H_{12}$, and 4-(η^6-$C_{10}H_8$)-4,1,12-*closo*-$RuC_2B_{10}H_{12}$ have been prepared and characterized [138]. The conformations of the naphthalene ligands with respect to the carborane cages in the three 13-vertex derivatives are in good agreement with those previously calculated for the analogous ferracarboranes 4-(η^6-$C_{10}H_8$)-4,1,*x*-*closo*-$FeC_2B_{10}H_{12}$.

The docosahedral metallacarboranes 4,4-$(PMe_2Ph)_2$-4,1,6-*closo*-$PtC_2B_{10}H_{12}$, 4,4-$(PMe_2Ph)_2$-4,1,10-*closo*-$PtC_2B_{10}H_{12}$, and $[N(PPh_3)_2]$[4,4-(1,5-C_8H_{12})-4,1,10-closo-$RhC_2B_{10}H_{12}$] have been synthesized [139]. Comparison with conformations previously established for six other $ML_2C_2B_{10}$ species of varying heteroatom patterns (4,1,2-MC_2B_{10}, 4,1,6-MC_2B_{10}, 4,1,10-MC_2B_{10}, and 4,1,12-MC_2B_{10}) reveals clear preferences. In all cases a qualitative understanding was afforded by simple MO arguments applied to the model heteroarene complexes $[(PH_3)_2PtC_2B_4H_6]^{2-}$ and $[(PH_3)_2PtCB_5H_6]^{3-}$. Moreover, DFT calculations on $[(PH_3)_2PtC_2B_4H_6]^{2-}$ in its various isomeric forms approximately reproduced the observed conformations in the 4,1,2-, 4,1,6-, and 4,1,10-MC_2B_{10} species. However, analogous calculations on $[(PH_3)_2PtCB_5H_6]^{3-}$ did not reproduce the conformation observed in the 4,1,12-MC_2B_{10} metallacarborane. DFT calculations on $(PH_3)_2PtC_2B_{10}H_{12}$ yielded good agreement with experimental conformations in all 4 structures.

Density functional theory employing the B3LYP functional with the 6-31G(d) basis set was used to optimize the structures of the 13-vertex metallacarboranes 4,1,6-$MC_2B_{10}H_{12}$ (M = Mn, Fe, Co, Ni) [140]. The polarizability and second-order nonlinear optical (NLO) coefficients were calculated by means of the finite-field (FF) method. In all metallacarboranes, smaller frontier molecular orbital energy gaps lead to larger second-order NLO coefficients. For the homogeneous metallacarboranes of different spin states, the dipole moments of high-spin state structures are larger than those of low-spin state structures. However, the polarizability and second-order NLO coefficients are not directly related to spin multiplicity. Therefore, the spin state has minor influence on the NLO properties.

Reactions of [*nido*-$(Me_2NCH_2CH_2)(R)C_2B_{10}H_{10}]^{2-}Na_2$ (R = H, $MeOCH_2CH_2$, $Me_2NCH_2CH_2$) with $ZrCl_4{\cdot}(THF)_2$ or $SmI_2(THF)_x$ gave metallacarboranes with unusual η^7-*arachno*-carboranyl ligands [141]. In these complexes heteroatom-containing pendant sidearms are both electronically and entropically necessary for the formation of such complexes with the central metal ions in the highest oxidation state. In view of the strong reducing power of the *nido*-$R_2C_2B_{10}H_{10}^{2-}$ dianion, a mechanism involving first the reaction of $[(Me_2NCH_2CH_2)(R'CH_2CH_2)$-$C_2B_{10}H_{10}]$ $Na_2(THF)_x$ with $ZrCl_4(THF)_2$ to give the divalent zirconacarborane intermediate $[\eta^1{:}\eta^1{:}\eta^6$-$(Me_2NCH_2CH_2)(R'CH_2$-$CH_2)C_2B_{10}H_{10}]Zr^{II}$ and the o-carborane 1-$Me_2NCH_2CH_2$-2-$R'CH_2CH_2$-1,2-$C_2B_{10}H_{10}$, followed by intramolecular electron transfer from the Zr(II) center to the *nido*-carborane. This proposal is supported by theoretical studies with the aid of B3LYP density functional theory calculations. The activation energy is calculated to be only 16.6 kcal/mol, indicating that this process can occur readily. Such a conversion involves intramolecular electron transfer in which the Zr(II) center formally loses two d electrons to the *nido*-carborane ligand, leading to the formation of the *arachno*-carborane ligand in the final product. The calculated reaction energy indicates that the intramolecular electron-transfer process is thermodynamically very favorable. The two pendant sidearms, which are classified as hard ligands in inorganic chemistry, are expected to contribute significantly to the driving force of this reaction because they stabilize a Zr(IV) instead of a Zr(II) metal center. The d^0 Zr(IV) center fulfills the bonding requirement with an η^7-*arachno*-carboranyl ligand.

The 13-vertex metallacarboranes $\{[(C_6H_5CH_2)_2C_2B_{10}H_{10}]M(THF)\}_2\{Na(THF)_3\}_2\cdot 2THF$ (M = Dy, Y, Er) and $[\{(C_6H_5CH_2)_2C_2B_{10}H_{10}\}M(THF)]_2\}[Li(THF)_4]_2$ (M = Y, Er) have been synthesized [142]. DFT calculations at the B3LYP level were performed on the model complex $[(\eta^7\text{-}C_2B_{10}H_{12})Y(H_2O)]_2^{2-}$ on the basis of the experimentally determined geometry. In the model complex, the benzyl substituents were replaced by hydrogen atoms and the coordinated THF ligand was simplified to H_2O for theoretical simplicity. Molecular orbital calculations indicate that the metal-carborane bonding is well delocalized and can be described as the orbital interactions between the metal five d orbitals and the cage five symmetry-adapted frontier orbitals. Only a d^0/f^n transition metal ion with the proper size is expected to be capable of being η^7-bound to an *arachno*-carboranyl ligand. Examination of the molecular orbitals obtained from the B3LYP calculations indicates that the five d orbitals of the yttrium atom are all significantly involved in the metal-$[\eta^7\text{-}C_2B_{10}H_{12}]$ bonding interactions. The relevant molecular orbitals are mainly found in the HOMO-LUMO region.

The 13-vertex stannadicarbaboranes 4,1,6-*closo*-$SnC_2B_{10}H_{12}$ and 1,6-Me_2-4,1,6-*closo*-$SnC_2B_{10}H_{10}$ have been synthesized and their fluxional processes studied using DFT [143]. In this connection the fluxional nature of 4,1,6-MC_2B_{10} metallacarboranes has been discussed in terms of a C_s transition state related to the ground state molecule by a double diamond–square rearrangement. The optimized 4,1,6-*closo*-$SnC_2B_{10}H_{12}$ structure corresponds very closely to that determined crystallographically for the dimethyl derivative 1,6-Me_2-4,1,6-*closo*-$SnC_2B_{10}H_{10}$ (the more accurate structure) with average and maximum disagreements on cage bond lengths of only 0.02 and 0.09 Å, respectively. Comparison with the transition state (optimized in C_s symmetry, one imaginary frequency) gives an activation energy of 6.0 kcal/mol. However, in other 4,1,6-$MC_2B_{10}H_{12}$ species with transition metal vertices, higher activation energies between 10 and 12 kcal/mol were consistently found. For 4-(C_6H_6)-4,1,6-$RuC_2B_{10}H_{12}$ fluxionality was found to occur through a double diamond-square-diamond (dsd) process via the C_s symmetric transition state originally suggested by Hawthorne and coworkers. Excellent agreement was found between E_{act} calculated for the 4-$(\eta^6\text{-}C_6H_6)$-4,1,6-$RuC_2B_{10}H_{12}$ and that estimated experimentally by ^1H NMR for the analogous p-cymene species [144].

The stannacarborane 1,2-μ-$(CH_2)_3$-4,1,2-*closo*-$SnC_2B_{10}H_{10}$ and its 2,2′-bipyridine (bipy), 1,10-phenanthroline (*o*-phen), and 4,4′-diphenyl-2,2′-bipyridine (Ph_2bipy) adducts, 1,2-μ-$(CH_2)_3$-4-(bipy)-4,1,2-*closo*-$SnC_2B_{10}H_{10}$, 1,2-μ-$(CH_2)_3$-4-(*o*-phen)-4,1,2-*closo*-$SnC_2B_{10}H_{10}$, and 1,2-μ-$(CH_2)_3$-4-(Ph_2bipy)-4,1,2-*closo*-$SnC_2B_{10}H_{10}$, together with the analogous compound 1,2-μ-{$C_6H_4(CH_2)_2$}-4-(bipy)-4,1,2-*closo*-$SnC_2B_{10}H_{10}$ have been prepared and characterized [145]. Both docosahedral and henicosahedral structures (Fig. 3.6) were found for the central SnC_2B_{10} unit. A method of quantifying the "percentage docosahedral character" of a *closo*-13-vertex cluster is described. Thus "carbons adjacent" compounds are all significantly less slipped than analogous "carbons apart" 4,1,6-SnC_2B_{10} species. This is an apparently surprising result in the context of earlier work on "carbons adjacent" and "carbons apart" icosahedral platinacarboranes. However, a computational study of "carbons adjacent" and "carbons apart" 12- and 13-vertex platinacarboranes reveals that only in one case, namely that of the "carbons adjacent" 12-vertex metallacarborane, is the degree of slipping determined by orbital effects. For all of the other systems studied (12-vertex "carbons apart" and both "carbons apart" and "carbons adjacent" 13-vertex species) orbital effects are less significant and the degree of slipping reflects the dominant role played by steric interactions.

3.3.12 14-Vertex Systems

The structures of metallaboranes with 14, 15, and 16 vertices are based on the Frank-Kasper deltahedra (Fig. 3.7). These are the only topologically possible deltahedra with only degree 5 and 6 vertices with no pair of degree 6 vertices sharing an edge. The icosahedra with only degree 5 vertices also trivially meets the requirement of a Frank-Kasper deltahedron.

A density functional theory B3LYP/6-31G(d) approach was employed to optimize the structures of 14-vertex bis-substituted carboranes and metallaboranes [146]. All of these structures are based on the 14-vertex Frank-Kasper deltahedron, namely the bicapped hexagonal antiprism (Fig. 3.7). The polarizabilities and

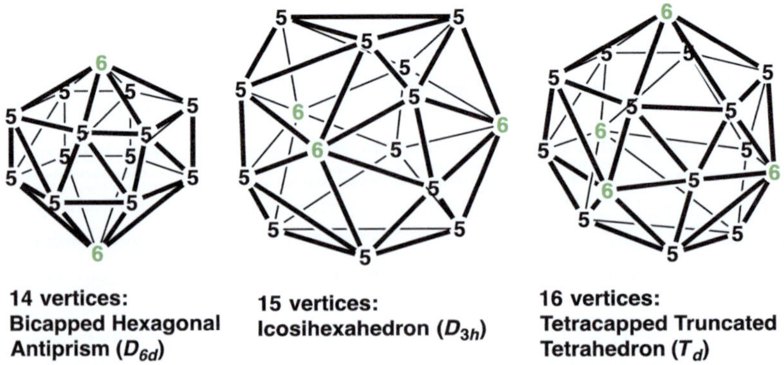

Fig. 3.7 The Frank-Kasper deltahedra having 14, 15, and 16 vertices

second-order hyperpolarizabilities were calculated by means of a finite-field (FF) method. The effective core potential basis set was used for the metal atoms in metallaboranes. However, the results were relatively insensitive to the basis set chosen and show that the bonding styles of carbon and metals are different in the 14-vertex disubstituted carboranes and metallaboranes. Introduction of metal atoms effectively enhances the nonlinear optical (NLO) coefficients of molecules. The metallaboranes, which in comparison to metal-free carboranes have lower HOMO-LUMO gaps, may exhibit semiconducting or even conducting properties.

The synthesis and structures of several 14-vertex MC_2B_{11} germa-, stanna-, and plumba-carboranes have been reported [147]. Single-crystal X-ray analyses reveal that they adopt a distorted bicapped hexagonal antiprism geometry, in which the p-block metal atom slips away from the above center of the C_2B_4 bonding face towards the boron side, leading to an η^4 bonding mode.

3.3.13 15- and 16-Vertex Systems

The 16-vertex closo-[$(Cp^*Rh)_3B_{12}H_{12}Rh\{Cp^*RhB_4H_9\}$, 15-vertex $(Cp^*Rh)_2B_{13}H_{13}$, 12-vertex $(Cp^*Rh)_2B_{10}H_n(OH)_m$, ($n=12$, $m=0$; $n=9$, $m=1$; $n=8$, $m=2$), and 10-vertex $(Cp^*Rh)_3B_7H_7$ and $(Cp^*Rh)_4B_6H_6$, have all been synthesized [148]. The 15-vertex rhodaborane cluster $(Cp^*Rh)_2B_{13}H_{13}$ exhibits icosihexahedron geometry (26 triangular faces) with three degree 6 vertices (Fig. 3.7). The 12-vertex rhodaborane clusters exhibit *isocloso* geometry, different from icosahedral geometry with the rhodium atoms at degree 6 vertices. The 10-vertex clusters $(Cp^*Rh)_3B_7H_7$ and $(Cp^*Rh)_4B_6H_6$ clusters have *isocloso* geometries (Fig. 3.2). Quantum-chemical calculations with DFT methods at the BP86 level of theory have been used to provide further insight into the electronic structure and stability of the optimized structures that are in satisfactory agreement with the experimental structure determinations by X-ray crystallography. The computed ^{11}B chemical shift values using gauge-including atomic orbital density functional theory [GIAO–DFT] method at the B3LYP/def2-TZVP level are consistent with the experimental values.

The identity of the first example of a 16-vertex *closo* metallaborane [$(Cp^*Rh)_3B_{12}H_{12}Rh\{Cp^*RhB_4H_9\}$] has been confirmed by multinuclear NMR spectroscopy, elemental analysis, infrared spectroscopy, and X-ray structural analysis [149]. The 16-vertex polyhedron corresponds to the Frank-Kasper polyhedron (Fig. 3.7) with a degree 6 Rh vertex bound to a hexagonal B_6 belt (B1-B6), below which lies an antiprismatic Rh_3B_3 belt and then a B_3 trigonal unit. Quantum-chemical calculations with DFT methods at the B3LYP (SDD, 6-31G*) level of theory have been used to provide further insight into the electronic structure and stability of this cluster using the unsubstituted Cp analogue. Satisfactory agreement is found between the computed bond lengths and those of the crystallographically characterized cluster. The DFT studies show the enhancement of thermodynamic stability of this 16-vertex tetrarhodaborane cluster over the simple metal-free *closo* borane $[B_{16}H_{16}]^{2-}$ as well as a still experimentally unknown $(CpRh)_4B_{12}H_{12}$ derivative of

$[B_{16}H_{16}]^{2-}$ in which CpRh units are located at the four degree 4 vertices of the 16-vertex Frank-Kasper deltahedron (Fig. 3.7). The three-dimensional aromatic character of this 16-vertex rhodaborane cluster is confirmed by a NICS(0) value of −16.1 ppm, which relates to the stability of boron deltahedra.

3.3.14 Oblatocloso *(Flattened) Polyhedral Metallaboranes*

The unusual dirhenaboranes $Cp_2Re_2B_nH_n$ ($n=6$–10) [21–23] have been the subject of recent theoretical studies. The experimentally observed *oblatocloso* deltahedra (Fig. 3.3) have been shown to be the lowest energy structures for these dirhenaboranes [150]. In some cases the energy differences between such oblate deltahedral structures and the next higher energy structures are quite considerable, that is, up to 25 kcal/mol for the nine-vertex $Cp_2Re_2B_7H_7$ structures. The higher energy $Cp_2Re_2B_{n-2}H_{n-2}$ structures, not yet known experimentally, are of the two types: (1) Most spherical (*closo*) deltahedra having unusually short 2.28–2.39 Å Re–Re edges with unusually high Wiberg bond indices suggesting formal multiple bonds on the deltahedral surface; (2) Deltahedra having one or two degree 3 vertices and 2.6–2.9 Å Re–Re edges. The latter deltahedra are derived from smaller deltahedra by capping Re_2B faces with the degree 3 vertices. In another theoretical study on these *oblatocloso* dirhenaboranes, the DFT-optimized Re–Re, Re–B and B–B distances were found to compare rather well with the bond lengths measured for the X-ray structures [151]. The rather good agreement between computed and experimental bond lengths in species of this sort gives confidence in the computed bond distances in other metallaboranes, including those containing hydrogen atoms which could not be properly located by X-ray techniques with certainty or in compounds exhibiting different possible isomers.

The analogous ditechnetaboranes $Cp_2Tc_2B_{n-2}H_{n-2}$ (n = 8, 9, 10, 11, 12) have also been studied by density functional theory [152]. As expected, their lowest energy structures are the flattened *oblatocloso* deltahedral structures (Fig. 3.3) similar to their rhenium analogues. These ditechnetaboranes typically have Tc–Tc distances ~0.05 Å shorter and Tc–Tc Wiberg bond indices 0.05 smaller than those in the analogous dirhenaboranes. Higher energy ditechnetaboranes are found with short Tc–Tc distances and *closo* or *isocloso* structures analogous to higher energy dirhenaboranes. These include 8-, 9-, 10-, and 12-vertex most spherical *closo* structures with formal Tc-Tc polyhedral surface quadruple bonds as well as 10- and 11-vertex *isocloso* structures with formal Tc ≡Tc polyhedral surface triple bonds.

3.3.15 Systems with Fused Metallaborane Polyhedra

The structure of the diruthenaborane $Cp^*_2Ru_2B_8H_{14}$ can be considered as two RuB_5 pentagonal pyramids that are fused together by sharing a B–B edge. DFT calculations on $Cp^*_2Ru_2(B_8H_{14})$ suggests a strong link with the pentalene complex

$Cp^*_2Fe_2(C_8H_6)$. Both compounds exhibit similar bonding modes and qualitatively related electronic structures [153]. There are however, differences that are related to the weaker B–B bonds in the B_8H_{14} ligand as compared to the C–C bonds in the pentalene ligand. Based on the calculations reported in this article, $Cp^*_2Ru_2(B_8H_{14})$ can be considered as being an isoelectronic metallaborane analogue of the pentalene complex $Cp^*_2Fe_2(C_8H_6)$ and other related 34-electron species. However, there are differences between the electronic structures of $Cp^*_2Ru_2B_8H_{14}$ and $Cp^*_2Fe_2C_8H_6$ which render $Cp^*_2Ru_2B_8H_{14}$ probably unique from the structural point of view. This contrasts with $Cp^*_2Fe_2C_8H_6$, which belongs to a series of pentalene complexes of ideal C_{2h} symmetry having electron counts varying between 32 and 38 and differing only by their metal hapticities. These differences originate from the different electronegativities of B and C.

The molybdenum compound $(\eta^5\text{-}C_5Me_5Mo)_3MoB_9H_{18}$, which has been structurally characterized by X-ray crystallography, represents a novel class of vertex-fused clusters in which a Mo atom has been fused in a perpendicular fashion between two molybdaborane clusters [154]. Electronic structure calculations employing density functional theory yield geometries in agreement with the structure determinations. One of the most interesting aspects of the structure is the range of Mo–Mo internuclear distances between bonding and nonbonding interactions. In an effort to understand the nature of various Mo–Mo interactions, studies using extended Huckel theory and DFT methods were carried out. The resulting model of the electronic structure rationalizes the factors behind the variations in Mo–Mo distances.

References

1. Williams RE (1971) Carboranes and boranes; polyhedra and polyhedral fragments. Inorg Chem 10:210–214
2. Williams RE (1992) The polyborane, carborane, carbocation continuum: architectural patterns. Chem Rev 92:177–207
3. King RB, Duijvestijn AJW (1990) The topological uniqueness of the deltahedra found in the boranes $B_nH_n^{2-}$ ($6 \leq n \leq 12$). Inorg Chim Acta 178:55–57
4. Wade K (1971) The structural significance of the number of skeletal bonding electron-pairs in carboranes, the higher boranes and borane anions, and various transition-metal carbonyl cluster compounds. Chem Commun 792–793
5. Mingos DMP (1972) Skeletal electron counting in cluster and ring compounds. Nat Phys Sci 236:99–102
6. Mingos DMP (1984) Polyhedral skeletal electron pair approach. Acc Chem Res 17:311–319
7. Aihara J (1978) Three-dimensional aromaticity of polyhedral boranes. J Am Chem Soc 100:3339–3342
8. King RB, Rouvray DH (1977) Chemical applications of group theory and topology. 7. A graph-theoretical interpretation of the bonding topology in polyhedral boranes, carboranes, and metal clusters. J Am Chem Soc 99:7834–7840
9. King RB (2001) Three-dimensional aromaticity in polyhedral boranes and related molecules. Chem Rev 101:1119–1152
10. Callahan KP, Hawthorne MF (1976) Ten years of metallocarboranes. Adv Organomet Chem 14:145–186

11. Grimes RN (1983) Role of metals in borane clusters. Acc Chem Res 16:22–26
12. Bould J, Kennedy JD, Thornton-Pett M (1992) Ten-vertex metallaborane chemistry. Aspects of the iridadecaborane *closo→isonido→isocloso* structural continuum. J Chem Soc Dalton Trans 563–576
13. Kennedy JD, Štibr B (1994) Polyhedral Metallaborane and Metallaheteroborane Chemistry. Aspects of Cluster Flexibility and Cluster Fluxionality. In: Kabalka GW (ed) Current topics in the chemistry of boron. Royal Society of Chemistry, Cambridge, pp 285–292
14. Kennedy JD (1998) Ch. 3. Disobedient Skeletons. In: Casanova J (ed) The borane-carborane-carbocation continuum. Wiley, New York, pp 85–116
15. Štibr B, Kennedy JD, Drdáková E, Thornton-Pett M (1994) Nine-vertex polyhedral iridamonocarbaborane chemistry. Products of thermolysis of [(CO)$(PPh_3)_2IrCB_7H_8$] and emerging alternative cluster-geometry patterns. J Chem Soc Dalton Trans 229–236
16. Kennedy JD (1986) Structure and bonding in recently isolated metallaboranes. Inorg Chem 25:111–112
17. Baker RT (1986) Hyper-closo metallaboranes. Inorg Chem 25:109–111
18. Johnston RL, Mingos DMP (1986) Molecular orbital calculations relevant to the hypercloso vs. iso-closo controversy in metallaboranes. Inorg Chem 25:3321–3323
19. Johnston RL, Mingos DMP, Sherwood P (1991) Bonding and electron counting in hypercloso metalloboranes and metallocarboranes. New J Chem 15:831–841
20. For a review of much of the relevant chemistry from Fehlner's group see Fehlner TP (2002) Metallaboranes of the earlier transition metals: relevance to the cluster electron counting rules. In: Shapiro PJ, Atwood DA (ed) Group 13 chemistry: from fundamentals to applications. American Chemical Society, Washington DC, pp 49–67
21. Ghosh S, Shang M, Li Y, Fehlner TP (2001) Synthesis of [$(Cp^*Re)_2B_nH_n$] n = 8–10: metal boride particles that stretch the cluster structure paradigms. Angew Chem Int Ed 40:1125–1128
22. Wadepohl H (2002) Hypoelectronic dimetallaboranes. Angew Chem Int Ed 41:4220–4223
23. Le Guennic B, Jiao H, Kahlal S, Saillard J-Y, Halet J-F, Ghosh S, Shang M, Beatty AM, Rheingold AL, Fehlner TP (2004) Synthesis and characterization of hypoelectronic rhenaboranes. Analysis of the geometric and electronic structures of species following neither borane nor metal cluster electron-counting paradigms. J Am Chem Soc 126:3203–3217
24. Farràs P, Juárez-Pérez EJ, Lepšík M, Luque R, Núñez R, Teixidor F (2012) Metallacarboranes and their interactions: theoretical insights and their applicability. Chem Soc Rev 41:3445–3463
25. Roy DK, Ghosh S, Halet J-F (2014) Beyond the icosahedron: the quest for high-nuclearity supraicosahedral metallaboranes. J Clust Sci 25:225–237
26. Becke AD (1993) Density-functional thermochemistry. III. The role of exact exchange. J Chem Phys 98:5648–5652
27. Lee C, Yang W, Parr RG (1988) Development of the Colle-Salvetti correlation-energy formula into a functional of the electron density. Phys Rev B 37:785–789
28. Becke AD (1988) Density-functional exchange-energy approximation with correct asymptotic behavior. Phys Rev A 38:3098–3100
29. Perdew JP (1986) Density-functional approximation for the correlation energy of the inhomogeneous electron gas. Phys Rev B 33:8822–8824
30. Perdew JP, Burke K, Ernzerhof M (1996) Generalized gradient approximation made simple. Phys Rev Lett 77:3865–3868
31. Lenthe E, Baerends EJ, Snijders JG (1993) Relativistic regular two component Hamiltonians. J Chem Phys 99:4597–4610
32. Andrae D, Haussermann U, Dolg M, Stoll H, Preuss H (1990) Energy-adjusted ab initio pseudopotentials for the second and third row transition elements. Theor Chim Acta 77:123–141
33. Møller C, Plesset MS (1934) Note on an approximation treatment for many-electron systems. Phys Rev 46:618–622

34. Cheeseman JR, Trucks GW, Keith TA, Frisch MJ (1996) A comparison of models for calculating nuclear magnetic resonance shielding tensors. J Chem Phys 104:5497–5509
35. Schleyer PR, Najafian K (1998) Stability and three-dimensional aromaticity of closo-monocarbaborane anions, $CB_{n-1}H_n^-$, and *closo*-dicarboranes, $C_2B_{n-2}H_n$. Inorg Chem 37:3454–3470
36. Chen Z, Wannere CS, Corminboeuf C, Puchta R, Schleyer PR (2005) Nucleus independent chemical shifts (NICS) as an aromaticity criterion. Chem Rev 105:3842–3512
37. Bader RFW (1990) Atoms in molecules: a quantum theory. Clarendon, Oxford
38. Becke AD, Edgecombe KE (1990) A simple measure of electron localization in atomic and molecular systems. J Chem Phys 92:5397–5403
39. Silvi B, Savin A (1994) Classification of chemical bonds based on topological analysis of electron localization functions. Nature 371:683–686
40. Savin A, Jepsen O, Flad J, Andersen OK, Preuss H, von Schnering HG (1992) Electron localization in solid-state structures of the elements – the diamond structure. Angew Chem Int Ed 31:187–188
41. Shameema O, Jemmis ED (2009) *Closo* versus *hypercloso* metallaboranes: a DFT study. Inorg Chem 48:7818–7827
42. Rabaâ H, Ghosh S, Sundholm D, Halet J-F, Saillard J-Y (2014) Addition and elimination reactions of H_2 in ruthenaborane clusters: a computational study. J Organomet Chem 761:1–9
43. Krishnamoorthy BS, Kahlal S, Ghosh S, Halet J-F (2013) Electronic, geometrical, and thermochemical studies on group-14 element-diruthenaborane cluster compounds: a theoretical investigation. Theor Chem Acc 132:1356
44. Muratov DV, Romanov AS, Petrovskii PV, Antipin MY, Siebert W, Kudinov AR (2012) A (diborole)cobalt complex with a C–H···B Bridge, $CpCo(1,3\text{-}C_3B_2Me_5H)$, and its thallium derivative: synthesis, structure, and bonding. Eur J Inorg Chem 4174–4182
45. Mebel AM, Musaev DG, Koga N, Morokuma K (1993) Metallaboranes with group 8 and 9 transition metals. Is accurate ab initio molecular orbital calculation of structure, stability and NMR chemical shifts possible? Bull Chem Soc Jpn 66:3259–3270
46. Anju RS, Roy DK, Mondal B, Ramkumar V, Ghosh S (2013) An early–late transition metal hybrid analogue of hexaborane(12). Organometallics 32:4618–4623
47. Brânzanic A, Lupan A, King RB (2014) Six-vertex hydrogen-rich $Cp_2M_2B_4H_8$ dimetallaboranes of the second- and third-row transition metals: effects of skeletal electron count on preferred polyhedra. Organometallics 33:6443–6451
48. Maguire JA, Fagner JS, Siriwardane U, Banewicz JJ, Hosmane NS (1980) Structural and bonding investigation of the donor-acceptor complex 1-Sn(Phenantroline)-2,3-$[Si(CH_3)_3]$2-2,3-$C_2B_4H_4$. Struct Chem 1:583–595
49. Siriwardane U, Maguire JA, Banewicz JJ, Hosmane NS (1989) Chemistry of C-trimethylsilyl-substituted main-group heterocarboranes. 5. Reactivity of stannacarboranes toward a tridentate Lewis base, 2,2′:6′,2″-terpyridine: synthetic, structural and bonding studies on the donor-acceptor complex 1-Sn-$[C_5H_{11}N_3]$-2-$(SiMe_3)$-3-(Me)-2,3- $C_2B_4H_4$ [R=$Si(CH_3)_3$ and CH_3]. Organometallics 8:2792–2800
50. Satpati P (2008) Structure and bonding of MCB_5H_7 and its sandwiched dimer CB_5H_6M–MCB_5H_6 (M = Si, Ge, Sn): Isomer stability and preference for slip distorted structure. J Organomet Chem 693:1159–1165
51. Barreto RD, Fehlner TP, Hosmane NS (1988) Quantum-chemical investigation of (2,2′-bpy) $SnB_4H_4(CCH_3)_2$ and $(CO)_3FeB_4H_4(CCH_3)_2$. Origin of observed structural distortions from idealized closo geometries. Inorg Chem 27:453–457
52. Fenske RF (1971) Molecular orbital theory for π-donor and π-acceptor complexes. Pure Appl Chem 27:61–72
53. Hall MB, Fenske RF (1972) Electronic structure and bonding in methyl- and perfluoromethyl (pentacarbonyl) manganese. Inorg Chem 11:768–775
54. Hosmane NS, Jia L, Zhang H, Maguire JA (1994) Chemistry of C-trimethylsilyl-substituted heterocarboranes. 15. Synthetic, spectroscopic, reactivity, and bonding studies on the

"carbons apart" closo-l-Sn-2-($SiMe_3$)-4-(R)-2,4-$C_2B_4H_4$: Crystal structures of the donor-acceptor complexes l-Sn(L)-2,4-$(SiMe_3)_2$-2,4-$C_2B_4H_4$ [R = $SiMe_3$, Me; L = 2,2'-$C_{10}H_8N_2$, 2,2'-$C_8H_6N_4$, or (η^5-C_5H_5)Fe(η^5-$C_5H_4CH_2(Me)_2N$)]. Organometallics 13:1411–1423
55. Ezhova MB, Zhang H, Maguire JA, Hosmane NS (1998) Experimental and theoretical studies on group 1 metallacarboranes: Synthesis, structure and ab initio calculations of the NMR chemical shifts of the 1-(THF)-1-(TMEDA)-1-Na-2,4-$(SiMe_3)_2$-2,4-$C_2B_4H_5$ and related carboranes. J Organomet Chem 550:409–422
56. Aldridge S, Hashimoto H, Kawamura K, Shang M, Fehlner TP (1998) Cluster expansion reactions of group 6 metallaboranes. Syntheses, crystal structures, and spectroscopic characterizations of $(Cp^*Cr)_2B_5H_9$, $(Cp^*Cr)_2B_4H_8Fe(CO)_3$, $(Cp^*Cr)_2B_4H_7Co(CO)_3$, and $(Cp^*Mo)_2B_5H_9Fe(CO)_3$. Inorg Chem 37:928–940
57. Sahoo S, Reddy KHK, Dhayal RS, Mobin SM, Ramkumar V, Jemmis ED, Ghosh S (2009) Chlorinated hypoelectronic dimetallaborane clusters: synthesis, characterization, and electronic structures of $(\eta^5\text{-}C_5Me_5W)_2B_5H_nCl_m$ (n=7, m=2 and n=8, m=1). Inorg Chem 48:6509–6516
58. de Biani FF, Corsini M, Zanello P, Yao H, Bluhm ME, Grimes RN (2004) Electronic properties of mononuclear, dinuclear, and polynuclear cobaltacarboranes: electrochemical and spectroelectrochemical studies. J Am Chem Soc 126:11360–11369
59. Qiu Y-Q, Liu X-D, Sun S-L, Fan M, Su Z-M, Wang R-S (2008) DFT study on second-order nonlinear optical properties of the derivatives of 7-vertex cobalt–carborane metallocenyl. J Mol Struct (THEOCHEM) 863:66–72
60. Wrackmeyer B, Schanz HJ (2004) Hexaethyl-2,4-dicarba-*nido*-hexaborane(8), deprotonation and complexation studied by NMR spectroscopy and density functional theory (DFT) calculations. Z Naturforsch 59b:685–691
61. Ma NN, Yang GC, Sun SL, Liu CG, Qiu YQ (2011) Computational study on second-order nonlinear optical (NLO) properties of a novel class of two-dimensional Λ- and W-shaped sandwich metallocarborane-containing chromophores. J Organomet Chem 696:2380–2387
62. Ma N, Liu C, Qiu Y, Sun S, Su Z (2011) Theoretical investigation on redox-switchable second-order nonlinear optical responses of push–pull $Cp^*CoEt_2C_2B_4H_3$-expanded (Metallo) porphyrins. J Comput Chem 33:211–219
63. Krishnamoorthy BS, Thakur A, Chakrahari KKV, Bose SK, Hamon P, Roisnel T, Kahlal S, Ghosh S, Halet J-F (2012) Theoretical and experimental investigations on hypoelectronic heterodimetallaboranes of group 6 transition metals. Inorg Chem 51:10375–10383
64. Singh AK, Sadrzadeh A, Yakobson BI (2010) Metallacarboranes: toward promising hydrogen storage metal organic frameworks. J Am Chem Soc 132:14126–14129
65. Geetharani K, Krishnamoorthy BS, Kahlal S, Mobin SM, Halet J-F, Ghosh S (2012) Synthesis and characterization of hypoelectronic tantalaboranes: comparison of the geometric and electronic structures of [$(Cp^*TaX)_2B_5H_{11}$] (X = Cl, Br, and I). Inorg Chem 51:10176–10184
66. Siebert W, Kudinov AR, Zanello P, Antipin MY, Scherban VV, Romanov AS, Muratov DV, Starikova ZA, Corsini M (2009) Synthesis of μ-diborolyl triple-decker complexes by electrophilic stacking. Similar bonding properties of anions [CpCo(1,3-$C_3B_2H_5$)]$^-$ and Cp- toward transition metals. Organometallics 28:2707–2715
67. Rochdi M, Saillard J-Y, Halet J-F, Ghosh S, Rabaâ H (2012) Can high-hydride content hypoelectronic rhenaborane clusters take up dihydrogen? A theoretical study. Polyhedron 43:31–35
68. Thakur A, Chakrahari KKV, Mondal B, Ghosh S (2013) Novel triple-decker sandwich complex with a six-membered [$B_3Co_2(\mu_4\text{-}Te)$] ring as the middle deck. Inorg Chem 52:2262–2264
69. Lupan A, King RB (2014) The buildup of eight-vertex tetrametallaborane clusters: bisdisphenoidal versus tetracapped tetrahedral structures. Eur J Inorg Chem 3614–3618
70. Bould J, Harrington RW, Clegg W, Kennedy JD (2012) Nine-vertex metallaborane chemistry. Preparation and characterization of [1,1,1-$(PMe_3)_2H$-*isocloso*-IrB_8H_7-8-X], where X = H or Cl. J Organomet Chem 721–722:155–163

71. Mingos DMP, Welch AJ (1980) Molecular-orbital studies on carbametallaboranes. Part 3. Tricapped trigonal prismatic carbaplatinaborane and related polyhedral molecules. Dalton Trans 1674–1681
72. Bould J, Clegg W, Kennedy JD (2006) Polyhedral iridaborane chemistry: elements of the 10-vertex *closo–isonido–isocloso* continuum. Molecular structures of $[(PPh_3)_2HIrB_9H_9(PPh_3)]$, $[(PPh_3)(Ph_2PC_6H_4)IrB_9H_7(PPh_3)]$, $[(PPh_3)(Ph_2PC_6H_4)HIrB_9H_6Cl(PPh_3)]$, $[(PPh_3)(Ph_2PC_6H_4)HIrB_9H_6(PPh_3)_2]$, and $[(PPh_3)(Ph_2PC_6H_4)HIrB_9H_{12}]$. Inorg Chim Acta 359:3723–3735
73. Bould J, Oro LA, Macías R, Kennedy JD, Londesborough MGS (2011) A DFT and crystallographic reinvestigation of the $[L_2RuC_2B_7H_9]$ and $[L_3RuC_2B_7H_9]$ '*hypercloso*' and *closo* systems. Polyhedron 30:2140–2145
74. Luaces S, Bould J, Macías R, Sancho R, Lahoza FJ, Oro LA (2012) Facile two-electron reduction of a *closo*-rhodathiadecaborane. Dalton Trans 41:11627–11634
75. Sharmila D, Ramalakshmi R, Chakrahari KK, Varghese B, Ghosh S (2014) Synthesis, characterization and crystal structure analysis of cobaltaborane and cobaltaheteroborane clusters. Dalton Trans 43:9976–9985
76. Macías R, Bould J, Holub J, Kennedy JD, Štíbr B, Thornton-Pett M (2007) Polyhedral metallaheteroborane chemistry. Synthesis, spectroscopy, structure and dynamics of eleven-vertex $\{RhNB_9\}$ and $\{PtCB_9\}$ metallaheteroboranes. Dalton Trans 2885–2897
77. Bould J, Cunchillos C, Lahoz FJ, Oro LA, Kennedy JD, Macías R (2010) New iridathiaboranes with reversible *isonido* ↔ *nido* cluster flexibility. Inorg Chem 49:7353–7361
78. Calvo B, Macías R, Artigas MJ, Lahoz FJ, Oro LA (2013) Reactions of 11-vertex rhodathiaboranes with HCl: synthesis and reactivity of new Cl-ligated clusters. Inorg Chem 52:211–221
79. Bould J, Macías R (2014) Do agostic interactions play a role in the stabilization of the *nido* structure of $[(PPh_3)_2RhSB_9H_{10}]$? J Organomet Chem 761:120–122
80. Calvo B, Roy B, Macias R, Artigas MJ, Lahozz FJ, Oro LA (2014) NH_3-promoted ligand lability in eleven-vertex rhodathiaboranes. Inorg Chem 53:12428–12436
81. Du S, Farley RD, Harvey JN, Jeffery JC, Kautz JA, Maher JP, McGrath TD, Murphy DM, Riis-Johannessen T, Stone FGA (2003) The seventeen- and eighteen-electron metallacarbaboranes $[1,1,1\text{-}(CO)_3\text{-}2\text{-}Ph\text{-}closo\text{-}1,2\text{-}MnCB_9H_9]^{n-}$ (n = 1, 2): a structurally characterized, redox-related pair. Chem Commun 1846–1847
82. Nafady A, Butterick R III, Calhorda MJ, Carroll PJ, Chong D, Geiger WE, Sneddon LG (2007) Hyperelectronic metal-carborane analogues of cymantrene ($MnCp(CO)_3$) anions: electronic and structural noninnocence of the tricarbadecaboranyl ligand. Organometallics 26:4471–4482
83. Muhammad S, Xu H, Liao Y, Kan Y, Su Z (2009) Quantum mechanical design and structure of the $Li@B_{10}H_{14}$ basket with a remarkably enhanced electro-optical response. J Am Chem Soc 131:11833–11840
84. González-Cardoso P, Stoica A-I, Farràs P, Pepiol A, Viñas C, Teixidor F (2010) Additive tuning of redox potential in metallacarboranes by sequential halogen substitution. Chem Eur J 16:6660–6665
85. Farràs P, Teixidor F, Rojo I, Kivekäs R, Sillanpää R, González-Cardoso P, Viñas C (2011) Relaxed but highly compact diansa metallacyclophanes. J Am Chem Soc 133:16537–16552
86. Farràs P, Viñas C, Teixidor F (2013) Preferential chlorination vertices in cobaltabisdicarbollide anions. Substitution rate correlation with site charges computed by the two atoms natural population analysis method (2a-NPA). J Organomet Chem 747:119–125
87. Farràs P, Olid-Britos D, Viñas C, Teixidor F (2011) Unprecedented B–H activation through Pd-catalysed B–C vinyl bond coupling on borane systems. Eur J Inorg Chem 2525–2532
88. Juárez-Pérez EJ, Viñas C, González-Campo A, Teixidor F, Sillanpää R, Kivekäs R, Núñez R (2008) Controlled direct synthesis of C-mono- and C-disubstituted derivatives of $[3,3'\text{-}Co(1,2\text{-}C_2B_9H_{11})_2]^-$ with organosilane groups: theoretical calculations compared with experimental results. Chem Eur J 14:4924–4938

89. Juárez-Pérez EJ, Viñas C, Teixidor F, Núñez R (2009) First example of the formation of a Si–C bond from an intramolecular Si–H···H–C dihydrogen interaction in a metallacarborane: a theoretical study. J Organomet Chem 694:1764–1770
90. Bühl M, Hnyk D, Macháček J (2005) Computational study of structures and properties of metallaboranes: cobalt bis(dicarbollide). Chem Eur J 11:4109–4120
91. Viñas C, Llop J, Teixidor F, Kivekäs R, Sillanpää R (2005) Restricted rotation in unbridged sandwich complexes: rotational behavior of *closo*-[Co(η^5-NC_4H_4)($C_2B_9H_{11}$)] derivatives. Chem Eur J 11:1933–1941
92. Juárez-Pérez EJ, Núñez R, Viñas C, Sillanpää R, Teixidor F (2010) The role of C–H···H–B interactions in establishing rotamer configurations in metallabis(dicarbollide) systems. Eur J Inorg Chem 2385–2392
93. Chevrot G, Schurhammer R, Wipff G (2007) Molecular dynamics study of dicarbollide anions in nitrobenzene solution and at its aqueous interface. Synergistic effect in the Eu(III) assisted extraction. Phys Chem Chem Phys 9:5928–5938
94. Coupez B, Wipff G (2004) The synergistic effect of cobalt-dicarbollide anions on the extraction of M^{3+} lanthanide cations by Calix[4]arenes: a molecular dynamics study at the water–'oil' interface. C R Chimie 7:1153–1164
95. Chevrot G, Schurhammer R, Wipff G (2006) Surfactant behavior of "ellipsoidal" dicarbollide anions: a molecular dynamics study. J Phys Chem B 110:9488–9498
96. Řezáčová P, Pokorná J, Brynda J, Kožíšek M, Cígler P, Lepšík M, Fanfrlík J, Řezáč J, Šašková KG, Sieglová I, Plešek J, Šícha V, Grüner B, Oberwinkler H, Sedláček J, Kräusslich HG, Hobza P, Král V, Konvalinka J (2009) Design of HIV protease inhibitors based on inorganic polyhedral metallacarboranes. J Med Chem 52:7132–7141
97. Barszcz B, Graja A, Ziolkovskiy DV, Starodub VA, Kravchenko AV (2010) Raman studies of TMTSF salt of cobalt bis(dicarbollide) anion. J Mol Struct 976:196–199
98. Kazheva ON, Alexandrov GG, Kravchenko AV, Kosenko ID, Lobanova IA, Sivaev IB, Filippov OA, Shubina ES, Bregadze VI, Starodub VA, Titov LV, Buravov LI, Dyachenko OA (2011) Molecular conductors with a 8-hydroxy cobalt bis(dicarbollide) anion. Inorg Chem 50:444–450
99. Bühl M, Hnyk D, Macháček J (2007) Computational studies of structures and properties of metallaboranes. Part 3: protonated iron bis(dicarbollide), [3-Fe-(1,2-$C_2B_9H_{11})_2H]^-$. Inorg Chem 46:1771–1777
100. Pennanen TO, Macháček J, Taubert S, Vaara J, Hnyk D (2010) Ferrocene-like iron bis(dicarbollide), [3-Fe^{III}-(1,2-$C_2B_9H_{11})_2]^-$. The first experimental and theoretical refinement of a paramagnetic ^{11}B NMR spectrum. Phys Chem Chem Phys 12:7018–7025
101. Núñez R, Tutusaus O, Teixidor F, Viñas C, Sillanpää R, Kivekäs R (2005) Highly stable neutral and positively charged dicarbollide sandwich complexes. Chem Eur J 11:5637–5647
102. Hawthorne MF, Zink JI, Skelton JM, Bayer MJ, Liu C, Livshits E, Baer R, Neuhauser D (2004) Electrical or photocontrol of the rotary motion of a metallacarborane. Science 303:1849–1851
103. Ma N-N, Li S-J, Yan L-K, Qiu Y-Q, Su Z-M (2014) Switchable NLO response induced by rotation of metallacarboranes [NiIII/IV($C_2B_9H_{11})_2]^{-/0}$ and C-, B-functionalized derivatives. Dalton Trans 43:5069–5075
104. Bühl M, Holub J, Hnyk D, Macháček J (2006) Computational studies of structures and properties of metallaboranes. 2. Transition-metal dicarbollide complexes. Organometallics 25:2173–2181
105. Perekalin DS, Kudinov AR (2005) Calculations of thermodynamic stability of $CpCoC_2B_9H_{11}$ cobaltacarborane isomers. Russ Chem Bull 54:1603–1605
106. Llop J, Viñas C, Teixidor F, Victori L, Kivekäs R, Sillanpää R (2001) C-C plasticity in boron chemistry: modulation of the Cc···Cc distance in mixed pyrrolyl/dicarbollide complexes. Organometallics 20:4024–4030
107. Herber RH, Kudinov AR, Zanello P, Nowik I, Perekalin DS, Meshcheryakov VI, Lyssenko KA, Corsini M, Fedi S (2006) Synthesis, structure, electrochemistry, and metal-atom dynamics of cyclopentadienyl ferracarboranes. Eur J Inorg Chem 1786–1795

108. Planas JG, Viñas C, Teixidor F, Comas-Vives A, Ujaque G, Lledós A, Light ME, Hursthouse MB (2005) Self-Assembly of mercaptane-metallacarborane complexes by an unconventional cooperative effect: a C–H···S–HH···H–B hydrogen/dihydrogen bond interaction. J Am Chem Soc 127:15976–15982
109. Mutseneck EV, Perekalin DS, Holub J, Starikova ZA, Petrovskii PV, Zanello P, Corsini M, Štíbr B, Kudinov AR (2006) (Tetramethylcyclobutadiene)cobalt complexes with phosphacarborane ligands. Organometallics 25:2419–2426
110. Alekseev LS, Safronov AV, Dolgushin FM, Korlyukov AA, Godovikov IA, Chizhevsky IT (2009) An unexpected cluster opening upon the formation of electronically unsaturated η^3-(cyclooctenyl)metallacarboranes of rhodium(III) and iridium(III) with sterically reduced $[(PhCH_2)_2C_2B_9H_9]^{2-}$ ligand. J Organomet Chem 694:1727–1735
111. Loginov DA, Starikova ZA, Corsini M, Zanello P, Kudinov AR (2013) (Cyclopentadienyl) metalladicarbollides 3-(η-C_5R_5)-3,1,2-$MC_2B_9H_{11}$ (M=Co, Rh, Ir): synthesis, electrochemistry, and bonding. J Organomet Chem 747:69–75
112. Webster CE, Hall MB (2003) De novo design in organometallic chemistry: stabilizing iridium(V). Coord Chem Rev 238–239:315–331
113. Loginov DA, Vinogradov MM, Starikova ZA, Petrovskii PV, Holub J, Kudinov AR (2010) The first metallacarborane triple-decker complexes with a bridging pentaphosphonyl ligand. Collect Czech Chem Commun 75:981–993
114. Kudinov AR, Zanello P, Herber RH, Loginov DA, Vinogradov MM, Vologzhanina AV, Starikova ZA, Corsini M, Giorgi G, Nowik I (2010) Ferracarborane benzene complexes $[(\eta\text{-}9\text{-}L\text{-}7,8\text{-}C_2B_9H_{10})Fe(\eta\text{-}C_6H_6)]^+$ (L = SMe_2, NMe_3): synthesis, reactivity, electrochemistry, Mössbauer effect studies, and bonding. Organometallics 29:2260–2271
115. Bould J, Kennedy JD (2014) An assessment of the intercarbon stretching phenomenon in C-substituted "pseudocloso" {3,1,2-RuC_2B_9} metalladicarbaboranes. J Organomet Chem 749:163–173
116. Fischer MJ, Jelliss PA, Phifer LM, Rath NP (2005) Halogenated rhenacarboranes: optoelectronic behavior of the iodinated rhenacarborane complex anion $[3,3,3\text{-}(CO)_3\text{-}8\text{-}I\text{-}closo\text{-}3,1,2\text{-}ReC_2B_9H_{10}]^-$. Inorg Chim Acta 358:1531–1544
117. Zou H-Y, Ma N-N, Sun S-L, Li X, Qiu Y-Q (2013) Structures and redox-switchable second-order nonlinear optics properties of N-legged piano stool shaped 12-vertex rhenacarborane half-sandwich complexes. J Organomet Chem 728:6–15
118. Fox MA, Howard JAK, Hughes AK, Malget JM, Yufit DS (2001) Synthesis of isomeric B-methylated tantalum carboranes, $(Me_2N)_3TaC_2B_9H_{10}Me$. Dalton Trans 2263–2269
119. Bould J, Baše T, Londesborough MGS, Oro LA, Macías R, Kennedy JD, Kubát P, Fuciman M, Polívka T, Lang K (2011) Reversible capture of small molecules on bimetallaborane clusters: synthesis, structural characterization, and photophysical aspects. Inorg Chem 50:7511–7523
120. Bould J, Londesborough MGS, Kennedy JD, Macías R, Winter REK, Císařová I, Kubát P, Lang K (2013) Isonitrile ligand effects on small-molecule-sequestering in bimetalladodecaborane clusters. J Organomet Chem 747:76–84
121. Mingos DMP, Forsyth MI, Welch AJ (1977) X-ray crystallographic and theoretical studies on 'slipped' metallacarboranes. J Chem Soc Chem Commun 605–607
122. Mingos DMP (1977) Molecular-orbital studies on carbametallaboranes. Part 1. Icosahedral carbaplatinaborane polyhedra. Dalton Trans 602–610
123. Calhorda MJ, Mingos DMP, Welch AJ (1982) Theoretical comparison of the 'slip' distortion and rotational barriers in comparable seven and twelve vertex carbaplatinaboranes. J Organomet Chem 228:309–320
124. Mingos DMP, Forsyth MI, Welch AJ (1977) Molecular and crystal structure of 3,3-bis(triethylphosphine)-1,2-dicarba-3-platinadodecaborane(11) and molecular-orbital analysis of the slip distortion in carbametallaboranes. Dalton Trans 1363–1374
125. Perekalin DS, Holub J, Golovanov DG, Lyssenko KA, Petrovskii PV, Štíbr B, Kudinov AR (2005) Ferra- and ruthenatricarbollides $CpFeC_3B_8H_{11}$ and $Cp^*RuC_3B_8H_{11}$. Organometallics 24:4387–4392

126. Perekalin DS, Glukhov IV, Štíbr B, Kudinov AR (2006) A restricted polyhedral rearrangement of an aminosubstituted 12-vertex ferratricarbollide. Inorg Chim Acta 359:3264–3268
127. Holub J, Grüner B, Perekalin DS, Golovanov DG, Lyssenko KA, Petrovskii PV, Kudinov AR, Štíbr B (2005) Synthesis and rearrangements of aminosubstituted ferra- and ruthenatricarbaboranes. Inorg Chem 44:1655–1659
128. Januszko A, Kaszynski P, Grüner B (2007) Liquid crystalline derivatives of bis(tricarbollide) Fe(II). Inorg Chem 46:6078–6082
129. Kiani FA, Hofmann M (2006) Ortho-, meta-, and para-directing influence of transition-metal fragments in 12-vertex *closo*-cyclopentadienylmetallaheteroboranes: additive nature of structural increments. Organometallics 25:485–490
130. McLellan R, Boag NM, Dodds K, Ellis D, Macgregor SA, McKay D, Masters SL, Noble-Eddy R, Platt NP, Rankin DWH, Robertson HE, Rosair GM, Welch AJ (2011) New chemistry of 1,2-*closo*-$P_2B_{10}H_{10}$ and 1,2-*closo*-$As_2B_{10}H_{10}$; in silico and gas electron diffraction structures, and new metalladiphospha- and metalladiarsaboranes. Dalton Trans 40:7181–7192
131. Hnyk D, Wann DA, Holub J, Bühl M, Robertson HE, Rankin DWH (2008) The gas-phase structure of 1-selena-*closo*-dodecaborane(11), 1-$SeB_{11}H_{11}$, determined by the concerted use of electron diffraction and computational methods. Dalton Trans 96–100
132. Lee J-D, Kim S-K, Kim T-J, Han W-S, Lee Y-J, Yoo D-H, Cheong M, Ko J, Kang SO (2008) Dicarbollylamine ligand as a tunable template for σ, σ- and π, σ-bonding modes: syntheses, structures, and theoretical studies of η5:η1-coordinated constrained-geometry group 13 metal complexes. J Am Chem Soc 130:9904–9917
133. Scott G, McAnaw A, McKay D, Boyd ASF, Ellis D, Rosair GM, Macgregor SA, Welch AJ, Laschi F, Rossib F, Zanello P (2010) Supraicosahedral indenyl cobaltacarboranes. Dalton Trans 39:5286–5300
134. McAnaw A, Lopez ME, Scott G, Ellis D, McKay D, Rosair GM, Welch AJ (2012) Untethered 4,1,2-MC_2B_{10} supraicosahedral metallacarboranes, their C, C′-dimethyl 4,1,6-, 4,1,8- and 4,1,12-MC_2B_{10} analogues, and DFT study of the (4,)1,2- to (4,)1,6-isomerisations of C_2B_{11} carboranes and MC_2B_{10} metallacarboranes. Dalton Trans 41:10957–10969
135. Burke A, Ellis D, Ferrer D, Ormsby DL, Rosair GM, Welch AJ (2005) Synthetic, spectroscopic, computational and structural studies of some 13-vertex ruthenacarboranes. Dalton Trans 1716–1721
136. Dustin DF, Dunks GB, Hawthorne MF (1973) Novel 13-vertex metallocarborane complexes formed by polyhedral expansion of 1,2-dicarba-*closo*-dodecaborane(12) (1,2-$B_{10}C_2H_{12}$). J Am Chem Soc 95:1109–1115
137. Ellis D, McKay D, Macgregor SA, Rosair GM, Welch AJ (2010) Room-temperature C-C bond cleavage of an arene by a metallacarborane. Angew Chem Int Ed 49:4943–4945
138. Scott G, Ellis D, Rosair GM, Welch AJ (2012) Icosahedral and supraicosahedral naphthalene ruthenacarboranes. J Organomet Chem 721–722:78–94
139. Dalby KJ, Ellis D, Erhardt S, McIntosh RD, Macgregor SA, Rae K, Rosair GM, Settels V, Welch AJ, Hodson BE, McGrath TD, Stone FGA (2007) The conformations of 13-vertex $ML_2C_2B_{10}$ metallacarboranes: experimental and computational studies. J Am Chem Soc 129:3302–3314
140. Nana M, Yongqing Q, Shiling S, Min F, Zhongmin S (2010) Density functional theory study on structure and nonlinear optical properties of 13-vertex metallacarborane. Acta Chim Sin 68:733–738
141. Cheung M-S, Chan H-S, Bi S, Lin Z, Xie Z (2005) Functional sidearm promoted electron-transfer reactions: a new route to metallacarboranes incorporating the η^7-*arachno*-carboranyl ligands. Organometallics 24:4333–4336
142. Chui K, Yang Q, Mak TCW, Lam WH, Lin Z, Xie Z (2000) Synthesis, structure, and bonding of d^0/f^n metallacarboranes incorporating the η^7-carboranyl ligand. J Am Chem Soc 122:5758–5764
143. Wilson NMM, Ellis D, Boyd ASF, Giles BT, Macgregor SA, Rosair GM, Welch AJ (2002) The first supraicosahedral p-block metallacarboranes. Chem Commun 464–465

144. Boyd ASF, Burke A, Ellis D, Ferrer D, Giles BT, Laguna MA, McIntosh R, Macgregor SA, Ormsby DL, Rosair GM, Schmidt F, Wilson NMM, Welch AJ (2003) Supraicosahedral (metalla)carboranes. Pure Appl Chem 75:1325–1333
145. Abram PD, McKay D, Ellis D, Macgregor SA, Rosair GM, Welch AJ (2010) Synthetic, structural and computational studies on adducts of the 4,1,2-SnC_2B_{10} supraicosahedral stannacarborane. Dalton Trans 39:2412–2422
146. Min F, Yongqing Q, Shiling S, Xiaodong L, Zhongmin S (2009) Density functional theory study on the polarizability and second-order hyperpolarizability of 14-vertex bis-substituted carborane and metallaborane. Acta Chim Sin 67:1171–1176
147. Zheng F, Xie Z (2014) Synthesis and structural characterization of 14-vertex germa-, stanna-, and plumba-carboranes. Dalton Trans 43:4986–4992
148. Roy DK, Mondal B, Shankhari P, Anju RS, Geetharani K, Mobin SM, Ghosh S (2013) Supraicosahedral polyhedra in metallaboranes: synthesis and structural characterization of 12-, 15-, and 16-vertex rhodaboranes. Inorg Chem 52:6705–6712
149. Roy DK, Bose SK, Anju RS, Mondal B, Ramkumar V, Ghosh S (2013) Boron beyond the icosahedral barrier: a 16-vertex metallaborane. Angew Chem Int Ed 52:3222–3226
150. Lupan A, King RB (2012) Hypoelectronic dirhenaboranes having eight to twelve vertices: Internal versus surface rhenium – rhenium bonding. Inorg Chem 51:7609–7616
151. Krishnamoorthy BS, Kahlal S, Le Guennic B, Saillard J-Y, Ghosh S, Halet JF (2012) Molecular transition-metal boron compounds. Any interest? Solid State Sci 14:1617–1623
152. Lupan A, King RB (2013) Comparison of hypoelectronic deltahedral ditechnetaboranes having eight to twelve vertices with their rhenium analogues: Examples of polyhedral surface metal–metal multiple bonds. Polyhedron 60:151–157
153. Boucher B, Ghosh S, Halet J-F, Kahlal S, Saillard J-Y (2012) Bonding and electronic structure of $Cp^*_2Ru_2(B_8H_{14})$, a metallaborane analogue of dinuclear pentalene complexes. J Organomet Chem 721–722:167–172
154. Dhayal RS, Sahoo S, Reddy KHK, Mobin SM, Jemmis ED, Ghosh S (2010) Vertex-fused metallaborane clusters: synthesis, characterization and electronic structure of $[(\eta^5\text{-}C_5Me_5Mo)_3MoB_9H_{18}]$. Inorg Chem 49:900–904

Chapter 4
Quantum Chemistry of Excited States in Polyhedral Boranes

Josep M. Oliva, Antonio Francés-Monerris, and Daniel Roca-Sanjuán

Abstract In this Chapter we describe the electronic structure of ground states and excited states of the two isomers of octadecaborane (22), *anti*- and *syn*-$B_{18}H_{22}$, and the new derivative of *anti*-$B_{18}H_{22}$, the polyhedral substituted borane 4,4′-$(HS)_2$-*anti*-$B_{18}H_{20}$. A theoretical interpretation is given on the fluorescence of the *anti*-$B_{18}H_{22}$ isomer, and the non-radiative decay of the *syn*-$B_{18}H_{22}$ isomer, an unsolved problem since 1962. For the new derivative of *anti*-$B_{18}H_{22}$, substitution of hydrogen atoms in positions 4 and 4′ by SH groups allows the tuning of the photophysical properties in 4,4′-$(HS)_2$-*anti*-$B_{18}H_{20}$, facilitating intersystem crossing from the excited singlet state to the triplet state.

4.1 Introduction

One of the foundations in quantum mechanics is Schrödinger's wave mechanics equation $\hat{H}\psi = E\psi$ [1, 2]. Solving this complex integro-differential equation allows us to obtain, always within approximations, the electronic structure of stationary states in a many-electron system. Almost a century later, the main developments within quantum chemistry have been impressive and widespread: Modification of the wave function Ψ in the Schrödinger's equation leads to the so-called wave-function methods (WFM), starting off with the molecular-orbital (MO) Hartree-Fock (HF) method (mean-field theory method) [3–13]. On the other hand, the application of the symmetric group S_N to quantum chemistry, and taking into account the formation of molecules from the wave functions of the constituent atoms, leads

J.M. Oliva (✉)
Instituto de Química-Física "Rocasolano" (CSIC), Serrano 119,
Madrid ES-28006, Spain
e-mail: J.M.Oliva@iqfr.csic.es

A. Francés-Monerris • D. Roca-Sanjuán
Institut de Ciència Molecular, Universitat de València,
P. O. Box 22085, València ES-46071, Spain
e-mail: Antonio.Frances@uv.es; Daniel.Roca@uv.es

D. Hnyk, M.L. McKee (eds.), *Boron*, Challenges and Advances in Computational Chemistry and Physics 20,
DOI 10.1007/978-3-319-22282-0_4

to the more complex but highly compact Valence-Bond (VB) methods [14–19]. Extensions of the HF method involve higher correlated wave functions, defined in a widespread spectrum of methods, such as Coupled-Cluster (CC) [20–25] and Configuration Interaction (CI) [26] and derived methods thereof. The full configuration interaction method (full CI) provides numerically exact solutions (with a complete basis set) to the non-relativistic Schrödinger equation; however, due to the exponential complexity, such method is not applicable using current digital computers, even for a system of more than, say, a few atoms heavier than hydrogen!

On the other hand, modification of the energy operator (Hamiltonian) in the Schrödinger equation leads to the so-called perturbative methods [27–31], which can also be combined with WFM. All these methods have been used with more or less success in quantum chemistry, as function of the size and type of the many-electron problem to be solved.

The appearance of Density-Functional Theory (DFT) [32, 33], based on functionals of the scalar electron density $\rho(\boldsymbol{r})$, and its implementation to quantum chemistry packages[1] [35, 36] by the end of the last century, have given an unprecedented push to the problem of correlation in many-electron systems. However, due to unpredictable behavior, we insist on being cautious when using DFT-based methods, depending on the many-electron problem to be solved.

The determination of higher-energy solutions of stationary states – the so-called excited states – in the Schrödinger equation is a very involved problem, and can be considered a sort of endeavor comparable to the quest for the *holy grail* – determination of all excited states at once. Spectroscopic accuracy in the determination of such states requires a tremendous intellectual and computational effort, with no clear-cut solution, at least in the short term and with the use of digital computers. In Sect. 4.3 we give a more detailed description of the electronic structure methods we use for the determination of excited states in polyhedral boranes.

Turning now to the world of the fifth element – boron – and more precisely, to the polyhedral borane clusters, a close look into the literature shows few references to studies of excited states in this type of compound, as compared to organic compounds [37, 38]. Excited-state studies were carried out in the polyhedral (substituted) borane structures B_4Cl_4, $B_6H_6^{2-}$, $B_9H_9^{2-}$, $B_{10}H_{10}^{2-}$, $B_{12}H_{12}^{2-}$, $B_{12}Cl_{12}^{2-}$ using CI computations with single and double excitations [39]. The main publications found thereafter in the literature deal mainly either with the polyhedral heteroborane acting as anion in a salt with a transition-metal cation [40, 41] or with photoluminescence and photoinduced electron transfer (PET) processes [38] with the polyhedral heteroborane as a ligand in a coordination (metal) complex [42–49].

In this Chapter we are mainly concerned with electronic excited states derived directly from borane and substituted borane polyhedral cages, with the targeted phenomenon and property stemming directly from the boron compound. Our group has contributed to study the quantum chemistry of excited states in polyhedral heteroboranes and substituted boranes with some works in the last decade [50–58], and two sections of this Chap. 4 (Sects. 4.4 and 4.5, *vide infra*) are the outcome

[1] See, for instance, the computational package Gaussian-09 [34].

of a collaboration between the group of the quantum chemistry of the excited state at the University of Valencia (Spain) and the group at the Institute of Inorganic Chemistry of the Academy of Sciences of the Czech Republic, v.v.i. [59–61]. The study of excited states in heteroborane and substituted borane polyhedral compounds turns out to be an unexplored and promising research field with potential applications in biosciences and nanosciences [62–72].

4.2 Photophysics and Photochemistry Processes

Photoinduced excited-state chemistry begins with the absorption (*Abs*) process in which the population is promoted from the ground to the excited states by means of radiation (see Fig. 4.1). If several wavelengths are used, the population is distributed among the states located at the range of energies of the light used to excite the system. Nevertheless, not all the excited states are equality populated. Selection rules based on multiplicity and symmetry considerations have been developed for predicting the permitted and forbidden electronic transitions. Thus, only those transitions that do not imply a change in the spin multiplicity are allowed, and only totally symmetric matrix elements yield nonzero probabilities at first order. For the allowed one-photon transitions, the probability can be estimated using Fermi's Golden Rule,

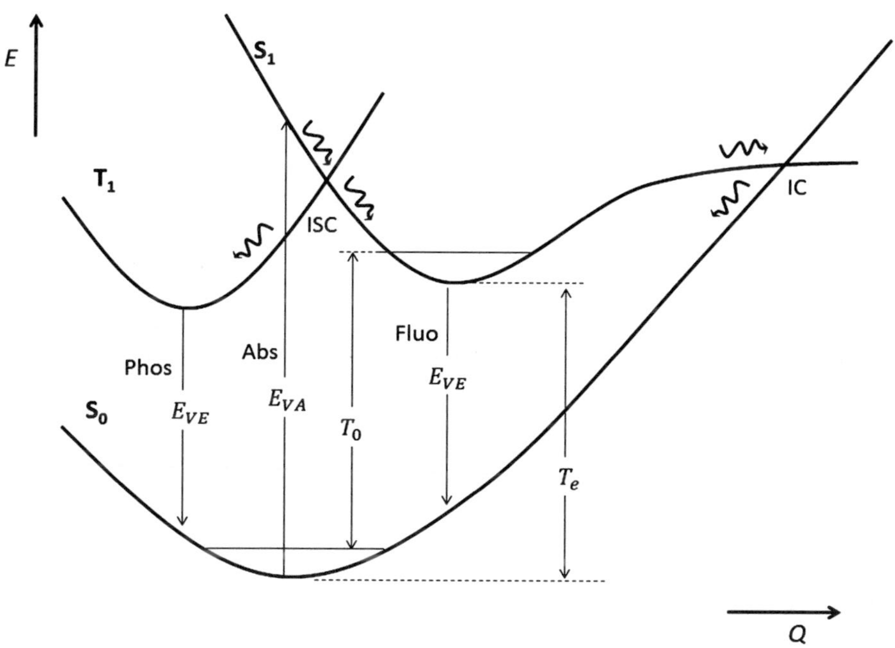

Fig. 4.1 Scheme of the main photophysical and photochemical molecular events discussed in the present Chapter

which establishes that this probability is proportional to the square of the transition dipole moment (*TDM*) between the initial and final electronic states: $TDM = <\varphi_1|\vec{d}|\varphi_2>$. Specifically, the transition probability is theoretically estimated by means of the oscillator strength (f) associated to each excited state via the equation

$$f = \frac{2}{3} E_{VA} TDM^2,$$

where E_{VA} is the vertical energy difference between the excited and ground electronic states (at the equilibrium geometry of the ground state; see Fig. 4.1). The electronic f can be compared with the experimental estimation based on the transition band height and width. In addition to the electronic properties, the vibrational modes also contribute to the strength of the transition (band intensity) through the so-called Franck-Condon factors, which are the vibrational overlap terms between the vibrational states belonging to the two electronic states: $\langle v_1 | v_2 \rangle$. The highest overlap, and thus the most probable transition, commonly occurs at the ground-state equilibrium structure, which is called the Franck-Condon (FC) region. Hence, the E_{VA} is usually compared with the experimental absorption band maximum. Still, the absorption can also take place at lower energies up to the energy difference between the ground and excited electronic state minima, which is called the adiabatic energy, T_e (see Fig. 4.1). In many cases, the computed T_e provides enough information to assign the band origin. Nevertheless, for a direct comparison with the experimental data, the zero-point vibrational energy (ZPVE) must be computed in both electronic states and added to T_e to obtain the vibrational band origin, T_0 (see Fig. 4.1).

Once the molecule is excited, distinct decay mechanisms can occur for deactivating the excess of energy provided to the molecule. We shall describe here only those processes not involving energy or charge transfer to other molecules. Radiative and non-radiative processes can be then distinguished. Within the radiative decay mechanisms, fluorescence (*Fluo*) or phosphorescence (*Phos*) may occur. Since the most probable states to be populated are the singlet excited states, *Phos* implies previous intersystem crossing (ISC) processes in which the population is transferred to the triplet manifold. Regarding the non-radiative decay mechanism to the ground state, internal conversions (IC) are usually involved (see Fig. 4.1). The dominant decay mechanism is the one without energy barriers along the corresponding deactivation paths and without non-efficient processes. A brief description of the theoretical characterization of these processes is given in the following. For further reading, we recommend the reviews from Refs. [73–79].

4.2.1 Emission: Fluorescence

Many excited organic and inorganic molecular systems undergo a vibrational relaxation along the singlet excited state hypersurfaces until a stable structure is reached, from which light emission takes place. As for the absorption process, the vertical transition, with an associated vertical emission energy of E_{VE} (see Fig. 4.1), accounts for the highest probability (band maximum). Thus, it may be compared with the band maximum, yet emission may take place at other energies until the energy difference T_0. From the computation of the *TDM*, the radiative lifetime (τ_{rad}) for this process (that is related to the probability of transition) can be estimated by using the Einstein coefficients for spontaneous emission (A_{21}) and the Strickler-Berg approximation:

$$A_{21} = \frac{1}{\tau_{rad}} = 2.142005 \times 10^{10}\, E_{VE}^{3} TDM^{2}$$

where τ_{rad} is measured in s. Typical values of τ_{rad} are in the range $10^{-9} - 10^{-6}$ s.

4.2.2 Non-radiative Decay

Certain molecules, however, do not show fluorescent bands in the experiments. And the time-resolved transient absorption technique gives rise to very short lifetime signals. In this case, the molecule relaxes the extra energy through IC processes that brings the system back to the ground-state potential energy hypersurface (PEH). From a theoretical standpoint, no stable structures (minima) are found on the excited-state surface along the decay paths and the molecule evolves from the FC region through a barrierless profile towards a crossing between the excited and ground state PEHs. This point of crossing is called a conical intersection (CI) and it enables the population transfer between surfaces. CIs are N-dimensional points (N is the total number of internal coordinates of the system) in which an energy degeneracy between the two states that crosses takes place among N-2 coordinates (intersecting space) [80–84]. The two dimensions that break the degeneracy form the branching space which is defined by the gradient differential vector and the non-adiabatic coupling vector. Usually, an energy minimization is performed in the intersection space to locate the minimum energy crossing point (MECP), which is expected to be the CI point of interest for the photochemistry of the molecule. However, the accessibility of this structure (that funnels the energy to the ground state) from the FC region or the stable structures on the excited-state surfaces must be analysed. For this purpose, minimum energy path (MEP) computations must be performed from the bright state (the one with the highest probability of population after absorption of light) at the FC geometry in order to determine the most probable decay path of the system after excitation [37, 85, 86]. The evolution of this computation towards a CI point guarantees the non-radiative decay through this crossing point. An alternative strategy to verify the accessibility of a CI from the FC region or an excited-state minimum is by means of linear interpolations of the internal coordinates. Barrierless profiles obtained in such manner indicate a favorable decay process through the CI.

4.2.3 Emission: Phosphorescence

Excited molecules may also decay to the ground state through a mechanism involving states of different spin multiplicities (usually singlet and triplet states). In this case, the initially populated singlet excited state, or subsequent singlet states reached along the relaxation path, transfer the population to a triplet state through an ISC process. The ISC takes place via a crossing region between the surfaces of the excited singlet and triplet states (see Fig. 4.1). Taking into account that the spin multiplicity is different for both states, the ISC process is much less favorable than the IC process. For the sake of comparison, the lifetimes for these processes are in the range $10^{-8} - 10^{-2}$ and $10^{-12} - 10^{-6}$ s, respectively. In order to have an efficient singlet-triplet crossing (STC) process, the spin and angular momentum must couple. This coupling, called spin-orbit coupling (SOC) can be determined as:

$$SOC_{lk} = \sqrt{\sum_{u} |\langle T_{l,u}|\hat{H}_{SO}|S_k\rangle|^2}, \; u = x, y, z$$

where $\hat{H}_{SO}$ is the spin-orbit Hamiltonian and T and S refer to the triplet and singlet states, respectively.

After the ISC process, once on the triplet surface, the molecule usually reaches an equilibrium structure (minimum) from which it emits phosphorescent light (see Fig. 4.1). Emission from the triplet excited state to the singlet ground state is also much less favorable than fluorescence emission. Hence, the lifetime for this radiative process is larger, in the range of $10^{-3} - 10^{2}$ s.

4.3 Computational Methods for Excited States

Excited-state chemistry is much more complex than that of the ground state, and more difficult to study with theoretical methods and computational strategies [87]. Among the reasons, three are specially relevant: (1) the abundance of near-degeneracies between states, (2) the involvement of excited states of distinct nature, and (3) the need for higher solutions of the electronic Hamiltonian. As a consequence of the first reason, the Born-Oppenheimer approach, which allows treating the nuclei and electrons separately, is not valid in many situations. In addition, multiconfigurational descriptions of the wave functions are required. Also, the PEHs possess a higher number of relevant singular points as compared to the ground state surfaces, for example, CIs and STCs. These new structures imply the computation of higher-order couplings, such as non-adiabatic, spin-orbit, or vibronic couplings. Regarding the nature of the excited states (reason 2), distinct types of orbitals may characterize the excitation process from the ground to the excited state (σ/σ^*, π/π^*, lone pairs). In addition, the radial extension of the orbitals involved in the excitation adds complexity to the study. Thus, excited states with different spatial extensions may appear: valence excited states (*e.g.* $n \rightarrow \pi^*$ or $\pi \rightarrow \pi^*$) and Rydberg states. In the latter, electron promotions take place to atomic-like orbitals with higher principal quantum numbers (*e.g.* $n \rightarrow 3s$ or $\pi \rightarrow 3p$). The molecular charge distribution also determines the nature of the excited states. Thus, the movement of electron density from one part of the system to another generates charge-transfer states, which otherwise are localized states. Finally, focusing on the technical aspects (reason 3), the ground-state techniques are difficult to implement in the excited state and, therefore, more complex algorithms for resolving the equations are needed.

In order to deal with the huge number of possibilities that arise in the chemistry of the excited states, the theoretical methods chosen for studying the photophysical and photochemical processes must be flexible enough to provide a balanced description, with similar precision, of the different types of states and PEH features. Among the quantum chemistry methods, the so-called multi-reference (MR) multiconfigurational (MC) methods are the most appropriate [37, 88, 89]. In these two-step methods, first, an optimal MC wavefunction is determined by building a CI linear expansion of determinants or spin-adapted configuration state functions (CSFs) and then finding the optimum CI expansion coefficients and the optimum form of the orbitals that minimize the energy according to the variational principle. Next, a MR approach is applied to compute the dynamic correlation effects. Three basic MR approaches are possible: configuration interactions (MRCI), coupled cluster (MRCC), and perturbation theory (MRPT). Since the MR-MC methods are based on a MC wavefunction, electronic-structure problems that are difficult to describe with single-reference methods, such as biradical structures, CIs or multiexcited states, are properly described with the MR-MC methods. In the second step, accurate energies and molecular properties are obtained by using one of the MR approaches. Hence, the two-step approach allows for a general and accurate characterization of the excited states. Finally, it is fair to mention here that many interesting developments in methodology have expanded the applicability of single-reference methods to the excited states (see, for example, Ref. [90]).

Among the MR-MC methods, CASPT2 [91–93] is probably one of the most useful and practical for studying spectroscopy and photochemistry. It is based on a multiconfigurational CASSCF wavefunction, which account for static correlation, that is, the electron correlation related to the presence of near-degeneracies. In the CASSCF procedure, a full CI expansion is performed, taking into account only those molecular orbitals that are more relevant for the description of the electronic problem. In the MR step of the approach, second-order perturbation theory is employed to include a large amount of dynamic electron correlation (the one related to the electron cusp or the fact that the probability of finding two electrons close in distance is low). The most relevant advantages of the CASPT2 method are three. First, it can be applied to study systems of medium and relatively large size. Second, it can be used to describe all type of states and degeneracies. Finally, it provides an accuracy of around 0.2–0.3 eV. Nevertheless, the CASPT2 method is not free of problems. We briefly introduce below the main problems and also how they are usually solved:

- The standard zeroth-order Hamiltonian overestimates the energy difference between closed and open-shell systems. In order to remove this overestimation, a modified zeroth-order Hamiltonian (H_0) has been introduced in which an ionization potential electron affinity (IPEA) shift is considered [94]. The recommended value is 0.25 au.

- Weakly perturbing states (intruder states) might cause problems. In this case, a penalty parameter (level shift) may be added in order to minimize their effects [95].
- The CASPT2 wavefunctions of different states are not orthogonal and therefore the interaction between states is not considered. In order to account for the coupling of several electronic states, a multistate treatment can be carried out by building an effective Hamiltonian matrix [96]. The diagonal and off-diagonal elements are the CASPT2 energies and the coupling up to second order of dynamical correlation energy, respectively. In order to obtain the so-called multistate (MS)-CASPT2 solutions, this matrix is made symmetric, assuming therefore that the off-diagonal terms are very similar and negligible. This assumption must be verified in any particular case since otherwise the MS may lead to nonphysical results for both energies and wavefunctions [97].

Finally, since the CASPT2 gradients are very time and CPU-demanding, a common strategy in theoretical studies, is to perform geometry optimizations at the CASSCF level and then obtain the CASPT2 energies on the optimized structures. This protocol is often known as CASPT2//CASSCF. Since the CASSCF wave function does not take into account dynamical correlation, it is also common to employ other methods such as DFT or MP2 in order to optimize the ground-state equilibrium geometry for obtaining the computed absorption spectra.

4.4 Strange Case of Dr Jekyll (*syn*-$B_{18}H_{22}$) and Mr Hyde (*anti*-$B_{18}H_{22}$)[2]

The comparison of the *syn*- and *anti*- isomers of octadecaborane (Fig. 4.2) is a clear example of how the arrangement of the atoms in a molecule determines the photophysical and photochemical properties of the compound. Both isomers are formed by two $\{B_{10}\}$ units sharing a common B-B edge. In the *syn*-, the two units are pointing to the same side of the edge, whereas in the *anti*-, they point to opposite sides. Different photophysical behavior was observed by Londesborough et al. [59] for these isomers. The differences can be rationalized studying the absorption and emission processes and the associated geometrical changes in both species. Here, we will review the most important results and the corresponding interpretation.

4.4.1 *Absorption*

The computed CASPT2 vertical electronic energies of the studied $B_{18}H_{22}$ molecules are shown in Table 4.1. The electronic states involved in the absorption spectrum of the systems in the UV–vis range are dominated by one-electron excitations within

[2] Using analogies with Robert Louis Stevenson's novella, we associate the *syn*-$B_{18}H_{22}$ isomer with quiet (non-radiative) Dr Jekyll, and the *anti*-$B_{18}H_{22}$ isomer with evil (fluorescent) Mr Hyde.

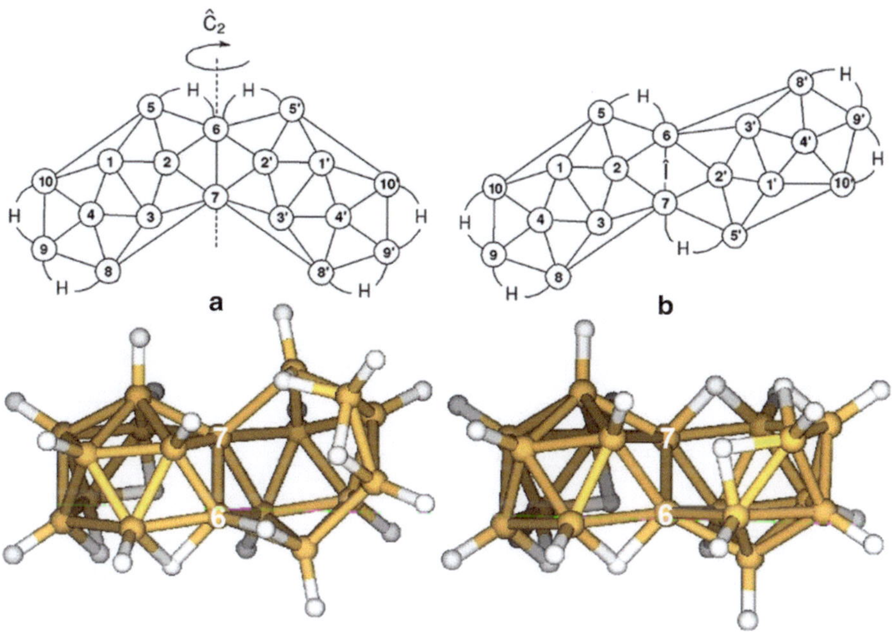

Fig. 4.2 Atom numbering and structure of (**a**) *syn*-$B_{18}H_{22}$ and (**b**) *anti*-$B_{18}H_{22}$ molecules

the four highest occupied and lowest unoccupied molecular orbitals (frontier orbitals): HOMO − 1, HOMO, LUMO, LUMO + 1 (see Ref. [59] for the contribution of the configurations in the electronic states). The *syn*-$B_{18}H_{22}$ isomer at the ground state equilibrium geometry belongs to point group C_2, consequently the state is labelled as 1^1A. Three relevant electronic singlet states were determined theoretically in the 4.0–6.0 eV range. Transitions from the ground state to the 1^1B state and to the 3^1B state can be ascribed to the two medium-intensity experimental bands recorded experimentally at 4.03 and 5.41 eV, respectively. The $1^1A \rightarrow 4^1A$ transition probably also contributes to the latter experimental band, according to its predicted energy (5.56 eV) and non-negligible oscillator strength (0.158). Two relevant triplet states were determined at the FC geometry, the 1^3A and the 1^3B states. Both lie below the low-lying singlet excited state. The energy difference between the 1^1B and the 1^3A states is 0.13 eV and they have a relatively strong SOC (1.2 cm^{-1}). Thus, in principle the population of the 1^3A state in the FC region would be possible through an ISC process from the 1^1B state. However, the shape of the singlet manifold leading to a barrierless CI with the ground state, a much faster and favorable process, will prevent population of the triplet state. The phenomenon will be reviewed in more detail in the next section.

The *anti*-$B_{18}H_{22}$ isomer also has three singlet excited states in the ~4.0–6.0 eV range, labelled as 1^1A_u (3.93 eV), 2^1A_u (4.99 eV) and 4^1A_u (5.85 eV), according to the C_i point group symmetry of the molecule at the ground state (1^1A_g) minimum. In contrast to the previous system, three maximum bands at 3.77, 4.56 and 5.76 were observed in the recorded UV–vis spectrum. They were mainly ascribed by

Table 4.1 Computed CASPT2 vertical absorption energies (E_{VA}), oscillator strengths (f), vertical emission energies (E_{VE}), electronic band origin energies (T_e) and natural radiative lifetimes (τ_{rad}) of the $B_{18}H_{22}$ isomers and derivatives [59]

State/Compound	E_{VA}/eV	f	E_{VE} /eV	T_e/eV	τ_{rad}/s
***syn*-$B_{18}H_{22}$**					
1^1B	4.26 (4.03)	0.338 (~0.6)	–	–	–
4^1A	5.56 (5.41)	0.158 (~0.5)	–	–	–
3^1B	5.76 (5.41)	0.251 (~0.5)	–	–	–
1^3B	3.76	–	2.67	3.31	0.96
1^3A	4.13	–	–	–	–
***anti*-$B_{18}H_{22}$**					
1^1A_u	3.93 (3.77)	0.265 (~0.4)	2.91 (3.05)	3.63 (3.41)	5.6×10^{-9} (1.2×10^{-8})
2^1A_u	4.99 (4.56)	0.051 (~0.2)	–	–	–
4^1A_u	5.85 (5.76)	0.884 (~1.0)	–	–	–
1^3A_u	3.48	–	2.49	3.04	0.51
2^3A_u	4.63	–	–	–	–
4,4′-$(SH)_2$-*anti*-$B_{18}H_{20}$					
1^1A_u	3.13 (3.25)	0.065	2.17 (2.31)	2.78	1.09×10^{-7} (0.94×10^{-9})
2^1A_g	3.27	–	–	–	–
2^1A_u	4.04 (3.92)	0.196	–	–	–
1^3A_u	3.02	–	1.97 (2.08)	2.61	1.69×10^{-3} (1.4×10^{-5})
$1^3A_u \rightarrow 2^3A_g$	–	0.210	–	–	–

Experimental values in brackets

Londesborough et al. [59] to the predicted electronic transitions from the ground state to the 1^1A_u, 2^1A_u, and 4^1A_u states, respectively. Computed oscillator strengths are also in good agreement with the experimental values (see Table 4.1). Energetics of the low-lying triplet states 1^3A_u and 2^3A_u are also shown in Table 4.1. Only the first one lies below the bright 1^1A_u state, and although the SOC values between the two states is relatively high, the energy difference of ~0.5 eV makes population transfer from the singlet to the triplet state unlikely. As the calculated SOC terms between these singlet and the triplet states are in general high, a more efficient ISC can be expected in other regions of the PEH where the energy difference is lower.

4.4.2 Emission

The characterization of the PEH of the bright excited state allows the prediction of emissive or non-emissive decay mechanisms of a molecule after the absorption of light. In general, the presence of a well-defined minimum before a relevant crossing with other electronic state in the excited-state hypersurface implies the emission of light by the compound, while a downward slope until a crossing with the ground

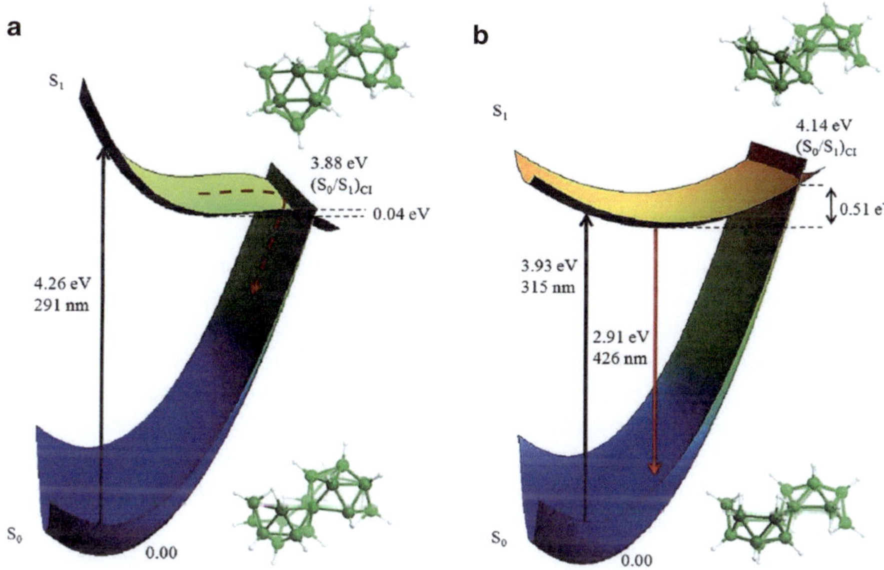

Fig. 4.3 Photophysics, PEHs and structures of the ground (S_0) and first excited (S_1) states of the (**a**) *syn*-$B_{18}H_{22}$ (*left*) and (**b**) *anti*-$B_{18}H_{22}$ systems (Reprinted with permission from Londesborough et al. [59]. Copyright 2012 American Chemical Society)

electronic state implies a fast or ultrafast non-emissive relaxation (see Sect. 4.2). Such difference is exactly what occurs in the comparison *syn*-$B_{18}H_{22}$ *vs.* *anti*-$B_{18}H_{22}$.

The photochemistry of both $B_{18}H_{22}$ isomers was studied by means of the CASPT2//CASSCF protocol [59]. The relevant PEHs for the two species are shown in Fig. 4.3. In the low-energy region of the spectrum the *syn*-$B_{18}H_{22}$ molecule has a 1^1B (S_1) bright state at 4.26 eV (*f*=0.338). The optimization of such excited state without symmetry requirements leads directly to a crossing structure with the ground state $(S_0/S_1)_{CI}$. Further optimization of the crossing point imposing the energy degeneracy between the S_0 and the S_1 states yields the minimum energy crossing point (MECP) $(S_0/S_1)_{CI,}$ placed adiabatically at 3.88 eV. The absence of significant barriers between the S_1 at the FC region and the CI optimized geometry was verified by performing a linear interpolation of internal coordinates between the two structures. The CI region is clearly below the excited-state energy at the FC geometry. Therefore, the system initially should have enough energy to reach the crossing area and decay to the ground state without emission of light, i.e., the excess of energy provided by the absorption of light is released through dissipation of heat and not by the emission of light. The failure to measure fluorescence or phosphorescence in the experiments is the consequence of the described scenario.

The barrierless energy profile towards the CI with the ground state is in general a favorable and thus fast process. Even though some triplet states were predicted at the studied energy ranges having significant SOC with the S_1 state, the presence of the fast non-emissive relaxation prevents an effective population of the triplets. This is supported by the fact that no experimental evidence of singlet oxygen generation or phosphorescence is observed in the *syn*-$B_{18}H_{22}$ complex [59].

The situation is different in the *anti*-$B_{18}H_{22}$ system. A minimum in the hypersurface of the bright 1^1A_u (S_1) state was predicted adiabatically at 3.63 eV. In addition, a $(S_0/S_1)_{CI}$ was calculated at 4.14 eV above the energy of the S_1 state (3.93 eV) at the FC region. To reach this point from the S_1 well, a barrier of 0.51 eV must be surmounted. In contrast to the *syn*-$B_{18}H_{22}$ isomer, strong fluorescence was observed in this case. By using the Strickler-Berg approximation, the natural radiative lifetime of the S_1 state (τ_{rad}) of 5.6×10^{-9} s was obtained. The experimental estimation of the same parameter (τ_F/φ_F) gave a value of $\tau_{rad} = 1.2 \times 10^{-8}$ s, reasonably in agreement with the theoretical result. Hence, in contrast to the previous isomer, after the absorption of light and population of the S_1 state the *anti*- system reaches the corresponding minimum, remaining there until the emission of a photon occurs, thus returning to the S_0 state. This radiative decay mechanism was confirmed experimentally as the most probable. In fact, an intense fluorescence of the compound was observed with a quantum yield (φ_F) of 0.97.

The population of the triplet states in the *anti*- isomer is hindered by the large energy difference (~0.50 eV) between the singlet and the triplet states at the studied FC and S_1 minimum regions. However, as the system approaches the $(S_0/S_1)_{CI}$ area, the energy gap decreases and the triplet states may be populated as a consequence. This can be the reason for the very weak phosphorescence observed in the system.

4.4.3 Geometrical Changes

The analysis of the geometry changes that take place after the excitation in a molecule can be very important to establish relationships among electronic structure, chemical functionalities and photochemical properties, and then help in the design of new materials with different photochemical behavior.

An analysis of the B-B distances in the optimized structures of both *syn*-$B_{18}H_{22}$ and *anti*-$B_{18}H_{22}$ molecules is thus useful to understand the different photochemistry of the isomers. To do that, Londesborough et al. [59] defined in their study the distance difference (Δd) of the B-B bonds as positive for elongations and negative for contractions. The two isomers show similar geometrical changes after excitation in the initial vibrational relaxation process along the S_1/T_1 surfaces, all of them within the ±0.15 Å range. On the contrary, the systems suffer important distortions in order to reach the CI points and, interestingly, clear differences between the *syn*- and *anti*-isomers can be found in this case (see Fig. 4.4). Large geometrical changes seem to be concentrated in a small number of bonds, in contrast to the minor changes distributed along the whole molecule in the initial relaxation. At the $(S_0/S_1)_{CI}$, the *syn*-$B_{18}H_{22}$ molecule has a largely elongated B_6-$B_{5'}$ bond of Δd ~ 1.2 Å, as compared to the S_0 equilibrium structure. On the other hand, three bonds of the *anti*-$B_{18}H_{22}$ system undergo large elongations: $B_{1'}$-$B_{10'}$ (Δd ~ 0.9 Å), B_2-$B_{5'}$ (Δd ~ 0.7 Å) and B_7-$B_{5'}$ (Δd ~ 0.55 Å).

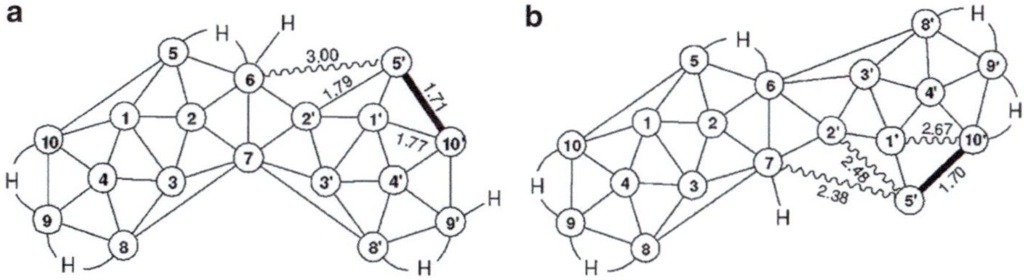

Fig. 4.4 Relevant B-B distances (Å) of (**a**) *syn*-$B_{18}H_{22}$ and (**b**) *anti*-$B_{18}H_{22}$ at the $(S_0/S_1)_{CI}$ point. Significant elongations are shown in *wavy lines*, and relevant contractions are shown as *heavy lines*

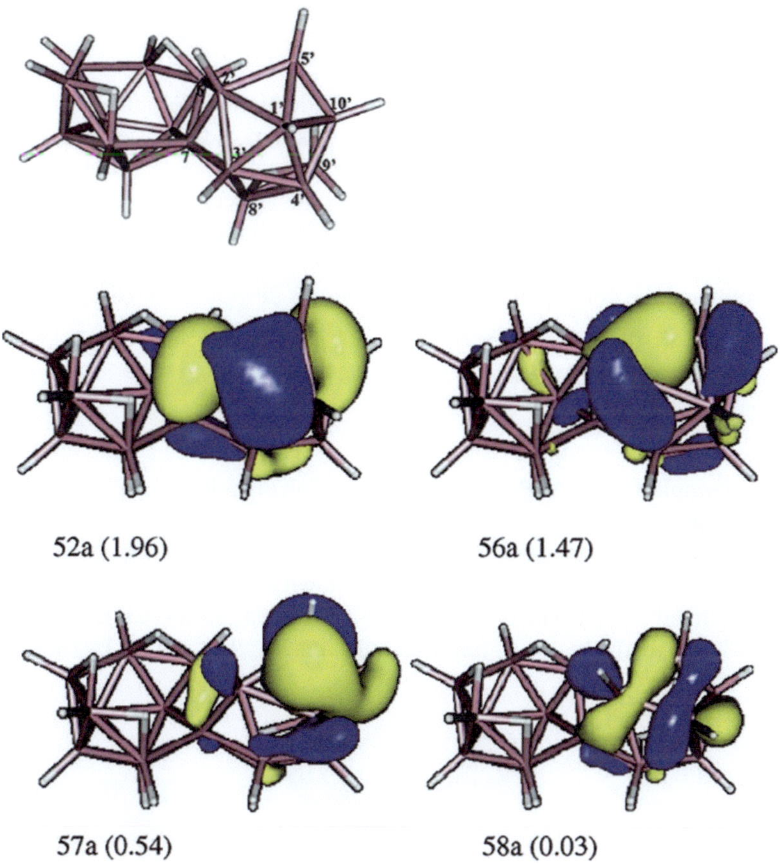

Fig. 4.5 Relevant CASSCF natural orbitals of the *syn*-$B_{18}H_{22}$ system at the $(S_0/S_1)_{CI}$ point. Occupation number within parenthesis (Reprinted with permission from Saurí et al. [60]. Copyright 2013 American Chemical Society)

The shape of the MOs is also helpful in the analysis of the geometrical changes. The relevant CASSCF MOs for the *syn*-isomer at the $(S_0/S_1)_{CI}$ geometry are shown in Fig. 4.5. They are mainly localized in the B_6-$B_{5'}$ connectivity and in the $B_{5'}$-$B_{10'}$

bond. The relevant loss of occupation in the 56a MO (occupation number of 1.47), with a strong σ bonding character of the B_6-$B_{5'}$ bond, is in clear agreement with the large elongation of the bond.

In summary, only one B-B bond is elongated (more than 1.0 Å) in the *syn-* compound while three B-B bonds are stretched (less than 1.0 Å) in the *anti-* isomer. This fact allow Londesborough et al. [59] to explain the existence of a S_1 minimum along the PEH in the *anti*-$B_{18}H_{22}$ case and the higher difficulty to reach the CI, as compared to the *syn*-isomer.

4.5 Excited States in the New Polyhedral Substituted Borane 4,4′-(SH)$_2$-*anti*-$B_{18}H_{20}$

In the present section we illustrate how the introduction of SH groups in the structure of a molecule lead to new photophysical properties of the system. As described above, the *syn-* and *anti*-$B_{18}H_{22}$ boron compounds do not have significant phosphorescence emission, whereas intense fluorescence is observed in the *anti-* complex ($\varphi_F = 0.97$). This situation changes substantially when the positions 4 and 4′ of the boron clusters are thiolated (see Figs. 4.2 and 4.6) [60]. The photophysics of this new system and the corresponding interpretation is described in the following.

4.5.1 *Absorption*

Table 4.1 compiles the CASPT2 vertical absorption energies of the 4,4′-(SH)$_2$-*anti*-$B_{18}H_{20}$ compound. Four relevant transitions from the ground state (1^1A_g) to the following excited states are predicted in the UV–vis region of the spectrum: 1^1A_u (3.13 eV), 2^1A_u (4.04 eV), 3^1A_u (4.83 eV, not shown in Table 4.1) and 4^1A_u (5.17 eV, not shown in Table 4.1). The computed oscillator strengths are 0.065, 0.196, 0.044 and 0.528, respectively. It is important to remark that the 1^1A_u (S_1) state is

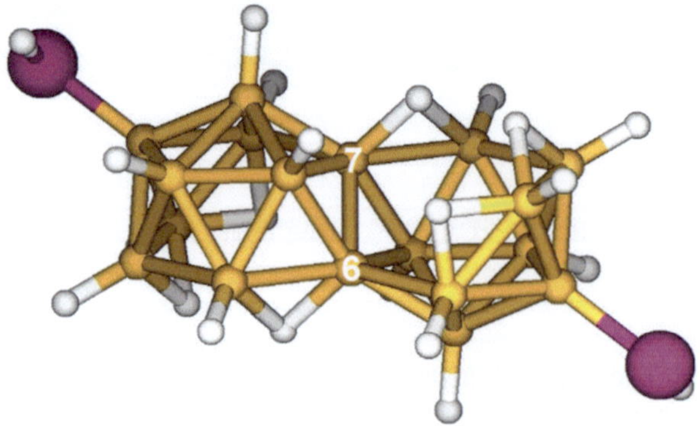

Fig. 4.6 DFT optimized structure of 4,4′-(SH)$_2$-*anti*-$B_{18}H_{20}$

dominated (85 %) by the one-electron promotion from the lone pair MO of the sulfur atoms to an anti-bonding MO delocalized over the boron cluster (see Ref. [60] for more information about the nature of the states). This process is obviously not possible in the non-substituted boranes due to the absence of thiol groups. According to the calculated energies and oscillator strengths, the two first theoretical values can be ascribed to the two experimental maximum bands recorded at 3.25 and 3.92 eV. The 4^1A_u state was safely assigned by Saurí et al. [60] to the intense experimental band centred at 5.02 eV. Nevertheless, a weak 3^1A_u state contribution to this band cannot be fully discarded.

In Table 4.1, the absorption properties of the thiolated system are compared with those of the non-substituted $B_{18}H_{22}$ complexes. As expected, the introduction of the SH functional groups modifies the electronic properties of the molecule, especially due to the presence of the lone pairs of the thiol group. This allows electronic transitions at lower absorption energies. Hence, the S_1 first excited state is placed at 4.26 eV in the *syn-* isomer, at 3.93 eV in the *anti-* isomer and at 3.13 eV in the case of the 4,4′-$(SH)_2$-*anti-* isomer. A significant difference is also found in the bright states of the systems. In the non-substituted complexes the bright state is S_1, while in the 4,4′-$(SH)_2$-*anti*-$B_{18}H_{20}$ it is the S_3 state.

Discrepancies were also found by Saurí et al. [60] in the triplet state population. The first triplet state 1^3A_u was predicted at 3.02 eV. Taking into account the low energy difference of 0.11 eV with the 2^1A_u (S_1) state and the significant SOC of 2.92 cm^{-1} between them, triplet population is then expected at the FC region of the system. This is clearly in contrast to the low triplet population probabilities of the *syn*-$B_{18}H_{22}$ and *anti*-$B_{18}H_{22}$ molecules.

4.5.2 *Emission*

The relevant PEHs computed for the 4,4′-$(SH)_2$-*anti*-$B_{18}H_{20}$ are shown in Fig. 4.7. In the following lines the most probable pathway (but not the only possible) determined by Saurí et al. [60] will be described. After absorption of light and population of the bright S_3 state, the authors assumed that the system decays to the S_1 state *via* vibrational relaxation and the corresponding IC processes. Then, as in the *anti*-$B_{18}H_{22}$ case, the system evolves towards the minimum of the S_1 predicted at 2.78 eV. The trapping of the molecule in this region allows two different processes: the radiative relaxation to the S_0 ground state, which explains the registered experimental fluorescence emission (see T_e values in brackets, Table 4.1), and the population of the T_1 triplet state. The latter was unlikely in the *anti*-$B_{18}H_{22}$ molecule due to the large energy difference between S_1 and T_1. However, in the present system the energy difference is significantly lower (0.14 eV) and the calculated SOC is higher (3.37 cm^{-1}). Thus, ISC processes are likely in the S_1 minimum in addition to the FC region discussed above. The competition between the fluorescence and the ISC processes is in agreement with the short fluorescence lifetime (0.94 ns) measured

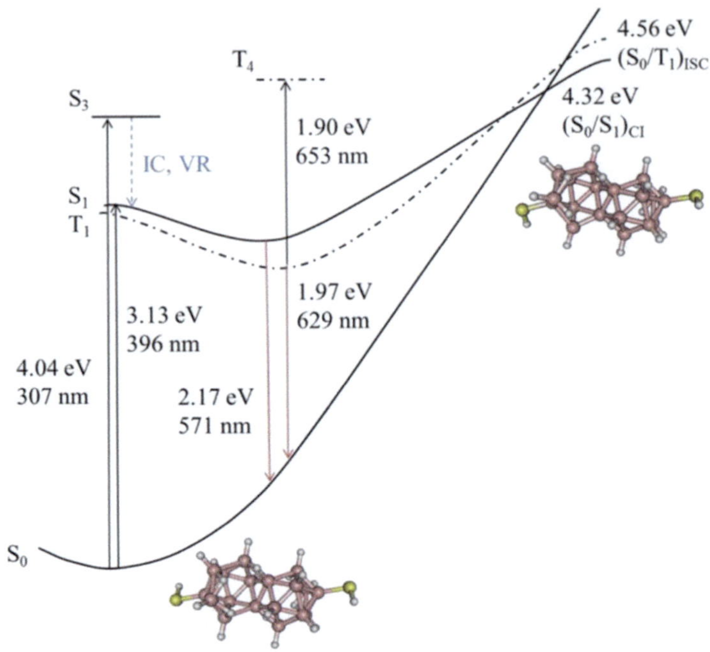

Fig. 4.7 Photophysics of the 4,4′-$(SH)_2$-*anti*-$B_{18}H_{20}$ system. The reaction coordinate is different for the S_1 (*solid line*) and the T_1 (*dashed line*) states (Reprinted with permission from Saurí et al. [60]. Copyright 2013 American Chemical Society)

experimentally. The observed transient absorption of long-lived species confirms the population of the triplet states in the 4,4′-$(SH)_2$-*anti*-$B_{18}H_{20}$ compound.

Once the triplet states are populated, phosphorescence emission becomes possible. Such emission was unambiguously confirmed experimentally, recording a phosphorescence band with a maximum at 596 nm (2.08 eV). The value is in clear agreement with the predicted 1.97 eV vertical emission energy for the 1^3A_u (T_1) state (see Table 4.1).

Regarding the efficiency of the non-radiative decay, it is worth noting that huge barriers between the S_1 minimum structures and the $(S_0/S_1)_{CI}$ (1.54 eV) or the $(S_0/T_1)_{ISC}$ (1.95 eV) were predicted at the CASPT2//CASSCF level. Hence, these points are not likely involved in the main photochemistry of the new 4,4′-$(SH)_2$-*anti*-$B_{18}H_{20}$ compound.

4.5.3 Geometrical Changes

As we did above for the photochemistry of the *syn*-$B_{18}H_{22}$ and *anti*-$B_{18}H_{22}$ compounds, we briefly review here the most important geometrical distortions that take place after the absorption of light. The changes in the atomic coordinates are analogous to the *anti*-$B_{18}H_{22}$ molecule, now also having the possibility of the SH groups rotations. In the singlet manifold, the greatest changes take place at the CI region, and they break the symmetry of the molecule. The B_8-B_9 bond suffers a strong elongation of Δd ~ 1.36 Å, the B_7-B_8 connectivity is shortened Δd ~ −0.26 Å and the B_1-B_4-SH dihedral angle varies from 9.8 to 90.0°. In the triplet manifold, the most important distortions were predicted at the STC region. The $B_{2'}$-$B_{5'}$ and the $B_{1'}$-$B_{10'}$ bonds are stretched Δd ~ 0.49 and Δd ~ 0.79 Å, respectively. On the contrary, the $B_{5'}$-$B_{10'}$ connectivity is contracted by Δd ~ −0.21 Å. Such strong geometrical distortions are the responsible of the energy increase of the molecule, causing the barriers for the non-emissive deactivation and allowing, in consequence, the fluorescence and phosphorescence phenomena of the molecule. The rest of molecular parameters remain approximately the same.

4.6 Conclusions

In the present Chapter, we review the absorption and emission properties of the *syn*-$B_{18}H_{22}$, *anti*-$B_{18}H_{22}$, and 4,4′-$(SH)_2$-*anti*-$B_{18}H_{20}$ boron complexes recorded experimentally and determined theoretically by Londesborough et al. [59] and Saurí et al. [60]. These compounds exhibit completely different photophysical behavior, which was rationalized on the basis of findings obtained with appropriate computational strategies and accurate theoretical methods. Figure 4.8 shows a schematic representation of the photophysics of the studied complexes, which shall be summarized here. First, a non-radiative relaxation is predicted for the *syn*-$B_{18}H_{22}$ complex according to the barrierless main decay path that characterizes the bright excited state of the molecule from the FC geometry towards the CI with the ground state. This crossing point funnels the population to the ground state. Ultrafast

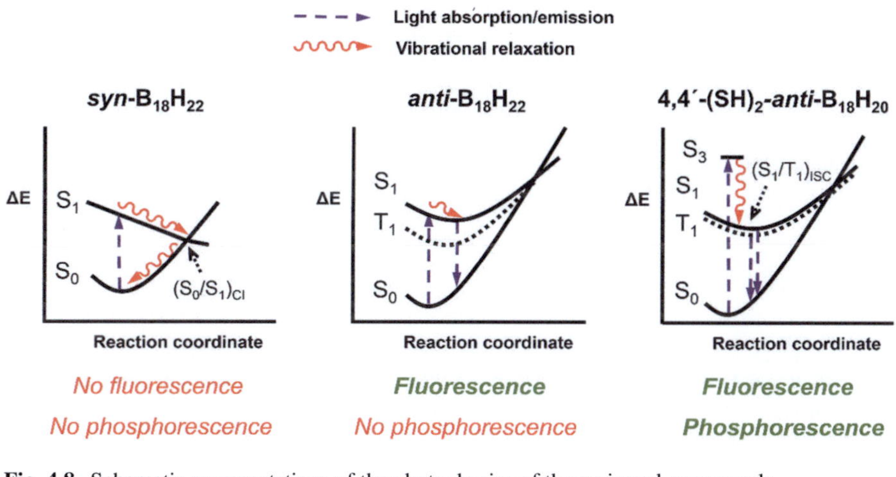

Fig. 4.8 Schematic representations of the photophysics of the reviewed compounds

non-emissive deactivation of the excited states is therefore the main photochemical feature of this compound. The scenario is totally different in the second compound, the *anti*-$B_{18}H_{22}$ isomer. A well-defined minimum on the PEH of the lowest-lying excited state traps the system in this area, favouring the fluorescence radiative mechanism. A relatively large energy gap between the singlet and triplet states in this area significantly reduces the population of the triplet state. Consequently, strong fluorescence and negligible phosphorescence is predicted for the *anti*-$B_{18}H_{22}$ system. The third compound is the new 4,4′-$(SH)_2$-*anti*-$B_{18}H_{20}$ complex. The thiol groups substituted at the 4 and 4′ positions of the complex tune the photophysics of the system, allowing the triplet population and thus increasing the efficiency of the phosphorescence emission process. Low-energy differences between the singlet and triplet states as well as significant SOCs at the excited state surface minimum explain the observed phosphorescence in the 4,4′-$(SH)_2$-*anti*-$B_{18}H_{20}$ complex. Therefore, fluorescence and phosphorescence processes compete in the deactivation of the excited states.

The studies reviewed in the present Chapter provide clear evidence of the important role of computational chemistry in the understanding of photophysical and photochemical phenomena. Even though direct comparisons with the experimental findings are still far from being possible due to the difficulties found to treat accurately the solvent effects and the large demands of time and CPU to evaluate dynamical aspects, the quantum chemistry of the excited state provides currently computational tools able to determine the main decay paths of excited molecules. As shown in the studies of this Chapter, differences in the absorption and emission spectra can be clearly interpreted. Thus, the analysis performed at the molecular level allows determining the structural and chemical origin of the different photoinduced behavior. Apart from the interpretation of experimental data, the quantum chemistry of the excited state is able to predict interesting properties in well-known molecules and also in new systems. Hence, the theoretical findings can motivate new ideas and encourage development of new experiments.

Acknowledgements We are grateful to Thomas Schmalz (Texas A&M University at Galveston, USA) for reading the manuscript. This research was supported by Projects CTQ2014-58624-P of the Spanish MEC/FEDER, GV2015-057 of the *Generalitat Valenciana*, and i-COOP-2013-B20040 from the Spanish National Research Council (CSIC). D.R.-S. thanks the "Juan de la Cierva" program of the Spanish MINECO (Ref. JCI-2012-13431). A.F.-M. thanks BES-2011-048326 FPI grant (MINECO).

References

1. Schrödinger E (1926) Quantisierung als Eigenwertproblem. Ann Phys (Berlin) 384:361–376
2. Schrödinger E (1982) Collected papers on wave mechanics, 3rd edn. Chelsea Publication Company, New York
3. Hartree DR (1928) The wave mechanics of an atom with a non-coulomb central field. Part I – Theory and methods. Proc Cambridge Phil Soc 24:89–110

4. Hartree DR (1928) The wave mechanics of an atom with a non-coulomb central field. Part II – Results and discussion. Proc Cambridge Phil Soc 24:111–132
5. Hartree DR (1928) The wave mechanics of an atom with a non-coulomb central field. Part III – Term values and intensities. Proc Cambridge Phil Soc 24:426–437
6. Hartree DR (1929) The wave mechanics of an atom with a non-coulomb central field. Part IV – Further results relating to terms of the optical spectrum. Proc Cambridge Phil Soc 25:310–314
7. Fock V (1930) Nähenungsmethode zur Lösung des quantenmechanischen Mehrkörperproblems. Z Phys 61:126–148
8. Fock V (1930) "Selfconsistent field" mit Austausch für Natrium. Z Phys 62:795–805
9. Slater JC (1960) Quantum theory of atomic structure. McGraw-Hill, New York
10. Roothaan CCJ (1951) New developments in molecular orbital theory. Rev Mod Phys 23:69–89
11. Hall GG (1951) The molecular orbital theory of chemical valency. VIII. A method of calculating ionization potentials. Proc Roy Soc A205:541–552
12. Pople JA, Nesbet RK (1954) Self-consistent orbitals for radicals. J Chem Phys 22:571–572
13. Roothaan CCJ (1960) Self-consistent field theory for open shells of electronic systems. Rev Mod Phys 32:179–185
14. Heitler W, London F (1927) Wechselwirkung neutraler Atome und homöopolare Bindung nach der Quantenmechanik. Z Phys 44:455–472
15. Coulson CA, Fischer I (1949) Notes on the molecular orbital treatment of the hydrogen molecule. Phil Mag 40:386–393
16. Pauling L (1960) The nature of the chemical bond. Cornell University Press, Ithaca
17. van Lenthe JH, Balint-Kurti GG (1983) The valence bond self-consistent field (VBSCF) method. J Chem Phys 78:5699–5713
18. Gerratt J, Cooper DL, Karadakov PB, Raimondi M (1997) Modern valence bond theory. Chem Soc Rev 26:87–100
19. Shaik S, Hiberty PC (2004) Valence bond theory, its history, fundamentals, and applications. A primer. Rev Comput Chem 20:1–100
20. Čížek J (1966) On the correlation problem in atomic and molecular systems. calculation of wavefunction components in Ursell-type expansion using quantum-field theoretical methods. J Chem Phys 45:4256–4266
21. Čížek J (1966) On the use of the cluster expansion and the technique of diagrams in calculations of correlation effects in atoms and molecules. Adv Chem Phys 14:35–89
22. Bartlett RJ (1981) Many-body perturbation theory and coupled cluster theory for electron correlation in molecules. Annu Rev Phys Chem 32:359–451
23. Paldus J (1981) Diagrammatic methods for many-Fermion systems (Lecture Notes ed.). University of Nijmegen, Njimegen, The Netherlands
24. Paldus J (2005) The beginnings of coupled-cluster theory: An eyewitness account. In: Dykstra CE, Frenking G, Kim KS, Scuseria GE (eds) Theory and applications of computational chemistry: the first forty years. Elsevier B.V, Amsterdam, pp 115–147
25. Shavitt I, Bartlett RJ (2009) Many-body methods in chemistry and physics. Cambridge Molecular Science, Cambridge
26. Sherrill CD, Schaefer III HF (1999) The configuration interaction method: advances in highly correlated approaches. In: Löwdin PO (ed) Advances in Quantum Chemistry, vol 34. Academic, San Diego, pp 143–269
27. Møller C, Plesset MS (1934) Note on an approximation treatment for many-electron systems. Phys Rev 46:618–622
28. Bartlett RJ, Silver DM (1976) Many-body perturbation theory applied to electron pair correlation energies. II. Closed-shell second-row diatomic hydrides. J Chem Phys 64:4578–4586
29. Bartlett RJ, Shavitt I (1977) Comparison of high-order many-body perturbation theory and configuration interaction for H_2O. Chem Phys Lett 50:190–198

30. Krishnan R, Frisch MJ, Pople JA (1980) Contribution of triple substitutions to the electron correlation energy in fourth order perturbation theory. J Chem Phys 72:4244–4245
31. Wilson S (1984) Electron correlation in molecules. Clarendon, Oxford
32. Hohenberg P, Kohn W (1964) Inhomogeneous electron gas. Phys Rev 136:B864–B871
33. Kohn W, Sham LJ (1965) Self-consistent equations including exchange and correlation effects. Phys Rev 140:A1133–A1138
34. Frisch MJ, Trucks GW, Schlegel HB, Scuseria GE, Robb MA, Cheeseman JR, Scalmani G, Barone V, Mennucci B, Petersson GA, Nakatsuji H, Caricato M, Li X, Hratchian HP, Izmaylov AF, Bloino J, Zheng G, Sonnenberg, JL, Hada M, Ehara M, Toyota K, Fukuda R, Hasegawa J, Ishida M, Nakajima T, Honda Y, Kitao O, Nakai H, Vreven T, Montgomery Jr JA, Peralta JE, Ogliaro F, Bearpark M, Heyd JJ, Brothers E, Kudin KN, Staroverov VN, Kobayashi R, Normand J, Raghavachari K, Rendell A, Burant, JC, Iyengar SS, Tomasi J, Cossi M, Rega N, Millam MJ, Klene M, Knox JE, Cross JB, Bakken V, Adamo C, Jaramillo J, Gomperts R, Stratmann RE, Yazyev O, Austin AJ, Cammi R, Pomelli C, Ochterski JW, Martin RL, Morokuma K, Zakrzewski VG, Voth GA, Salvador P, Dannenberg JJ, Dapprich S, Daniels AD, Farkas Ö, Foresman JB, Ortiz JV, Cioslowski J, Fox DJ (2009) Gaussian-09. Gaussian, Inc., Wallingford CT
35. Parr RG, Yang W (1989) Density-functional theory of atoms and molecules. Oxford University Press, New York
36. Marques MAL, Ullrich CA, Nogueira F, Rubio A, Burke K, Gross EKU (eds) (2006) Time-dependent density functional theory. Springer, Berlin Heidelberg
37. Klessinger M, Michl J (1995) Excited states and photochemistry of organic molecules. Wiley-VCH, New York
38. Ramamurthy V, Schanze KS (1998) Organic and inorganic photochemistry. Marcel Dekker, New York
39. Armstrong DR, Perkins PG, Stewart JJP (1973) Calculation of the electronic structure of boranes by the self-consistent molecular orbital method. Part III. Excited states of cage species. J Chem Soc Dalton Trans 21:2277–2280
40. Kunkely H, Vogler A (2005) Luminescence of silver 7, 8, 9, 10, 11, 12-hexabromo-closo-1-carbododecaborate. Inorg Chem Commun 8:992–993
41. Kunkely H, Vogler A (2007) Excited state properties of $Tl_2B_{12}H_{12}$. Metal-centered photoluminescence. Inor Chim Acta 360:679–680
42. Kwon S, Wee K, Cho Y-J et al (2014) Carborane dyads for photoinduced electron transfer: Photophysical studies on carbazole and phenyl-o-carborane molecular assemblies. Chem Eur J 20:5953–5960
43. Weber L, Kahlert J, Brockhinke R et al (2012) Luminescence properties of C-diazaborolyl-ortho-carboranes as donor-acceptor systems. Chem Eur J 18:8347–8357
44. Wee K-R, Cho Y-J, Song J-K et al (2013) Multiple photoluminescence from 1, 2-dinaphthyl-ortho-carborane. Angew Chem Int Ed 52:9682–9685
45. Base K, Grinstaff MW (1998) Generation of an unprecedented excited state oxidant in a coordinately unsaturated platinum complex. Inorg Chem 37:1432–1433
46. Hong E, Yang H, Kim Y, Jeoung SC, Do Y (2001) Mechano- and electroluminescence of a dissymmetric Hafnium carborane complex. Adv Mater 13:1094–1096
47. Calhorda MJ, Crespo O, Gimeno MC, Jones PG, Laguna A, López-de-Luzuriaga JM, Perez JL, Ramon MA, Veiros LF (2000) Synthesis, structure, luminescence and theoretical studies of tetranuclear Gold clusters with phosphinocarborane ligands. Inorg Chem 39:4280–4285
48. Crespo O, Gimeno MC, Jones PG, Laguna A, López-de-Luzuriaga JM, Monge M, Perez JL, Ramon MA (2003) Luminescent nido-carborane-diphosphine anions $[(PR_2)_2C_2B_9H_{10}]^-$ (R = Ph, ^{i}Pr). Modification of their luminescence properties upon formation of three-coordinate Gold(I) complexes. Inorg Chem 42:2061–2068
49. Bae H-J, Chung J, Kim H et al (2014) Deep red phosphorescence of cyclometalated iridium complexes by o-carborane substitution. Inorg Chem 53:128–138

50. Oliva JM, Serrano-Andrés L (2006) A computational study of the lowest singlet and triplet states of neutral and dianionic 1,2-substituted icosahedral and octahedral o-carboranes. J Comput Chem 27:524–535
51. Serrano-Andrés L, Oliva JM (2006) Photochemical window mechanism for controlled atom release in carborane endohedral boxes: theoretical evidence. Chem Phys Lett 432:235–239
52. Serrano-Andrés L, Klein DJ, Schleyer PR, Oliva JM (2008) What electronic structures and geometries of carborane mono- and ortho-, meta-, and para-diradicals are preferred? J Chem Theory Comput 4:1338–1347
53. Oliva JM, Klein DJ, Schleyer PR, Serrano-Andrés L (2009) Design of carborane molecular architectures with electronic structure computations: From endohedral and polyradical systems to multidimensional networks. Pure Appl Chem 81:719–729
54. Oliva JM, Serrano-Andrés L, Havlas Z, Michl J (2009) On the electronic structure of a dianion, a radical anion, and a neutral biradical $(HB)_{11}C\text{-}C{\equiv}C\text{-}C(BH)_{11}$ carborane dimer. J Mol Struct – THEOCHEM 912:13–20
55. Crespo O, Gimeno MC, Laguna A, Ospino I, Aullón G, Oliva JM (2009) Organometallic gold complexes of carborane. Theoretical comparative analysis of ortho, meta and para derivatives and luminescence studies. Dalton Trans 19:3807–3813
56. Oliva JM (2012) Energy landscapes in boron chemistry: Bottom-top approach towards design of novel molecular architectures. Adv Quantum Chem 64:105–119
57. Oliva JM, Alcoba DR, Lain L, Torre A (2013) Electronic structure studies of diradicals derived from closo-carboranes. Theor Chem Acc 132(1–6):1329
58. Oliva JM, Alcoba DR, Oña OB, Torre A, Lain L, Michl J (2015) Toward (car)borane-based molecular magnets. Theor Chem Acc 134:9. 8 p
59. Londesborough MGS, Hnyk D, Bould J et al (2012) Distinct photophysics of the isomers of $B_{18}H_{22}$ explained. Inorg Chem 51:1471–1479
60. Saurí V, Oliva JM, Hnyk D et al (2013) Tuning the photophysical properties of *anti*-$B_{18}H_{22}$: efficient intersystem crossing between excited singlet and triplet states in new 4,4′-(HS)2-anti-$B_{18}H_{20}$. Inorg Chem 52:9266–9274
61. Vicenta Saurí Peris (2013) Theoretical study of the molecular bases that control photochemical processes with biological and nanotechnological interest. PhD Thesis, Universitat de València
62. Cerdán L, Braborec J, Garcia-Moreno I, Costela A, Londesborough MGS (2015) A borane laser. Nature Commun 6:5958. 7 p
63. Hosmane NS (2011) Boron science. CRC Press, Boca Raton
64. Sibaev IB, Bregadze VI (2009) Polyhedral boron hydrides in use: current status and perspectives. Nova Science, New York
65. Sibaev IB, Bregadze VI (2009) Polyhedral boranes for medical applications: current status and perspectives. Eur J Inorg Chem 1433–1450
66. Hosmane NS, Maguire JA, Zhu Y, Takagaki M (2011) Boron and gadolinium neutron capture therapy for cancer treatment. World Scientific, Singapore
67. Grimes RN (2011) Carboranes, 2nd edn. Academic, London
68. Goddard WA III, Brenner DW, Lyshevski SE, Iafrate GJ (2003) Handbook of nanoscience, engineering and technology. CRC Pres, Boca Ratón
69. Pichierri F (2007) Polyhedral heteroborane clusters for nanotechnology. In: Mansoori GA, George TF, Assoufid L, Zhang G (eds) Molecular Building Blocks for Nanotechnology. Top Appl Phys 109:256–274
70. Fanfrlík J, Přáda A, Padělková Z, Pecina A, Macháček J, Lepšík M, Holub J, Růžička A, Hnyk D, Hobza P (2014) The dominant role of chalcogen bonding in the crystal packing of 2D/3D aromatics. Angew Chem Int Ed 53:10139–10142
71. Kennedy DC, Duguay DR, Tay LL, Richeson DR, Pezacki JP (2007) SERS detection and boron delivery to cancer cells using carborane labelled nanoparticles. Chem Commun 6750–6752
72. Barry NPE, Pitto-Barry A, Romero-Canelón I, Tran J, Soldevila-Barreda JJ, Hands-Portman I, Smith CJ, Kirby N, Dove AP, O'Reilly RK, Sadler PJ (2014) Precious metal carborane

polymer nanoparticles: characterization of micellar formulations and anticancer activity. Faraday Discuss 175:229–240
73. Serrano-Andrés L, Merchán M (2004) Spectroscopy: applications. In: Schleyer PR et al (eds) Encyclopedia of computational chemistry. Wiley, Chichester
74. Serrano-Andrés L, Merchán M (2008) Photostability and photoreactivity in biomolecules: Quantum chemistry of nucleic acid base monomers and dimers. In: Shukla MK, Leszczynski J (eds) Radiation induced molecular phenomena in nucleic acid: a comprehensive theoretical and experimental analysis. Chapter 16:435–472. Springer, The Netherlands
75. Serrano-Andrés L, Roca-Sanjuán D, Olaso-González G (2011) Recent trends in computational photochemistry. In: Albini A (ed) Photochemistry, vol 38, Series: specialist periodical reports. Royal Society, London, pp 10–36
76. González L, Escudero D, Serrano-Andrés L (2012) Progress and challenges in the calculation of electronic excited states. Chem Phys Chem 13:28–51
77. Serrano-Andrés L, Serrano-Pérez JJ (2012) Calculation of excited states: molecular photophysics and photochemistry on display. In: Handbook of computational chemistry. Springer, Berlin, pp 483–560
78. Giussani A, Segarra-Martí J, Roca-Sanjuán D, Merchán M (2015) Excitation of nucleobases from a computational perspective I: Reaction paths. Top Curr Chem 355:57–98
79. Roca-Sanjuán D, Fernández Galván I, Lindh R, Liu YJ (2015) Recent method developments and applications in computational photochemistry, chemiluminescence, and bioluminescence. In: Albini A (ed) Photochemistry, vol 42, Series: specialist periodical reports. Royal Society, London, pp 11–42
80. Teller E (1973) The crossing of potential surfaces. J Phys Chem 41:109–116
81. Herzberg G, Longuet-Higgins LC (1963) Intersection of potential energy surfaces in polyatomic molecules. Faraday Discuss 35:77–82
82. Robb MA, Bernardi F, Olivucci M (1996) Conical intersections as a mechanistic feature of organic-photochemistry. Pure Appl Chem 67:783–789
83. Domcke W, Yarkony DR, Köppel H (eds) (2004) Conical intersections: Electronic structure, dynamics, and spectroscopy. World Scientific, Singapore
84. Bearpark MJ, Robb MA (2007) Conical intersection species as reactive intermediates. Wiley, Hoboken
85. Garavelli M, Bernardi F, Cembran A (2005) In: Olivucci M (ed) Theoretical and computational chemistry. Elsevier, Amsterdam, p 191
86. Michl J (2005) Foreword. In: Olivucci M (ed) Computational photochemistry. Elsevier, Amsterdam, p ix
87. Serrano-Andrés L, Merchán M (2005) Quantum chemistry of the excited state. J Mol Struc – Theochem 729:109–118
88. Roos BO, Andersson K, Fülscher MP, Malmqvist PA, Serrano-Andrés L, Pierloot K, Merchán M (1996). New methods in computational quantum mechanics. In: Prigogine I, Rice S (eds) Multiconfigurational perturbation theory – applications in electronic spectroscopy. Adv Chem Phys 93:219–331. John Wiley, New York
89. Merchán M, Serrano-Andrés L (2005) Ab initio methods for excited states. In: Olivucci M (ed) Computational photochemistry. Elsevier, Amsterdam
90. Gozem S, Krylov AI, Olivucci M (2013) Conical intersection and potential energy surface features of a model retinal chromophore: Comparison of EOM-CC and multireference methods. J Chem Theory Comput 9:284–292
91. Andersson K, Malmqvist PÅ, Roos BO, Sadlej AJ, Wolinski K (1990) Second-order perturbation theory with a CASSCF reference function. J Phys Chem 94:5483–5488
92. Andersson K, Malmqvist PÅ, Roos BO (1992) Second-order perturbation theory with a complete active space self-consistent field reference function. J Chem Phys 96:1218–1226
93. Roca-Sanjuán D, Aquilante F, Lindh R (2012) Multiconfiguration second-order perturbation theory approach to strong electron correlation in chemistry and photochemistry. Wiley Interdisciplinary Reviews. Comput Mol Sci 2:585–603

94. Ghigo G, Roos BO, Malmqvist PÅ (2004) A modified definition of the zeroth-order Hamiltonian in multiconfigurational perturbation theory (CASPT2). Chem Phys Lett 396:142–149
95. Forsberg N, Malmqvist PÅ (1997) Multiconfiguration perturbation theory with imaginary level shift. Chem Phys Lett 274:196–204
96. Finley J, Malmqvist PÅ, Roos BO, Serrano-Andrés L (1998) The multi-state CASPT2 method. Chem Phys Lett 288:299–306
97. Serrano-Andrés L, Merchán M, Lindh R (2005) Computation of conical intersections by using perturbation techniques. J Chem Phys 122:104107-1-10

Chapter 5
Deconvoluting the Reaction Path from $B_{10}H_{14}$ Plus BH_4^- to $B_{12}H_{12}^{2-}$. Can Theory Make a Contribution?

Michael L. McKee

Abstract Reactions of boron hydrides can involve many competing pathways and intermediates. One example is the reaction of the borohydride anion BH_4^- with decaborane(14) $B_{10}H_{14}$ to form the dodecaborohydride dianion ($B_{12}H_{12}^{2-}$). Presumably, the reaction involves the addition of two BH_4^- anions and the elimination of five hydrogen molecules. Gas phase optimizations at the B3LYP/6-31G(d) level followed by implicit solvation modeling with CPCM (Conductor-like polarized model) are used to provide insights into the reaction mechanism.

5.1 Introduction

The field of boron chemistry is very broad [1–7]. In contrast to organic chemistry, reaction mechanisms can be very complex because there is often little energy difference between various intermediates and transition states [8–13]. Nevertheless, great progress has been made understating the reaction mechanisms of boron hydrides and related carboranes [14–21]. For example, the carboranes $C_2B_nH_{n+2}$, n = 5–10 are often formed with the carbon atoms adjacent to each other but can rearrange to more thermodynamically stable carboranes where the carbon atoms are further apart. The mechanism may involve different rearrangement steps such as the diamond-square-diamond (DSD) step. In the DSD step two fused triangles go through a square transition state and continue to two fused triangles where the originally non-adjacent vertices become bonded in the product. One or more DSD steps can occur concurrently. When two DSD steps occur at the same time, the step can be view as "local bond rotation" [17]. Other mechanistic steps are terminal-bridge-terminal hydrogen migration as well as a hinge or flip step. Some rearrangements can be viewed as nonclassical-classical-nonclassical (or *closo-nido-closo*) steps where the bonding

M.L. McKee (✉)
Department of Chemistry and Biochemistry, Auburn University, Auburn, AL 36949, USA
e-mail: mckeeml@auburn.edu

D. Hnyk, M.L. McKee (eds.), *Boron*, Challenges and Advances in Computational Chemistry and Physics 20,
DOI 10.1007/978-3-319-22282-0_5

environment around a boron center changes from three-center two-electron to two-center two-electron with an empty orbital on one boron atom.

Decaborane(14) $B_{10}H_{14}$ is a stable boron hydride that serves as the starting material for the synthesis of many boron-containing compounds [22–26]. An alkali metal borohydride can react with $B_{10}H_{14}$ to form the dodecaborohydride dianion plus five hydrogen molecules in refluxing diglyme (160 °C) [2]. There is some evidence for the $B_{11}H_{14}^-$ anion as an intermediate, but the mechanism is still a complete mystery. Other boron hydride dianions can possibly be formed, such as $B_{11}H_{11}^{2-}$ or $B_{10}H_{10}^{2-}$ (Eqs. 5.1, 5.2 and 5.3). It is interesting

$$B_{10}H_{14} + 2BH_4^- \rightarrow B_{12}H_{12}^{2-} + 5H_2 \tag{5.1}$$

$$B_{10}H_{14} + 2BH_4^- \rightarrow B_{11}H_{11}^{2-} + BH_3 + 4H_2 \tag{5.2}$$

$$B_{10}H_{14} + 2BH_4^- \rightarrow B_{10}H_{10}^{2-} + 2BH_3 + 3H_2 \tag{5.3}$$

to note that Štíbr and co-workers have recently shown [27] that the $C_2B_8H_{12} \rightarrow C_2H_8H_{10}$ *nido* → *closo* transformation can take place. This process is isoelectronic to a $B_{10}H_{14} \rightarrow B_{10}H_{10}^{2-}$ transformation (see Eq. 5.3).

The reaction steps with the largest activation barriers involve the loss of molecular hydrogen. All other steps have much lower barriers. In mapping out the potential energy surface, a search was made for the lowest-energy barrier for H_2 loss. From these transition states, the connected reactant and product structures were found. These intermediates were then connected together (via other transition states) in a time-consuming process that involved exploring many possibilities. In comparing different steps in the reaction mechanism, a comparison of free energy barriers in solution alone is not sufficient to decide on the preferred pathway. In many cases, there is competition between a unimolecular step and a bimolecular step. While it is likely that the H_2-elimination steps are irreversible, the addition/elimination of BH_4^- and or BH_3 may be reversible. The concentration of BH_4^- will depend on the initial reaction conditions. In contrast, any BH_3 will be coordinated to the solvent and its concentration may be dependent on many different competing reactions. For five reaction steps, no enthalpic barrier could be located for the addition of BH_3 to an intermediate (see steps with "No TS" label in Fig. 5.1). The most likely path through the reaction steps is shown by the outlined arrows in Fig. 5.1.

5.2 Computational Details

Geometries were optimized at the B3LYP/6-31G(d) level using the Gaussian 09 program system [28]. While the 6-31G(d) basis set is small by modern standards, the basis set allows extensive searching of the potential energy surface while at the same time providing reasonable geometries and relative energies of structures in a variety of different bonding environments. Vibrational frequencies are used to

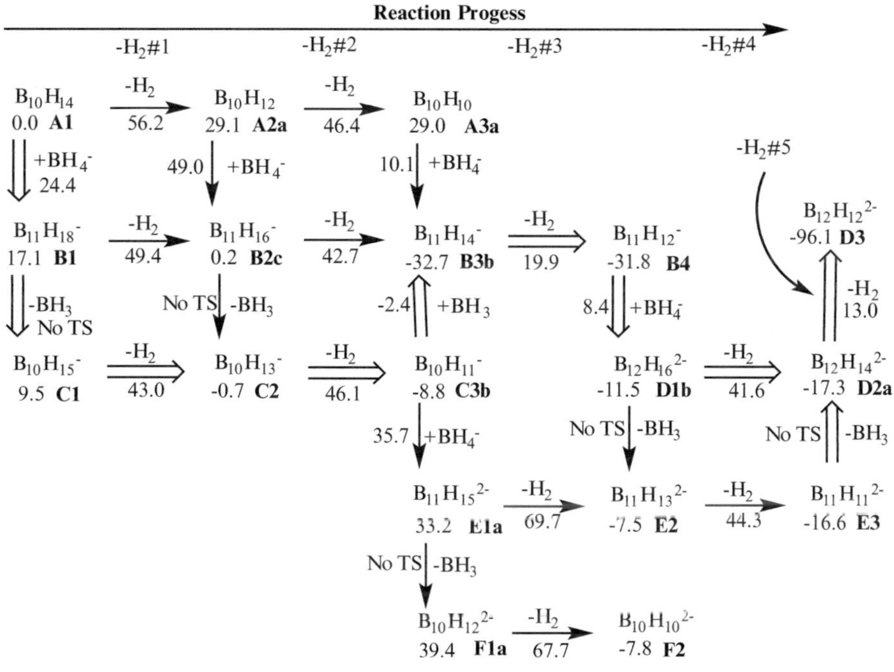

Fig. 5.1 Summary of all reaction steps in the reaction of $B_{10}H_{14}+2\ BH_4^-$ to $B_{12}H_{12}^{2-}+5\ H_2$. The numbers below each species are the free energy in solvent at 433 K relative to $B_{10}H_{14}+2BH_4^-$ in kcal/mol. The numbers beside arrows are free energies of transition states relative to the same reference. As the reaction proceeds, five successive hydrogen molecules are lost (H_2#1 – H_2#5)

compute zero-point and thermal corrections to 298 K. In order to estimate free energies in refluxing diglyme, the TΔS term was computed at 433 K (160 °C). Implicit solvation was modeled with the CPCM method with Pauling radii [29, 30]. The Pauling radii were found to give more accurate solution free energies for the boron hydride dianions at the B3LYP/6–311+G(2d,p) level of theory [28]. Free energies at 160 °C in diethyl ether were computed using Eq. 5.4.

$$\Delta G(\text{diglyme}, 433\,\text{K}) \approx \Delta G(\text{g}, 433\,\text{K}) + \Delta G(\text{solv}, \text{DEE} = \text{diethyl ether}) \quad (5.4)$$

The experimental solvent was diglyme (2-methoxyethyl ether, $RCH_2CH_2OCH_2CH_2R$, R = OCH_3) with a boiling point of 162 °C and a dielectric constant of 7.30. Because solvent parameters were lacking for diglyme, the CPCM method used parameters for diethyl ether ($RCH_2CH_2OCH_2CH_2R$, R = H) which has a dielectric constant of 4.24.

Most transition states were located in the gas phase; solvation effects were included via single-point energy calculations. However, locating three transition states required using explicit solvation in the geometry optimization. These transition states were the addition of BH_4^- to a boron hydride anion. In the gas phase, the barrier is quite high due to the electrostatic repulsion of the two fragments. In solution, the repulsion is screened by the solvent. In addition, the solvation free energy of the dianion product is greater than the sum of the solvation free energies of the two reactants.

5.3 Results and Discussions

The broad variations in the proposed mechanism will be presented first followed by a closer analysis of the various steps (Eqs. 5.5a, 5.5b, 5.5c, 5.6a, 5.6b, and 5.6c). In the scheme below (see Fig. 5.1), the various H_2 molecules released are numbered from #1 to #5.

$$B_{10}H_{14} \rightarrow B_{10}H_{12} + H_2(\#1) \quad (5.5a)$$

$$B_{10}H_{12} + BH_4^- \rightarrow B_{11}H_{16}^- \rightarrow B_{11}H_{14}^- + H_2(\#2) \quad (5.5b)$$

$$B_{10}H_{14} + BH_4^- \rightarrow B_{11}H_{18}^- \rightarrow B_{11}H_{14}^- + H_2(\#1) + H_2(\#2) \quad (5.5c)$$

$$B_{11}H_{14}^- + BH_4^- \rightarrow B_{12}H_{18}^{2-} \rightarrow B_{12}H_{12}^{2-} + H_2(\#3) + H_2(\#4) + H_2(\#5) \quad (5.6a)$$

$$B_{11}H_{14}^- \rightarrow B_{11}H_{12}^- + H_2(\#3) \quad (5.6b)$$

$$B_{11}H_{12}^- + BH_4^- \rightarrow B_{12}H_{16}^{2-} \rightarrow B_{12}H_{12}^{2-} + H_2(\#4) + H_2(\#5) \quad (5.6c)$$

The reaction scheme is quite complex with a large number of variations. In the actual reaction, the $B_{12}H_{12}^{2-}$ reaction product is likely formed via multiple pathways, where the amount of product depends on the bottlenecks along the various pathways. The intermediates which are considered in this study are listed in Table 5.1. The various steps in the pathway will be differentiated by letters and numbers (see Scheme 5.1). Steps going from "**1**" to "**2**" to "**3**" will involve the elimination of H_2. Steps from "**A**" to "**B**" to "**C**", etc. will involve the addition of BH_4^- or the loss of BH_3. A small letter "**a**", "**b**", etc. designates isomers/conformers of the same species. For example, the reaction **A1** → **A2** (through transition state **TSA1/A2**) is the loss of H_2 from $B_{10}H_{14}$ to form $B_{10}H_{12}$. Likewise, the **A1** → **B1** step (through transition state **TSA1/B1**) is the addition of BH_4^- to form $B_{11}H_{18}^-$. Some of the pathways involve rearrangements such as **B3a/B3c/B3d/B3e/B3f** which involves rearrangements on the $B_{11}H_{14}^-$ potential energy surface. Despite considerable effort, it cannot be guaranteed that the lowest-energy pathway has been found. The overall scheme is presented in Fig. 5.1, and thermodynamic properties are presented in Table 5.2. The suggested lower-energy pathways are shown by outlined arrows in Fig. 5.1. The species with enhanced thermodynamic stability are $B_{10}H_{14}$, $B_{10}H_{13}^-$, $B_{11}H_{14}^-$, $B_{10}H_{12}^{2-}$, $B_{10}H_{10}^{2-}$, $B_{11}H_{11}^{2-}$, and $B_{12}H_{12}^{2-}$. In the discussion below, various parts of

Table 5.1 Notation (and Literature References) of various species with the number of isomers considered in parentheses

	Ref.		Ref.		Ref.		Ref.
A1 $B_{10}H_{14}$		**A2**(5) $B_{10}H_{12}$	[31]	**A3**(2) $B_{10}H_{10}$	[32]		
B1 $B_{11}H_{18}^-$		**B2**(4) $B_{10}H_{16}^-$		**B3**(6) $B_{11}H_{14}^-$	[33–39]	**B4** $B_{11}H_{12}^-$	[40]
C1 $B_{10}H_{15}^-$		**C2** $B_{10}H_{13}^-$	[41–44]	**C3** $B_{10}H_{11}^-$	[45, 46]		
D1(2) $B_{12}H_{16}^{2-}$		**D2**(4) $B_{12}H_{14}^{2-}$		**D3** $B_{12}H_{12}^{2-}$	[23]		
E1(3) $B_{11}H_{15}^{2-}$		**E2** $B_{11}H_{13}^{2-}$		**E3** $B_{11}H_{11}^{2-}$	[24]		
F1(2) $B_{10}H_{12}^{2-}$	[47, 48]	**F2** $B_{10}H_{10}^{2-}$					

(a) On going from **A1** to **A2**, or **B3** to **B4**, etc., a hydrogen molecule is eliminated
(b) Species with enhanced stability are $B_{10}H_{14}$, $B_{10}H_{13}^-$, $B_{10}H_{12}^{2-}$, $B_{11}H_{14}^-$, $B_{10}H_{10}^{2-}$, $B_{11}H_{11}^{2-}$, and $B_{12}H_{12}^{2-}$

Scheme 5.1 Steps involving loss of BH_4^{2-} or addition of BH_3

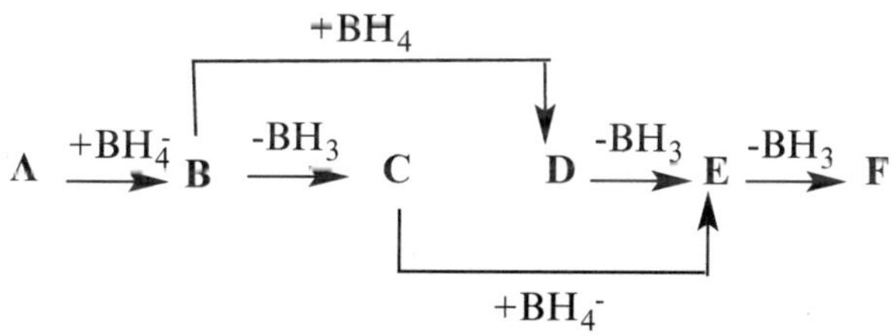

the reaction pathways are illustrated (see Scheme 5.2 for a guide to which reaction steps are illustrated in which figures). The reaction steps with the largest activation barriers all involve elimination of H_2.

The initial step is either elimination of H_2 from $B_{10}H_{14}$ (**A1** → **A2**) or addition of BH_4^- to $B_{10}H_{14}$ (**A1** → **B1**). If the first step is elimination of H_2, the product is $B_{10}H_{12}$ which can be viewed as a doubly protonated *closo* $B_{10}H_{10}^{2-}$ boron hydride dianion (Fig. 5.2). The symmetrical elimination of two bridging hydrogen atoms (**A1** → **A2b**) is very similar to the loss of H_2 from B_4H_{10} where the lowest energy transition state has C_s symmetry [15]. However, in a slightly lower energy process, two nonadjacent bridging hydrogen atoms can be lost to form an intermediate (**A1** → **A2a**) in a non-spontaneous reaction (ΔG(DEE, 433 K) = 29.1 kcal/mol). A transition state, which converts bridge and terminal hydrogen atoms, produces another intermediate (**A2a** → **A2b**). The indirect formation of **A2b** from **A1** has a 56.2 kcal/mol free energy barrier, 2.1 kcal/mol lower than the direct pathway. From **A2b**, a lower barrier (**TSA2b/A2e**) exists to **A2e** which corresponds to the lowest-energy form of doubly protonated $B_{10}H_{10}^{2-}$. The lowest-energy transition state for loss of H_2 from $B_{10}H_{12}$ requires some rearrangement of $B_{10}H_{12}$; specifically, **A2b** → **A2c** → **A2d** → **A3a** + H_2 where **A2d** is a $B_{10}H_{10}/H_2$ complex.

Loss of H_2 from $B_{10}H_{14}$ has a very high free energy barrier, which is consistent with the known stability of $B_{10}H_{14}$. The lowest-energy transition state (**TSA1/B1**, Fig. 5.3) for addition of BH_4^- to $B_{10}H_{14}$ involves the approach of BH_4^- to one of the

Table 5.2 Solvation Free Energies in Diethyl ether (DEE), Relative Energies, Relative Enthalpies (0 and 298 K), Relative Free Energies (g, 298 K), and Relative Free Energies (DEE, 433 K) in kcal/mol Relative to $B_{10}H_{14}$ plus 2 BH_4^-

	ΔG(sol)	ΔE	ΔH (0 K)	ΔH (298 K)	ΔG(*g*) (298 K)	ΔG(DEE) (433 K)
H_2	1.00					
BH_4^-	−47.77					
BH_3	1.07					
A1 $B_{10}H_{14}$	7.29	0.0	0.0	0.0	0.0	0.0
TSA1/A2b	7.52	67.0	60.3	61.3	58.1	58.3
A2b	7.78	50.3	40.7	43.3	27.8	29.3
TSA2b/A2c	7.93	54.1	43.9	46.1	31.8	33.5
A2c	9.66	53.8	45.5	47.7	33.3	36.7
TSA2c/B2a	−22.38	36.6	29.9	31.7	29.9	49.0
TSA1/A2a	6.12	64.0	58.6	59.0	57.3	56.2
A2a $B_{10}H_{12}$	9.21	47.6	38.7	41.1	26.2	29.1
TSA2a/A2b	9.03	67.0	55.6	58.1	42.9	45.6
TSA2b/A2e	8.34	51.3	41.0	43.1	29.0	31.0
A2e	9.96	38.3	29.2	31.6	16.9	20.6
TSA2c/A2d	10.06	58.9	48.5	50.7	36.6	40.4
A2d	4.67	53.0	43.6	45.8	31.8	30.2
TSA2d/A3a	5.36	72.3	59.8	62.5	47.3	46.4
A3a $B_{10}H_{10}$	6.54	66.9	51.6	56.0	27.8	29.0
TSA3a/A3b	9.58	78.9	61.4	65.9	37.2	41.5
A3b	10.96	60.5	45.4	49.4	23.3	29.0
TSA1/B1	−27.89	3.5	2.6	2.7	11.8	24.4
B1 $B_{11}H_{18}^-$	−12.02	−23.3	−21.7	−21.6	−11.4	17.1
C1 $B_{10}H_{15}^-$	−16.45	−13.5	−14.4	−13.9	−15.6	9.5
TSC1/C2	−18.42	25.3	21.0	21.5	19.8	43.0
C2 $B_{10}H_{13}^-$	−18.63	−6.3	−12.7	−10.7	−24.6	−0.7
TSC2/C3a	−17.65	44.9	34.0	36.7	21.2	46.1
TSB1/B2a	−14.38	14.3	12.5	12.6	23.3	49.4
B2a	−13.97	−18.0	−21.7	−20.2	−21.1	6.4
TSB2a/B2b	−14.39	−3.6	−9.2	−7.6	−8.6	18.5
B2b	−14.42	−23.2	−26.5	−25.0	−25.5	1.6
TSB2b/B2c	−13.48	−13.3	−17.4	−16.3	−15.7	12.3
B2c $B_{11}H_{16}^-$	−15.54	−26.5	−28.9	−28.1	−25.7	0.2
TSB2c/B2d	−15.91	−21.9	−25.0	−24.5	−21.4	4.2
B2d	−14.48	−24.0	−26.1	−25.4	−22.8	4.2
TSB2d/B3a	−15.99	20.2	14.3	15.3	17.2	42.7
B3a	−16.26	−7.3	−16.6	−14.3	−24.0	2.3
TSB3a/B3b	−16.88	−5.1	−15.7	−13.4	−22.8	2.8
B3b $B_{10}H_{14}^-$	−17.2	−43.5	−51.1	−49.1	−58.0	−32.7
TSA3a/B3c	−14.66	4.1	−10.2	−8.0	−17.7	10.1

(continued)

Table 5.2 (continued)

	ΔG(sol)	ΔE	ΔH (0 K)	ΔH (298 K)	ΔG(*g*) (298 K)	ΔG(DEE) (433 K)
B3c	−12.98	−9.6	−20.8	−17.3	−31.9	−2.4
TSB3c/B3d	−13.03	9.5	−3.6	−0.1	−14.3	15.2
B3d	−12.81	1.8	−9.9	−6.3	−20.9	8.8
TSB3d/B3e	−14.02	9.0	−4.1	−0.4	−16.1	12.4
B3e	−13.03	−1.8	−13.3	−9.8	−24.1	5.3
TSB3e/B3f	−14.10	26.5	14.7	17.9	5.3	33.7
B3f	−16.27	−7.3	−16.6	−14.3	−23.9	2.3
TSB3f/B3b	−16.88	−5.1	−15.7	−13.4	−22.8	2.8
TSB3b/B4	−17.06	13.6	1.8	4.2	−5.5	19.9
B4 $B_{11}H_{12}^{-}$	−16.17	−27.6	−41.3	−37.5	−59.1	−31.8
TSB4/D1a	−100.92	36.4	23.7	27.1	18.0	8.4
D1a	−101.15	20.3	9.5	12.7	4.4	−5.5
TSD1a/D1b	−101.78	24.9	12.7	15.9	7.9	−2.6
D1b $B_{12}H_{16}^{2-}$	−101.72	15.5	4.7	8.0	−1.0	−11.5
TSD1b/D2a	−100.81	73.0	57.0	60.7	51.1	41.6
D2a $B_{12}H_{14}^{2-}$	−101.89	26.5	9.0	14.5	−7.7	−17.3
TSD2a/D2b	−103.11	42.0	21.3	29.1	9.3	−1.6
D2b	−104.52	40.4	23.8	28.6	9.1	−3.2
TSD2b/D2c	−107.06	51.7	34.1	38.3	20.6	5.8
D2c	−107.46	43.3	27.0	31.3	13.5	−1.7
TSD2c/D2d	−107.36	44.2	27.0	31.0	13.9	−1.2
D2d	−107.34	31.7	16.3	20.4	2.9	−12.2
TSD2d/D3	−110.51	63.1	44.9	49.2	31.2	13.0
D3 $B_{12}H_{12}^{2-}$	−111.14	−36.0	−55.2	−49.9	−78.3	−96.1
C3a $B_{10}H_{11}^{-}$	−18.17	10.8	−3.2	0.7	−26.1	−0.7
TSC3a/C3b	−18.10	18.5	3.6	7.6	−19.3	6.2
C3b	−18.41	3.7	−11.0	−7.0	−33.9	−8.8
TSC3b/E1a	−103.21	70.6	58.3	61.8	47.5	35.7
E1a	−101.74	66.4	54.7	58.4	43.6	33.2
TSE1a/E1b	−101.85	71.3	58.0	61.8	46.5	35.9
E1b $B_{11}H_{15}^{2-}$	−101.89	68.5	56.7	60.5	45.0	34.4
TSE1b/E1c	−105.42	77.5	66.4	69.5	56.9	42.8
E1c	−108.04	67.7	57.9	60.8	49.4	32.7
TSE1c/E2	−108.22	109.0	95.6	98.8	86.6	69.7
E2 $B_{11}H_{13}^{2-}$	−113.68	45.1	31.4	35.1	13.9	−7.5
TSE2/E3	−112.48	102.3	82.8	87.1	64.5	44.3
E3 $B_{11}H_{11}^{2-}$	−114.35	53.7	33.2	39.1	4.4	−16.6
F1a $B_{10}H_{12}^{2-}$	−115.00	99.6	85.4	89.6	62.0	39.4
TSF1a/F1b	−114.74	102.9	87.6	91.7	64.2	41.9
F1b	−115.35	98.1	83.8	88.1	60.2	37.2
TSF1a/F1c	−115.32	108.2	92.4	96.6	68.8	45.8

(continued)

Table 5.2 (continued)

	ΔG(sol)	ΔE	ΔH (0 K)	ΔH (298 K)	ΔG(*g*) (298 K)	ΔG(DEE) (433 K)
F1c	−115.56	96.7	83.2	87.2	60.7	37.5
TSF1b/F2	−117.86	133.5	116.5	120.7	93.2	67.7
F2 $B_{10}H_{10}^{2-}$	−119.00	68.5	49.7	55.2	17.8	−7.8

(a) Energy at the B3LYP/6-311+G(2d,p)//B3LYP/6-31G(d) level
(b) Solvation at the CPCM with parameters for diethyl ether (DEE) as a substitute for diglyme with special keywords"cav, rep, disp" to compute solute-solvent dispersion interaction energy, solute-solvent repulsion energy, and solute cavitation energy, respectively
(c) Zero-point energy and heat capacity (integrated to 298 K) and entropy at the B3LYP/6-31G(d) level
(d) Enthalpies in kcal/mol (relative to $B_{10}H_{14}+2BH_4^-$) at 298 K computed with zero-point and thermal corrections
(e) Free energies in kcal/mol (relative to $B_{10}H_{14}+2BH_4^-$) from ΔH(298 K) – TΔS where T=433 K and solvation free energy computed at the CPCM(diethylether)/B3LPY/6-311+G(2d,p) level

Scheme 5.2 Guide to which figure describes which reaction sequence

Reaction Sequence	
A1→A2	Figure 2
A1→B1→B2	Figure 3
A2→A3	Figure 4
A2→B2	Figure 5
B2→B3	Figure 6
C1→C2	Figure 7
B3→B4→D1	Figure 8
D1→D2→D3	Figure 9
E2→E3	Figure 10
E1→F2	Figure 11

boron atoms. In the transition state, the H atom of BH_4^- is 2.17 Å from a boron center; in the product (**B1**), the terminal hydrogen is converted into a bridging hydrogen (Fig. 5.3). **B1** can either lose H_2 to form $B_{11}H_{16}^-$ (**B2a**) or lose BH_3 to form $B_{10}H_{15}^-$ (**C1**).

The neutral $B_{10}H_{10}$ is an example of a *hypercloso* boron hydride [49]. **A3a** and **A3b** have almost the same energy (Fig. 5.4). However, **A3a** is very interesting in that one boron atom does not have a terminal hydrogen. Thus, this species is well set up to add BH_4^- (**TSA3a/A3b**) to form $B_{11}H_{14}^-$. In the reaction of $B_{10}H_{14}$ plus

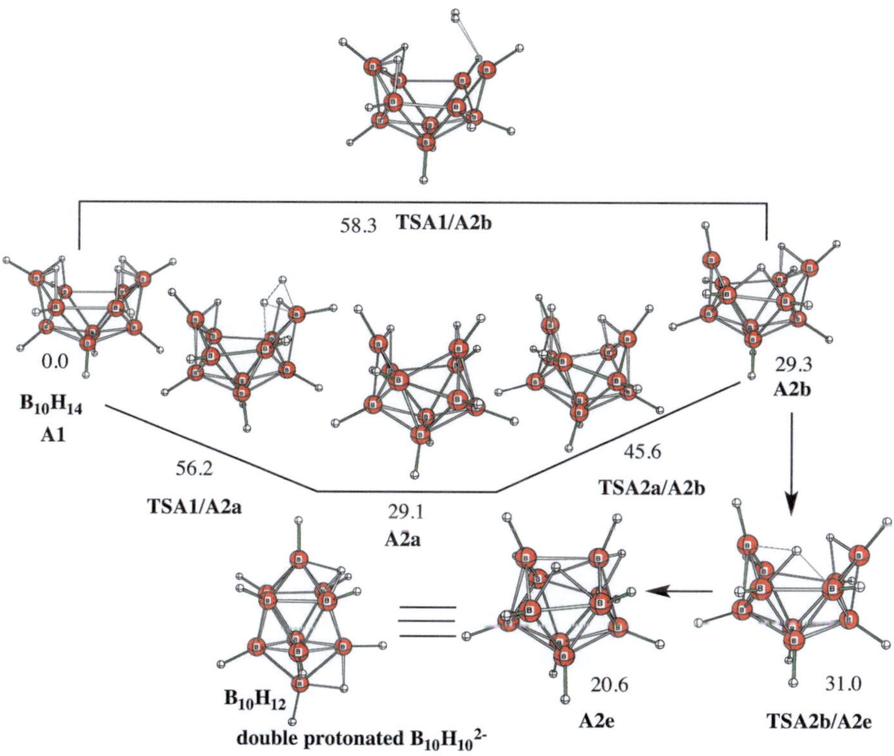

Fig. 5.2 Reaction pathway from $B_{10}H_{14}$ to $B_{10}H_{12}+H_2$ (**A1 → A2**)

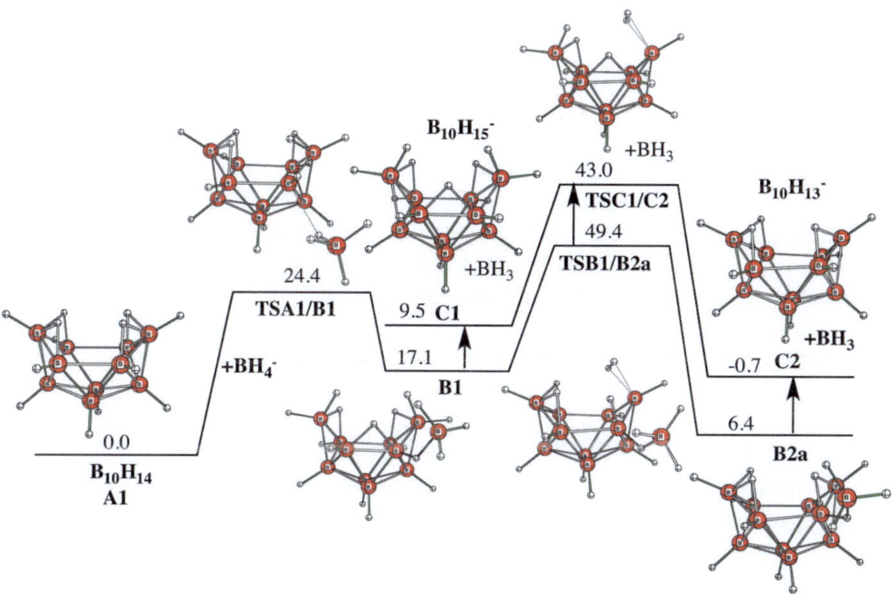

Fig. 5.3 Reaction pathway from $B_{10}H_{14}+BH_4^-$ to $B_{10}H_{18}^-+H_2$ and loss of H_2 from $B_{10}H_{18}^-$ to $B_{10}H_{16}^-$ (**A1 → B1 → B2**)

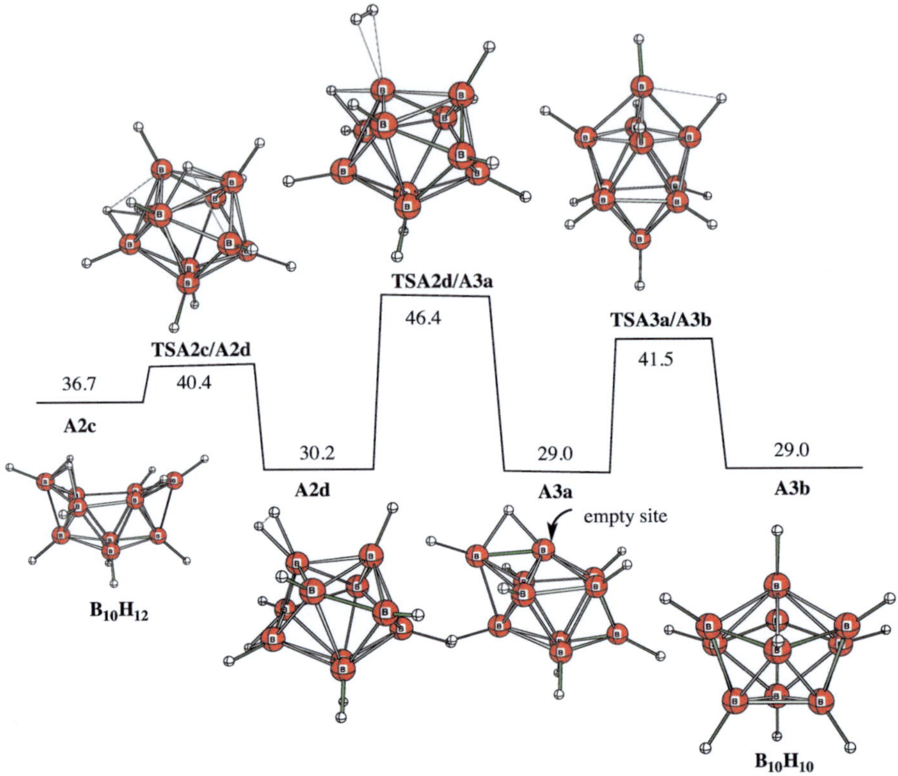

Fig. 5.4 Reaction pathway from $B_{10}H_{12}$ to $B_{10}H_{10} + H_2$ (**A2→A3**)

BH_4^-, this pathway is not visited due to the very high barrier for H_2 loss from $B_{10}H_{14}$ compared to the barrier for addition of BH_4^- (**TSA1/B1**). $B_{10}H_{12}$ (**A2b**) can rearrange to a species where one boron center lacks a terminal hydrogen (**A2c**, Fig. 5.5). This species can add BH_4^- (**TSA2c/B2a**) where the adding H atom of BH_4^- is 2.58 Å from the boron center. The free energy barrier for **A2c→TSA2c/B2a** is only 12.3 kcal/mol but it must be added to the already high free energy of **A2c** to give a free energy relative to reactants ($B_{10}H_{14}/2BH_4^-$) of 49.0 kcal/mol (Fig. 5.5).

In steps, the external BH_3 of **B2a** can fuse into the molecular framework (**B2a/B2b/B2c/B2d**) with little change in stability (Fig. 5.6). **B2d** then loses H_2 to form **B3a** and then **B3b** (Fig. 5.6). $B_{11}H_{14}^-$ (**B3b**) is a well-studied species [33–39] which is known to be double bridged in the open five-membered ring. The preferred route to $B_{11}H_{14}^-$ (**B3b**) is to add BH_4^- to $B_{10}H_{14}$ and lose BH_3 to form $B_{10}H_{15}^-$ (**C1**). **C1** can lose H_2 to form **C2** and lose another H_2 to form $B_{10}H_{11}^-$ (**C3a**) (Fig. 5.7). $B_{11}H_{14}^-$ (**B3b**) can then be formed from **C3b** by the addition of BH_3 which occurs with a very small activation barrier. The transition state free energy for this BH_3-addition step is estimated as the energy of **B3c** (not shown) which is a higher-energy isomer of $B_{11}H_{14}^-$ with a BH_3 coordinated to a hydrogen atom.

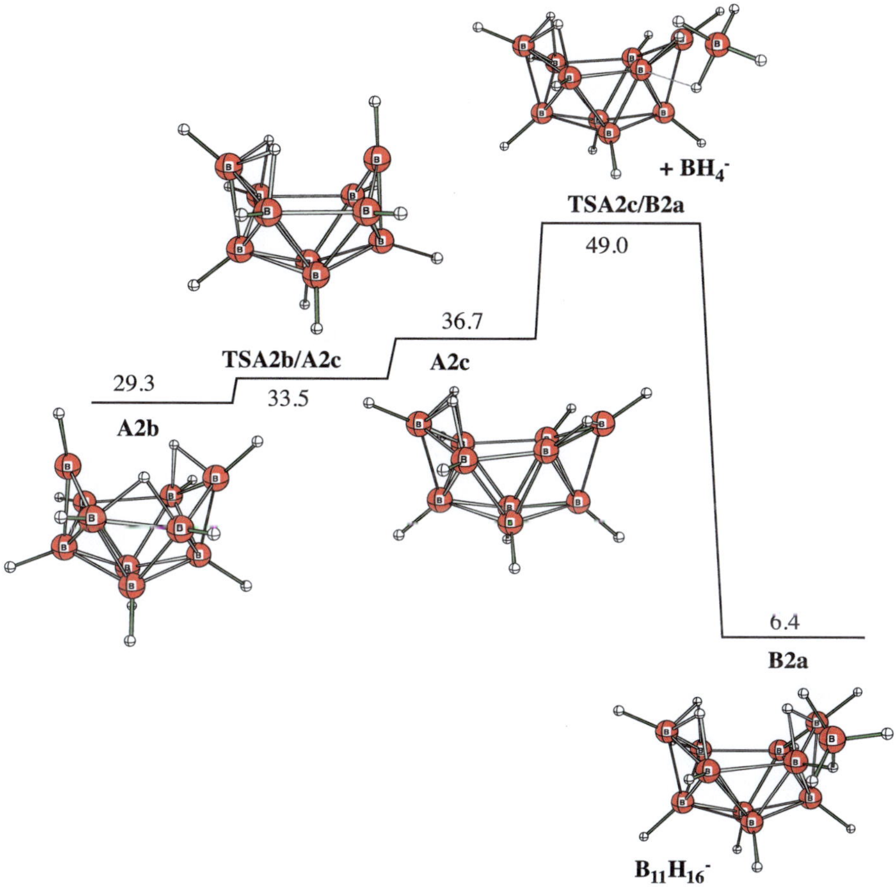

Fig. 5.5 Reaction pathway from $B_{10}H_{12}+BH_4^-$ to $B_{11}H_{16}^-$ (**A2** → **B2**)

From $B_{11}H_{14}^-$ (**B3b**) another H_2 is lost (via **TSB3b/B4**) to form B4 ($B_{11}H_{12}^-$, Fig. 5.8). The addition of BH_4^- to $B_{11}H_{12}^-$ (via **TSB4/D1a**) has a much larger free energy barrier than other BH_4^- addition steps and the H atom of BH_4^- approaches a boron center much more closely (1.87 Å) in the transition state (Fig. 5.8). **D1b** is formed from **D1a** (Fig. 5.8), which then eliminates the fourth H_2 molecule to form **D2a** (via **TSD1b/D2a**, Fig. 5.9). **D2a**, in a series of steps, incorporates the exterior BH_3 into the cage framework (**D2a** → **D2b** → **D2c** → **D2d**). The last H_2 elimination step has a small free energy barrier (**TSD2d/D3**, 25.2 kcal/mol, Fig. 5.9). No doubt the loss of H_2 is facilitated by the very large increase in stability of the $B_{12}H_{12}^{2-}$ (**D3**) product.

Figure. 5.10 illustrates the elimination of H_2 from $B_{11}H_{13}^{2-}$ (**E2**) to form $B_{11}H_{11}^{2-}$ (**E3**). This pathway may also be active in the $B_{10}H_{14}$ plus BH_4^- reaction. Indeed **E3** can add BH_3 without barrier to form **D2a**, which can then easily form $B_{12}H_{12}^{2-}$ (**D3**).

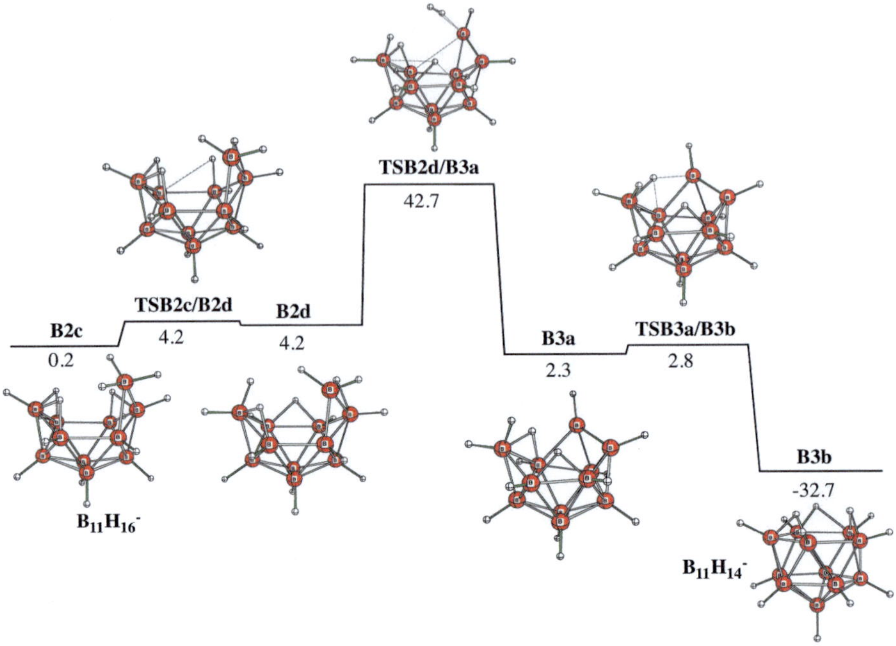

Fig. 5.6 Reaction pathway from $B_{11}H_{16}^-$ to $B_{11}H_{14}^- + H_2$ (**B2** → **B3**)

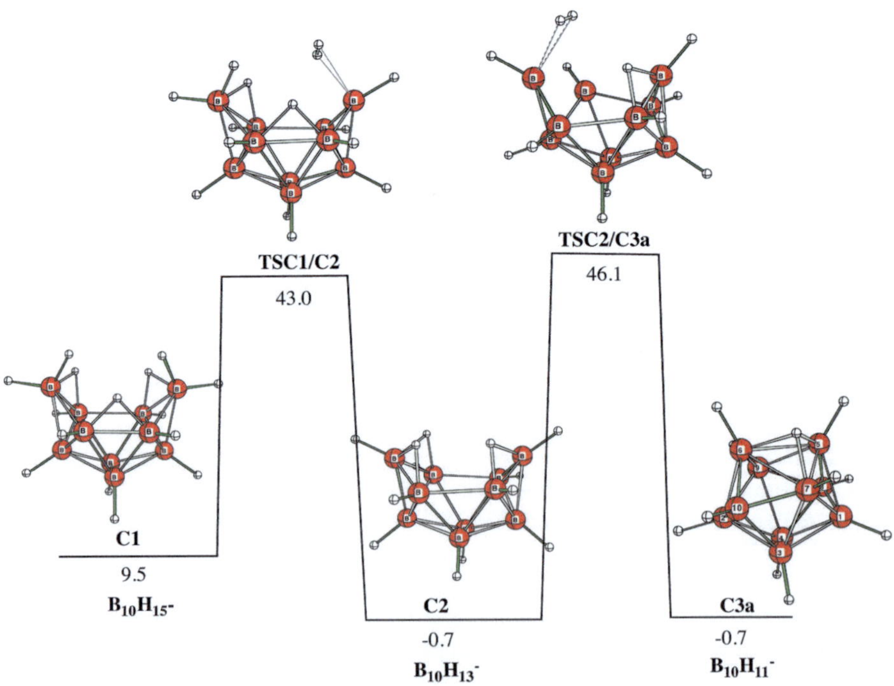

Fig. 5.7 Reaction pathway from $B_{10}H_{15}^-$ to $B_{10}H_{13}^- + H_2$ (**C1** → **C2**)

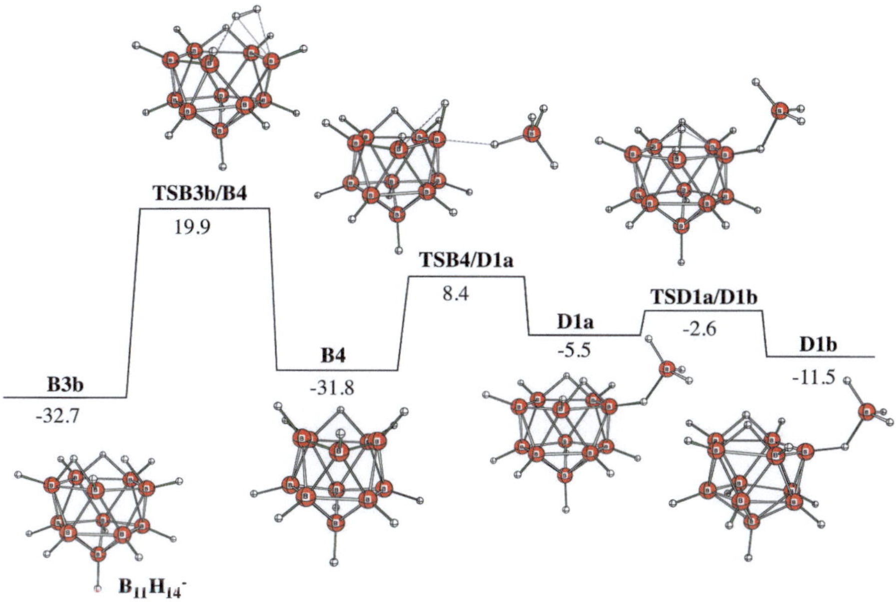

Fig. 5.8 Reaction pathway from $B_{11}H_{14}^-$ + BH_4^- to $B_{12}H_{16}^{2-}$ + H_2 (**B3 → B4 → D1**)

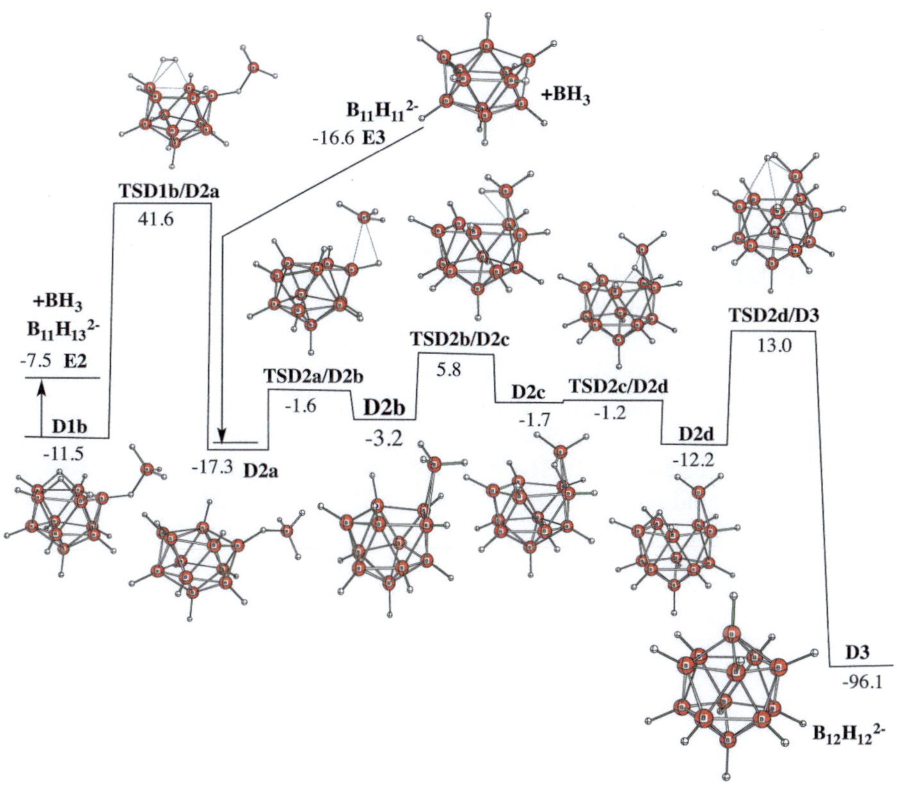

Fig. 5.9 Reaction pathway from $B_{12}H_{16}^{2-}$ to $B_{12}H_{14}^{2-}$ + H_2 and loss of H_2 to $B_{12}H_{12}^{2-}$ (**D1 → D2 → D3**)

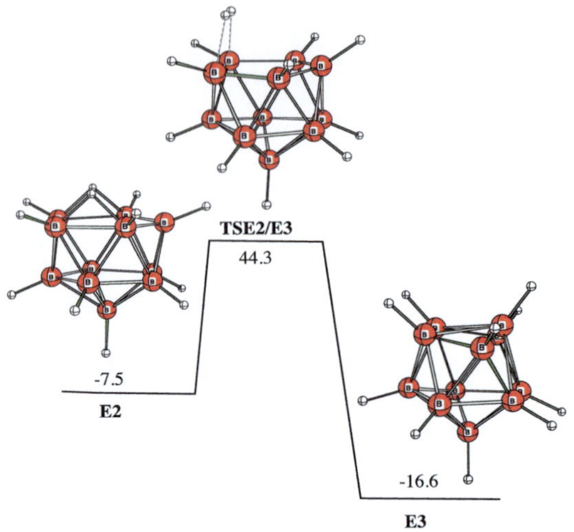

Fig. 5.10 Reaction pathway from $B_{11}H_{13}^{2-}$ to $B_{11}H_{11}^{2-} + H_2$ (**E2** → **E3**)

Volkov reported [24] that the $B_{11}H_{11}^{2-}$ dianion and an adduct of BH_3 could be transformed into $B_{12}H_{12}^{2-}$ at 130 °C (Eq. 5.7).

$$B_{11}H_{11}^{2-} + BH_3(NEt_3) \rightarrow B_{12}H_{12}^{2-} + H_2 \tag{5.7}$$

Figure. 5.11 illustrates the formation of $B_{10}H_{12}^{2-}$ (**F1a**) by elimination of BH_3 from $B_{11}H_{15}^{2-}$ (**E1a**). The elimination of H_2 from $B_{10}H_{12}^{2-}$ takes place from **F1b** which is reached by **TSF1a/F1b**. The free energy of the H_2-elimation transition state (**TSF1b/F2**) is very high (67.7 kcal/mol) which blocks formation of $B_{10}H_{10}^{2-}$ (**F2**) in the reaction even though the species is a well-known stable boron hydride dianion. The experimentally known $B_{10}H_{12}^{2-}$ structure [47, 48] is **F1c** which is reached from **F1a** via **TSF1a/F1c**. The solvation free energies of **F1b** and **F1c** are within 0.3 kcal/mol of each other (**F1b** (37.2), **F1c** (37.5 kcal/mol)).

Depending on the reaction conditions (temperature, counterion, etc.), different products of the reaction of $B_{10}H_{14}$ with BH_4^- have been reported. For example, Volkov and Paetzold reported [38] formation of $B_{11}H_{14}^-$ or $B_{10}H_{13}^-$ based on NMR evidence (Eqs. 5.8 and 5.9). This suggests that the reaction to $B_{12}H_{12}^{2-}$ may go through $B_{10}H_{13}^-$ (**C2**) and $B_{11}H_{14}^-$ (**B3**) as intermediates.

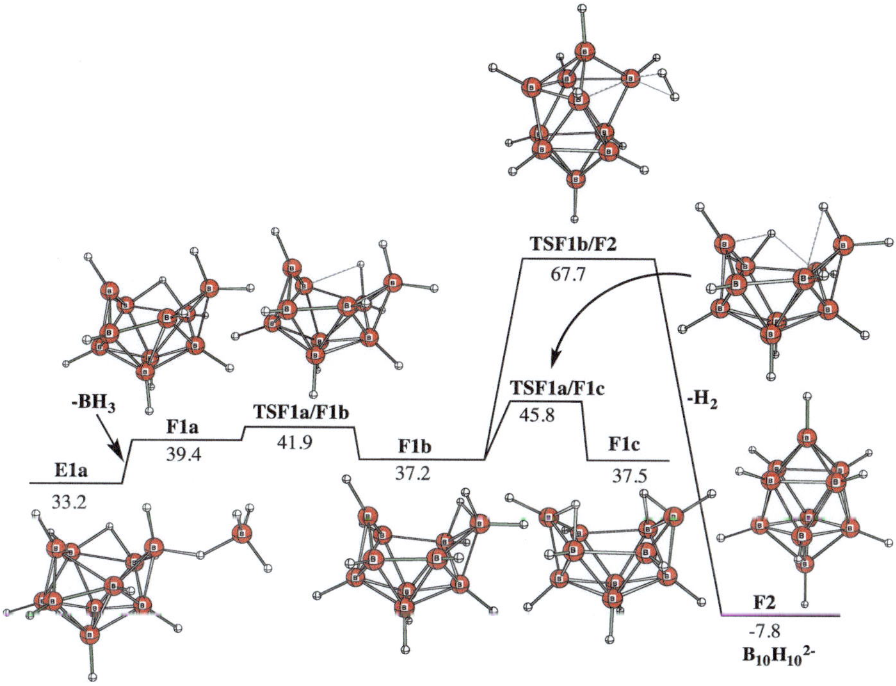

Fig. 5.11 Reaction pathway from $B_{11}H_{15}^{2-}$ to $B_{10}H_{10}^{2-}+H_2$ (**E1**$\rightarrow$**F2**)

$$B_{10}H_{14} + BH_4^- \rightarrow B_{11}H_{14}^- + 2H_2 \tag{5.8}$$

$$B_{10}H_{14} + BH_4^- \rightarrow B_{10}H_{13}^- + H_2 + 0.5\,B_2H_6 \tag{5.9}$$

5.4 Conclusions

The above calculations suggest a particular sequence of reactions that describe a plausible mechanism for the formation of $B_{12}H_{12}^{2-}$ from $B_{10}H_{14}$ and BH_4^-, subject to the limited sampling of chemical space. In fact, it is very likely that many competing paths are simultaneously active. The progress of the reaction is likely controlled by the H_2 elimination steps since these reaction steps have very high activation barriers and the reverse barriers are bimolecular reactions.

The investigation of the reaction mechanisms involving boron hydrides and carboranes present significant difficulties due to the large number of intermediates which are often not identified. Computational chemistry can contribute by mapping out pathways to the known intermediates and products that have greater stability. If branching ratios are in agreement with known products, that provides support for the existence of the unknown intermediates. The reaction of $B_{10}H_{14}$ with BH_4^- can form $B_{12}H_{12}^{2-}$ in boiling diglyme and long reaction times. The pathways explored in the present study presents a reasonable mechanism for the formation of $B_{12}H_{12}^{2-}$. The mechanism goes through $B_{10}H_{13}^-$ and $B_{11}H_{14}^-$ intermediates which are known to be formed in the reaction under different conditions.

Acknowledgments The Alabama Supercomputer Center is acknowledged for a generous allocation of computer time.

References

1. Todd LJ (1970) The chemistry of polyhedral borane ions. In: Brotherton RJ, Steinberg H (eds) Progress in boron chemistry, vol 2, 1st edn. Pergamon Press, New York, pp 1–35
2. Greenwood NN, Earnshaw A (1984) Chemistry of the elements. Pergamon Press, New York
3. King RB (2001) Three-dimensional aromaticity in polyhedral boranes and related molecules. Chem Rev 101(5):1119–1152
4. Chen Z, King RB (2005) Spherical aromaticity: recent work on fullerenes, polyhedral boranes, and related structures. Chem Rev 105(10):3613–3642
5. Fox MA, Wade K (2003) Evolving patterns in boron cluster chemistry. Pure Appl Chem 75(9):1315–1323
6. Grimes RN (2004) Boron clusters come of age. J Chem Educ 81(5):658–672
7. Shameema O, Jemmis ED (2011) Computational studies: boranes. In: Solomon EI, Scott RA, King RB (eds) Encyclopedia of inorganic and bioinorganic chemistry. Wiley, Chichester, pp 539–550
8. Workman DB, Squires RR (1988) Hydride binding energies of boranes. Inorg Chem 27(11):1846–1848
9. McKee ML (1990) Estimation of heats of formation of boron hydrides from ab initio energies. J Phys Chem 94(1):435–440
10. McKee ML (1994) AB initio study of nine- and ten-vertex *nido* and *arachno* boranes and heteroboranes. Inorg Chem 33(26):6213–6218
11. Cheng MF, Ho HO, Lam CS, Li WK (2002) Heats of formation for the boron hydrides: a Gaussian-3 study. Chem Phys Lett 356(1):109–119
12. Dixon DA, Gutowski MJ (2005) Thermodynamic properties of molecular borane amines and the $[BH_4^-][NH_4^+]$ salt for chemical hydrogen storage systems from ab initio electronic structure theory. J Phys Chem A 109(23):5129–5135
13. Nguyen MT, Matus MH, Dixon DA (2007) Heats of formation of boron hydride anions and dianions and their ammonium salts $[B_nH_m^{y-}][NH_4^+]_y$ with y=1-2. Inorg Chem 46(18):7561–7570
14. Greenwood NN, Greatrex R (1987) Kinetics and mechanism of the thermolysis and photolysis of binary boranes. Pure & App Chem 59(7):857–868
15. McKee ML (1990) Theoretical study of hydrogen (H_2) elimination from boron hydrides B_4H_{10}, B_5H_{11}, and B_6H_{12} and of boron trihydride (BH_3) elimination from B_4H_{10}. J Am Chem Soc 112(19):6753–6759
16. McKee ML (1996) Theoretical study of the reaction of acetylene with B_4H_8. A proposed mechanism of carborane formation. 2. J Am Chem Soc 118(2):421–428
17. McKee ML (1988) Ab initio study of the mechanisms of rearrangements in $C_2B_4H_6$ and $C_2B_5H_7$. J Am Chem Soc 110(16):5317–5321
18. Sayin H, McKee ML (2007) Carboranes and baskets from reaction of B_4H_{10} with allene. Inorg Chem 46(7):2883–2891

19. Wales DJ (2005) Electronic structure of clusters. In: King RB (ed) Encyclopedia of inorganic chemistry, 2nd edn. Wiley, Chichester, pp 1506–1525
20. Sugden IJ, Plant DF, Bell RG (2013) Thermal rearrangement mechanisms in icosahedral carboranes and metallocarboranes. Chem Commun 49:975–977
21. McKay D, Macgregor SA, Welch AJ (2015) Isomerisation of *nido*-$[C_2B_{10}H_{12}]^{2-}$ dianions: unprecedented rearrangements and new structural motifs in carborane cluster chemistry. Chem Sci 6:3117–3128
22. Dunks GB, Barker K, Hedaya E, Hefner C, Palmer-Ordonez K, Remec P (1981) Simplified synthesis of $B_{10}H_{14}$ from $NaBH_4$ via $B_{11}H_{14}^-$ Ion. Inorg Chem 20(6):1692–1697
23. Sivaev IB, Bregadze VI, Sjöberg S (2002) Chemistry of *closo*-dodecaborate anion $[B_{12}H_{12}]^{2-}$: a review. Collect Czech Chem C 67:679–727
24. Volkov O (1999) Undecaborates $M_2[B_{11}H_{11}]$: facile synthesis, crystal structure, and reactions. Z Anorg Allg Chem 625:1193–1201
25. Miller HC, Miller NE, Muetterties EL (1964) Chemistry of boranes. XX. Syntheses of polyhedral boranes. Inorg Chem 3(10):1456–1463
26. He L, Li HW, Hwang SJ, Akiba E (2014) Facile solvent-free synthesis of anhydrous alkali metal dodecaborate $M_2B_{12}H_{12}$ (M=Li, Na, K). J Phys Chem C 118(12):6084–6089
27. Bakardjiev M, Štíbr B, Holub J, Padělková Z, Růži ka A (2015) Simple synthesis, halogenation, and rearrangement of *closo*-1,6-$C_2B_8H_{10}$. Organometallics 34(2):450–450
28. Frisch MJ, Trucks GW, Schlegel HB, Scuseria GE, Robb MA, Cheeseman JR, Scalmani G, Barone V, Mennucci B, Petersson GA, Nakatsuji H, Caricato M, Li X, Hratchian HP, Izmaylov AF, Bloino J, Zheng G, Sonnenberg JL, Hada M, Ehara M, Toyota K, Fukuda R, Hasegawa J, Ishida M, Nakajima T, Honda Y, Kitao O, Nakai H, Vreven T, Montgomery JA Jr, Peralta JE, Ogliaro F, Bearpark M, Heyd JJ, Brothers E, Kudin KN, Staroverov VN, Kobayashi R, Normand J Raghavachari K, Rendell A, Burant JC, Iyengar SS, Tomasi J, Cossi M, Rega N, Millam NJ, Klene M, Knox JE, Cross JB, Bakken V, Adamo C, Jaramillo J, Gomperts R, Stratmann RE, Yazyev O, Austin AJ, Cammi R, Pomelli C, Ochterski JW, Martin RL, Morokuma K, Zakrzewski VG, Voth GA, Salvador P, Dannenberg JJ, Dapprich S, Daniels AD, Farkas Ö, Foresman JB, Ortiz JV, Cioslowski J, Fox DJ (2009) Gaussian09, Revision A.2. Gaussian, Inc., Wallingford
29. Lee TB, McKee ML (2011) Dissolution thermochemistry of alkali metal dianion salts (M_2X_1, M=Li^+, Na^+, and K^+ with X=CO_3^{2-}, SO_4^{2-}, $C_8H_8^{2-}$, and $B_{12}H_{12}^{2-}$). Inorg Chem 50(22):11412–11422
30. Lee TB, McKee ML (2012) Redox energetics of hypercloso boron hydrides B_nH_n (n=6–13) and $B_{12}X_{12}$ (X=F, Cl, OH, and CH_3). Inorg Chem 51(7):4205–4214
31. Kochnev VK, Avdeeva VV, Malinina EA, Kuznetsov NT (2013) Theoretical study of dodecahydro-*closo*-decaborane $B_{10}H_{12}$, the diprotonated boron cluster $B_{10}H_{10}^{2-}$. Russ J Inorg Chem 58(7):793–799
32. Kochnev VK, Avdeeva VV, Goeva LV, Malinina EA, Kuznetsov NT (2014) Theoretical study of molecular hydrogen elimination from the undecahydrodecaborate monoanion $[B_{10}H_{11}]^-$. Exopolyhedral substitution intermediates: $[B_{10}H_9]^-$ monoanion and neutral $[B_{10}H_{10}]$ cluster. Russ J Inorg Chem 59(7):706–712
33. Adams RM, Siedle AR, Grant J (1964) Convenient preparation of the dodecahydrododecaborate Ion. Inorg Chem 3(3):461
34. Gaines DF, Bridges AN, Hayashi RK (1994) Synthesis of *nido*-$B_{11}H_{14}^-$ and alkyl derivatives via systematic cage enlargement of the decaborane(14) system: crystal structure of 7-Thx-$B_{11}H_{13}^-$. Inorg Chem 33(7):1243–1244
35. Maitre P, Eisenstein O, Michos D, Luo XL, Siedle AR, Wisnieski L, Zilm KW, Crabtree RH (1993) Borate anion ($B_{11}H_{14}^-$): a *nido* cage with No H...H interaction. J Am Chem Soc 115(17):7747–7751
36. Kiani FA, Hofmann M (2004) Structural increment system for 11-vertex *nido*-boranes and carboranes. Inorg Chem 43(26):8561–8571
37. Volkov O, Radacki K, Thomas RL, Rath NP, Barton LJ (2005) A new look at the nido-undecaborate system. J Organomet Chem 690(11):2736–2744

38. Volkov O, Paetzold P (2003) The chemistry of the undecaborates. J Organomet Chem 680(1-2):301–311
39. Titov LV, Gavrilova LA, Petrovskii PV (2008) Synthesis and some properties of $Nd(B_{11}H_{14})_3.4Dg$ (Dg is diglyme). Russ J Inorg Chem 53(4):565–567
40. Volkov O, Radacki K, Paetzold P, Zheng X (2001) Dodecahydro-*closo*-undecaborate $[B_{11}H_{12}]^-$. Z Anorg Allg Chem 627(6):1185–1191
41. Smith DE, Rupp EB, Shriver DF (1967) The mechanisms of electrolytic reduction for decaborane (14), $B_{10}H_{14}$, in an aprotic solvent. II. The second reduction step and the reduction step and the reduction of decaborane (13)ate (-1), $B_{10}H_{13}^-$1a. J Am Chem Soc 89(22):5568–5573
42. Graybill BM, Pitochelli AR, Hawthorne MF (1962) The preparation and reactions of $B_{10}H_{13}$ (ligand) anions. Inorg Chem 1(3):622–626
43. Yoon CW, Kusari U, Sneddon LG (2008) Computational studies of the reactions of $B_{10}H13^-$ with alkynes and olefins: pathways for dehydrogenative alkyne-insertion and olefin-hydroboration reactions. Inorg Chem 47(20):9216–9227
44. Sneddon LG, Huffman JC, Schaeffer RO, Streib WE (1972) Structure of the $B_{10}H_{13}^-$ Ion. J Chem Soc Chem Commun 474–475
45. Shore SG, Hamilton EJM, Bridges AN, Bausch J, Krause-Bauer JA, Dou D, Liu J, Liu S, Du B, Hall H, Meyers EA, Vermillion KE (2003) The solid state structure of $[B_{10}H_{11}]^-$ and its dynamic NMR spectra in solution. Inorg Chem 42(4):1175–1186
46. Kochnev VK, Avdeeva VV, Goeva LV, Malinina EA, Kuznetsov NT (2012) The undecahydrodecaborate anion $B_{10}H_{11}^-$ as the starting reagent in exopolyhedral substitution and complexation: theoretical and experimental prerequisites. Russ J Inorg Chem 57(3):331–336
47. Bridges AN, Gaines DF (1995) The dianion of decaborane(14), nido-dodecahydrodecaborate (2-), $[B_{10}H_{12}^{2-}]$ and its solution behavior. Inorg Chem 34(18):4523–4524
48. Hofmann M, Schleyer PR (1998) Structures of *arachno*- and *hypho*-B10 clusters and stability of their possible Lewis base adducts ($[B_{10}H_{12}]^{2-}$, $[B_{10}H_{12}.L]^{2-}$, $[B_{10}H_{12}.2L]^{2-}$, $[B_{10}H_{13}]^-$, $[B_{10}H_{13}.L]^-$, $[B_{10}H_{12}.2L]$). an ab initio/IGLO/NMR investigation. Inorg Chem 37(21):5557–5565
49. McKee ML, Wang ZX, Schleyer PR (2000) Ab initio study of the hypercloso boron hydrides B_nH_n and $B_nH_n^-$. Exceptional stability of neutral $B_{13}H_{13}$. J Am Chem Soc 122(19):4781–4793

Chapter 6
Big Borane Assemblies, Macropolyhedral Species and Related Chemistry

John D. Kennedy

Abstract Structural and behavioral chemistries based on carbon hydrides are extensive. It can be argued that chemistries based on boron hydrides are in principle similarly extensive. Molecular chemistry based on boron-hydrides is characterised by cluster formation, and has been dominated by work on single-cluster compounds. For an extensive 'big molecule' chemistry based on boron hydrides – one that rivals the extent of chemistry based on carbon hydrides – a chemistry that is based on the intimate fusion of single-cluster borane-based entities, to generate so-called 'macropolyhedral' species, has developed. In contrast to carbon-based chemistry, and thence with no natural feedstocks available, boron-containing cluster chemistry is entirely a human-made creation, and so the area necessarily progresses by exploratory experimental chemistry augmented more recently by computational approaches. This chapter attempts to offer a perspective on aspects of the field of larger borane-based molecular compounds from the approach of preparative and pragmatic bench science, and, in accord with the general theme of this volume, points out areas in which calculational chemistry has played a role and in which useful future roles can be envisaged. Following from the initial elucidation – now more than about thirty years ago – of most of the basic binary boron-hydride macropolyhedrals, an emphasis is placed on subsequent work which has largely been concerned with metallaboranes, thiaboranes and metallathiaboranes, as well as very intimately fused globular 'megaloborane' entities.

It is often upheld that, of all the elements, only carbon uniquely has the ability to form extensive series of stable hydrides. The basic carbon hydride structural units – alkanes, alkenes and alkynes and aromatic rings – can and do join together to result in countless numbers and varieties of what are essentially extended stable networks

A perspective on aspects of larger-boranes chemistry from the point of view of preparative and pragmatic bench chemistry.

J.D. Kennedy (✉)
School of Chemistry of the University of Leeds, Leeds LS2 9JT, England, UK

Institute of Inorganic Chemistry of the Academy of Sciences of the Czech Republic, v.v.i. CZ – 250 68, Husinec-Řež, The Czech Republic
e-mail: j.d.kennedy@leeds.ac.uk

D. Hnyk, M.L. McKee (eds.), *Boron*, Challenges and Advances in Computational Chemistry and Physics 20,
DOI 10.1007/978-3-319-22282-0_6

of atoms joined by two-electron two-center *sigma*-bonds involving sp^3, sp^2 or *sp* hybridized carbon atoms, generally augmented in the last two cases by *pi*-bonding. Electron-rich main-group elements of the nitrogen and oxygen groups also can and also do become involved, overall resulting in a vast and fascinating variety of carbon-based chemistry – organic chemistry. Utilizing this carbon-based chemistry, and basic inorganic units such as phosphates, and inorganic phenomena such as hydrogen-bonding and the flexible oxidation states of metallic elements, life on Earth in all its varieties and complexities has evolved; one consequence of this is that Nature has revealed not only the variety of structuralities and functionalities available in carbon-hydride-based chemistry, but it also provides feedstocks for the examination of these phenomena, as well as suggesting ways in which intelligent humans can develop or exploit them.

Not so generally appreciated is that that boron, carbon's preceding nearest-neighbor in the Periodic Table, can also form extensive series of stable hydrides – the boranes. As with the carbon hydrides, other main-group non-metallic elements can be involved, to yield heteroboranes, for example $SB_{11}H_{11}$ and the very stable and very well-known trio of $C_2B_{10}H_{12}$ isomers, and transition-element and main-group metallic centers can also be incorporated to generate metallaboranes and metallaheteroboranes. The basic structural units that characterize the resulting boron-containing cluster compounds are based on open or closed triangulated deltahedral single clusters and cages, generally containing three or more boron atoms. The cluster vertices are defined by the positions of the boron and the hetero atoms, with the hydrogen atoms generally bound around the cluster periphery. There is a tremendous variety of boranes, heteroboranes, and metallaboranes known which have cluster sizes of between 2 and 13 vertices. In contrast to the two-electron two-center sigma-bonding networks that characterize much of carbon hydride chemistry, the formation of the cluster compounds of the boron hydrides is governed and characterized by multicenter bonding.

This last is well demonstrated by the simplest borane, B_2H_6. A simplistic representation of the bonding (schematic **1**) shows that the boron atoms are held together by two two-electron three-center BHB bonds. Addition of two electrons enables the resulting $[B_2H_6]^{2-}$ dianion to adopt an ethane-like structure; now only one two-electron bond holds the boron atoms together (schematic **2**); interestingly, therefore, addition of electrons reduces the bonding between the boron atoms by removing a bonding orbital from the cluster proper. This is a general polyhedral borane characteristic.

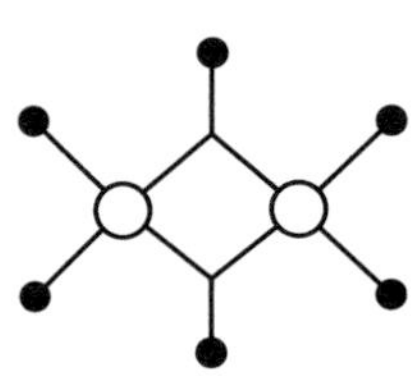

1

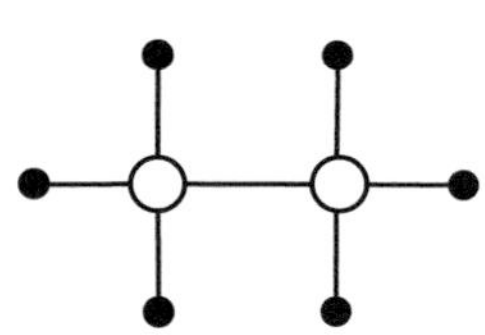

2

With higher boranes the multicenter bonding schemes become more complex. A 'localized' bonding scheme, as pioneered by Lipscomb, often used for the *arachno* decaborane dianion $[B_{10}H_{14}]^{2-}$, is given in structure **3**. Here in the B(2,5,6,7) and B(4,8,9,10) regions 'partial three-center bonding' is invoked because individual localized structures **3 A** and **3 B** do not mimic the two-fold symmetry of the molecule, and so a partial three-center bonding as in **3** is invoked to account for this, being regarded as the hybrid of the two canonical forms **3 A** and **3 B**.

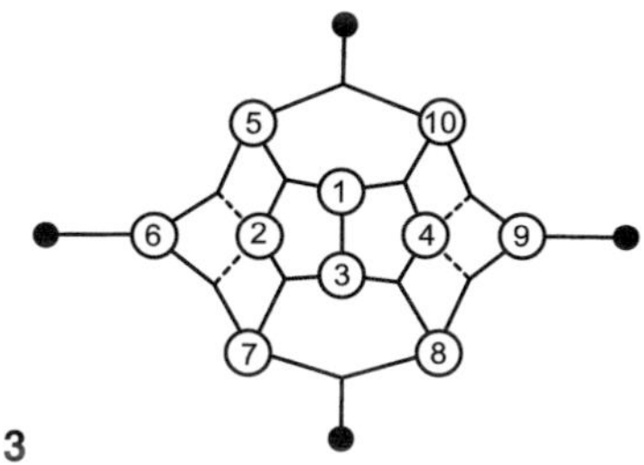

3

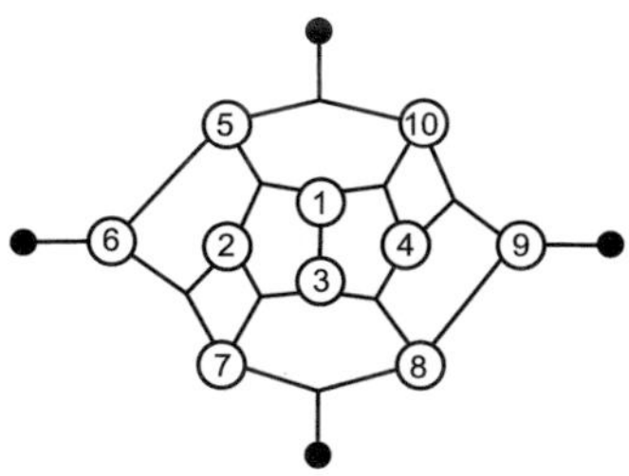

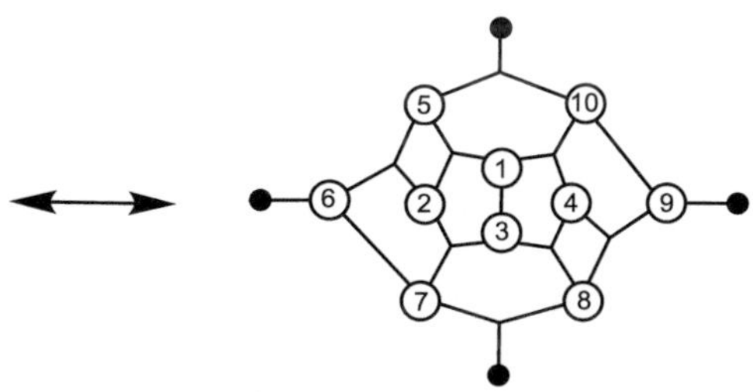

3 A **3 B**

With more symmetrical species the partial three center bonding approach is subsumed by schemes that are necessarily more delocalized. For example, for *nido*-B_5H_9 the localized two- and three-center bonding of structure **4 A** does not reflect the molecular C_4v square-pyramidal geometry; the average of four canonical forms is then better represented by **4 B** in which the hatched circle represents a symmetrical six-electron quasi-aromatic system linking the BH(1) vertex to the four basal boron atoms.

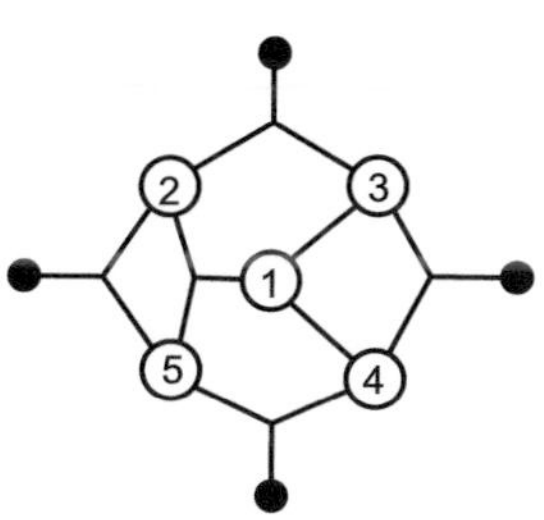

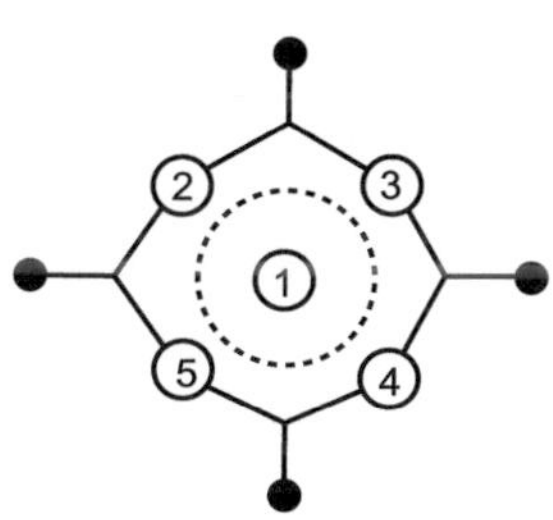

4 A **4 B**

There are increasing difficulties in preparing boron-containing single cages with increasingly more than 12 vertices: beyond 15 vertices or so, syntheses become very difficult. For the potential extent of borane-based cluster chemistry to even approach a rivalling of the extent of carbon-based chemistry, a big borane chemistry needs to be developed. Such a development implies that means must be sought to link single boron-containing clusters together. This can be accomplished by simply linking individual cluster units together using simple two-center *sigma* bonds (schematic **5**) or more elaborate *sigma*-bonded linkages of types familiar from carbon-based chemistry (*e.g.* schematic **6**), but this cannot significantly change the known chemistry of the single clusters.

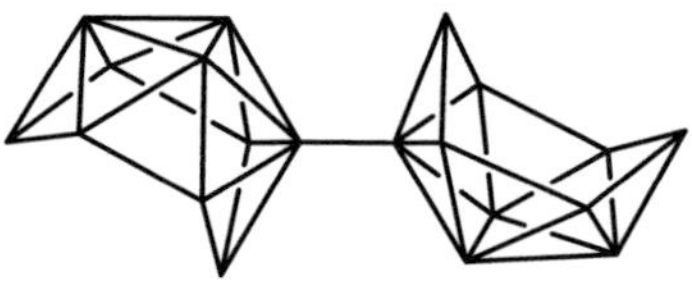

5

6

More intellectually and synthetically challenging, and necessary to develop the chemistry into truly novel discovery areas with unprecedented and unpredicted properties and behavior, is to consider molecular assemblies in which individual single clusters may be regarded as being more intimately joined together, *i.e.* fused such that the covalent multicenter bonding characteristics typical of the boranes, as mentioned above for the *nido*-B_5H_9 and the [*arachno*-$B_{10}H_{14}]^{2-}$ anion, extend through and are part of the intercluster conjunction. Such fused cluster species have become known as 'macropolyhedral' species, a term first coined in the early 1980s [1, 2]. The individual clusters are thence no longer separate entities; rather, they can be regarded as 'fused' with one, two, three, or more atoms held in common at the point of fusion. Unlike carbon-hydride chemistry, for which Nature has been most helpful not only in revealing what types of larger carbon-based structures may be available, but also in pointing the way to new targets, boron-containing cluster chemistry is a completely man-made creation, so that experimentation is required to reveal the types of intercluster fusion modes and macropolyhedral cluster assemblies that may be available. In this quest, theoretical calculations, both to substantiate and augment experimental results, and to explore potential new avenues based on reasonable extrapolations from known structural and behavioral motifs, have a currently increasingly important role (see other chapters in this Volume).

An initial perception by Williams [3, 4] recognised that the structures of polyhedral boron-containing species could be largely interpreted in terms of patterns set by *(a)* a set of *closo* compounds $[B_nH_n]^{2-}$ that had skeletons that are closed deltahedral polyhedra of highest symmetry, *(b)* a set of *nido* compounds B_nH_{n+4} with skeletons derived from the *closo* polyhedra by the removal of one vertex of highest connectivity, and *(c)* a set of *arachno* compounds B_nH_{n+6} with skeletons derived by the removal of a further vertex from a *nido* skeleton. Wade [5, 6] thence perceived that the number of skeletal electrons in these sets can be correlated with the structures as follows: the n-vertex *closo* deltahedral boranes are particularly stable if they contain $2n+2$ skeletal electrons, the n-vertex *nido* boranes B_nH_{n+4} retain the $2(n+1)+2$ skeletal electrons of the $(n+1)$-vertex deltahedra from which they are derived and thus are $2n+4$ skeletal-electron systems, and the n-vertex *arachno* boranes B_nH_{n+6} are correspondingly $2n+6$ skeletal-electron systems. These considerations, together with considerations involving the incorporation of non-boron atoms within the clusters [6–9], are often called 'Wade's Rules', and are dealt with from alternative perspectives in Chaps. 2, 4 and 7 of this volume. The elegant Williams/Wade geometrical-structure/electron-counting paradigm that results is illustrated in Fig. 6.1.

It should be emphasised that this paradigm is a very useful structure-electron correlation, and not one of the Laws of Science [10]. This is not always appreciated. Even at the nascence of the recognition of these correlations between geometrical structures and their formal electron counts it was noted that many aspects of the correlations may well result from fortunate coincidence, and also that species can disobey the rules when constituent atoms other than boron do not mimic the behavior of boron [8, 9]. Some boron hydride species themselves simply do not obey the 'Rules', for example formally *nido* B_8H_{12} which has an *arachno*-shaped skeleton [11, 12]. Others, such as the closed-cluster [*closo*-$B_8H_8]^{2-}$ and [*closo*-$B_{11}H_{11}]^{2-}$ anions [13–17], are so readily fluxional *via* open-faced intermediates that such intermediates are of very similar stabilities to the closed structure itself, and it is almost by chance that it is the closed structure we observe in the solid state. As the field has developed, particularly into metallaborane chemistry, new sets of structures can be factored out to start to generate new sets of correlations, such as the so-called *isocloso – isonido – isoarachno* sequence (Fig. 6.2) [18, 19]. These '*iso*' structures are subject to ongoing experimentation [20] and often useful, but not always, computational analysis (again see Chap. 2 for a more comprehensive survey).

Perceptions have arisen that *isocloso* species may often have quite different electronic structures from their ostensibly related conventional *closo* equivalents, but that 'intermediate' *isonido* species are often distortions of *closo*, in which the original *closo* electronic structure of the true *closo* skeleton is not very significantly perturbed. This introduces the mutually related concepts of cluster flexibility and cluster fragility, derived from experimental observations (se Fig. 6.3) [10, 21]. By 'flexibility' it is meant that a given cluster geometry, that might in the general case be associated with a particular structural type among Figs. 6.1 and 6.2, may be subject to geometrical distortions either by change of constituents or substituents, or by distortions induced by polar solvent effects or intermolecular solid-state hydrogen-bonding, dihydrogen-bonding, and 'non-bonding' interactions and Van der Waals packing effects. In terms of this concept of 'flexibility', then such changes will not be associated with significant changes in cluster MO structure, although the geometries may in some cases deviate considerably from the presumed ideal geometry. By contrast, it may be that such distorting effects are such as to induce a flip into different electronic type and/or a different geometrical structural type: if so, then the cluster may be said to exhibit cluster 'fragility'. These effects will some-

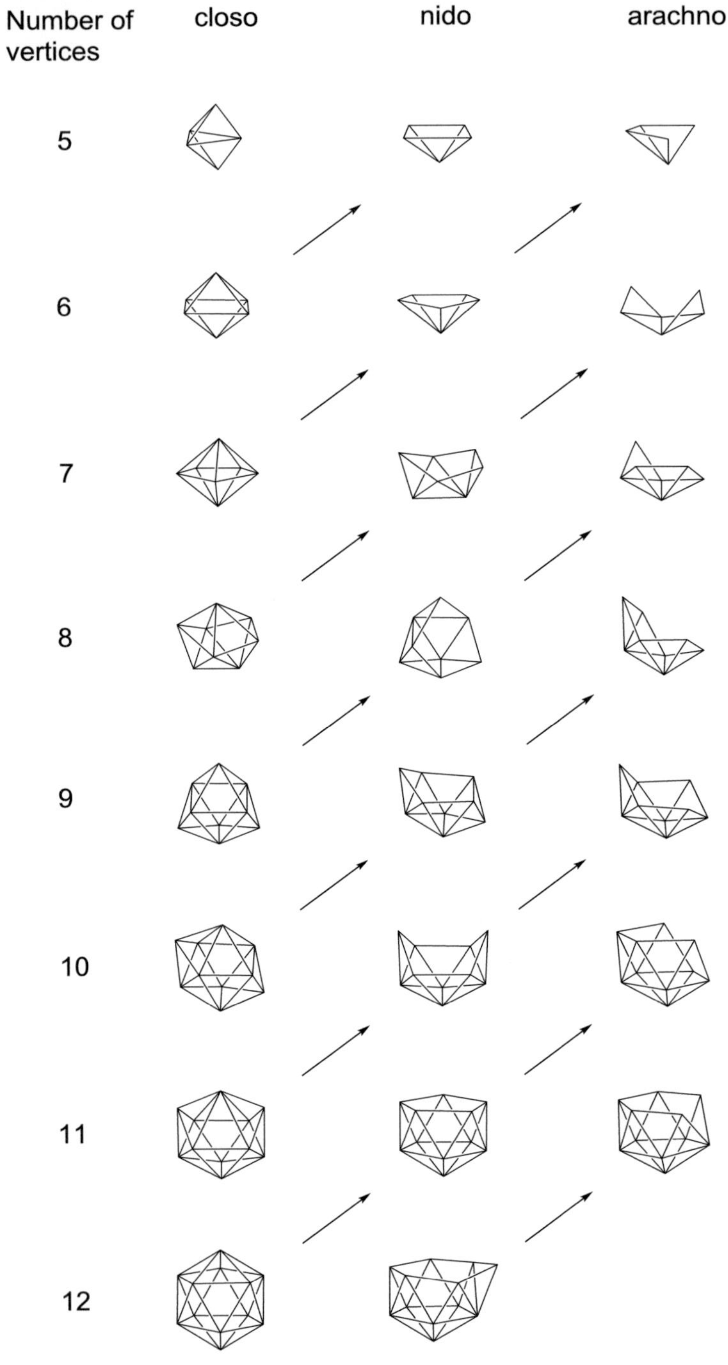

Fig. 6.1 Representation of the classical Williams/Wade structural patterns. The $(n-1)$-vertex geometries are derived from the n-vertex *closo* geometries by the removal of one vertex of high connectivity, and the $(n-2)$-vertex *arachno* geometries by the removal of a second, adjacent, vertex. Horizontal progressions *closo* → *nido* → *arachno* for the same number of vertices are achieved by the successive addition of pairs of electrons. In the 'Wade's Rules' formalism, the n-vertex *closo*, *nido* and *arachno* structures have respectively $(2n+2)$, $(2n+4)$ and $(2n+6)$ cluster-bonding electrons associated with them

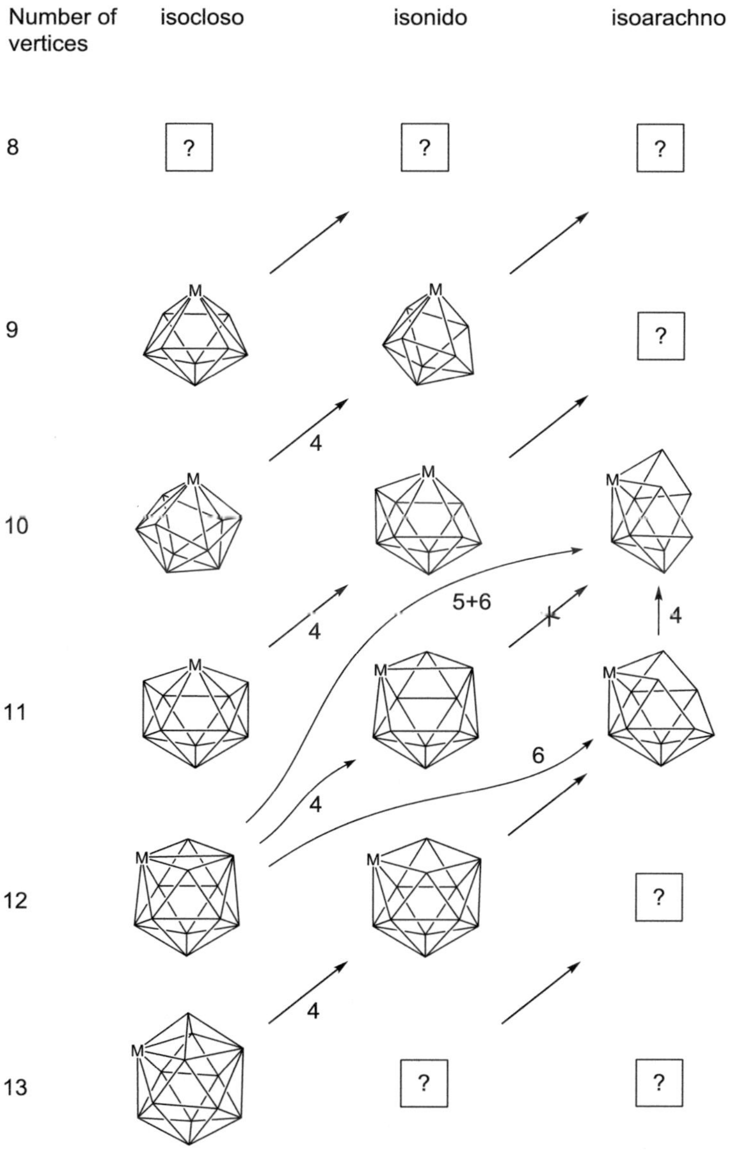

Fig. 6.2 A pattern of behaviour that can be devised for structural relationships among the emerging '*isocloso*', '*isonido*' and '*isoarachno*' boron-containing cluster systems (see also Ref. [18]). Removal of a vertex of low connectivity from an n-vertex *isocloso* geometry generates an $(n-1)$-vertex *isonido* geometry. Removal of two adjacent vertices generates a corresponding '*isoarachno*' geometry. Numbers against the arrows indicate the cluster connectivities of the removed vertices. Question marks indicate current gaps in experimental evidence, and point the way to computational assessments of viability of isolation

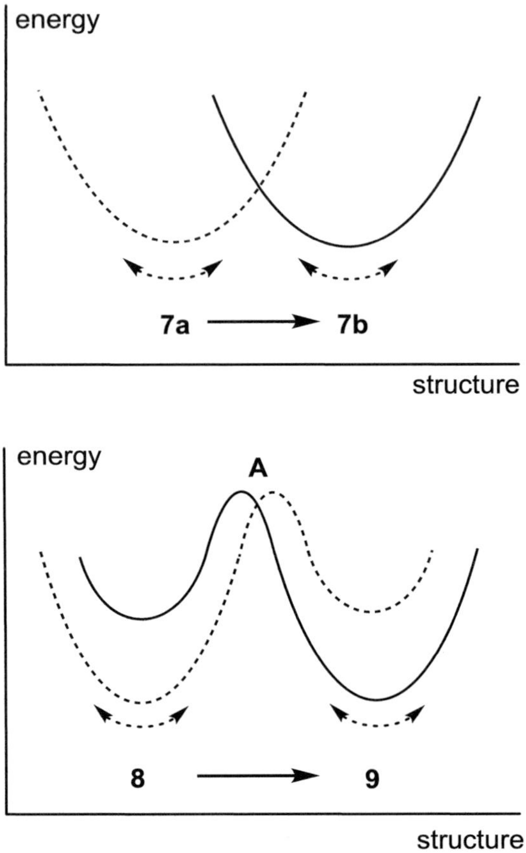

Fig. 6.3 The *upper diagram* illustrates the phenomenon of cluster flexibility, in which a change in molecular circumstances induces changes in dimensions within an overall cluster type **7**. The *lower diagram* illustrates cluster fragility, in which a change in molecular circumstances of a cluster type **8** induces a catastrophic flip into an alternative cluster type **9**. A presumption here, not always valid, would be that **8** and **9** will be fundamentally different electronically. In both cases, if the energetic minima are shallow, then additional cluster flexibility about the minima will be induced by perturbations such as crystal packing forces and solution effects (*hatched arrows*). This can lead to discrepancies between calculated 'gas-phase' geometries and those observed experimentally. With approximately equal energetics for **8** and **9**, and a low activation barrier **A**, fluxionality may result in solution, making the 'flexibility' and 'fragility' categories of behaviour difficult to distinguish (See also Ref. [19])

times mean that observed molecular structures may differ significantly from calculated 'gas-phase' structures, which some workers occasionally find disconcerting, and sometimes puzzling.

Computational work is a useful adjunct in this context to examine the progression among cluster types as their geometries are systematically distorted; here calculated nuclear magnetic shieldings can be of great help [21, 22]. Because these shieldings depend intimately on the molecular orbital structure, then significant changes in bonding within a cluster will be manifested in substantial nuclear shielding changes – these can be related to experimental 'checkpoints' from the experimental NMR behavior of sample compounds under consideration, even if the at-present-formidable task of elucidating which specific MO changes are associated with dominant contributions to the nuclear shielding changes [23] has not yet generally been addressed (the implicit challenge for computational chemistry here is in fact potentially very useful). NMR chemical shifts of course are an experimental measure of nuclear magnetic shielding. A general point is that computational work must be guided by reality, and in that regard experimental checkpoints, or reasonable extrapolations from experimental results, are vital.

All the above factors can operate when clusters fuse to form macropolyhedral species and can complicate interpretations. In particular, the bonding requirements at the point of intercluster fusion may often perturb the electronic structure of the individual subcluster units so that they exhibit changes that would be ascribable to cluster flexibility and cluster fragility if they were observed in single cluster species. This can be exacerbated by substituent effects. For example, 19-vertex $[(PMe_2Ph)_2PdB_{18}H_{20}]$ (schematic **10**) features a *nido* 10-vertex subcluster for which *nido*-$B_{10}H_{14}$ is the single-cluster borane model. The B(5)-B(10) 'gunwale' connectivity in the $B_{10}H_{14}$ single-cluster model itself is 'long' at *ca* 1.97 Å, its length indicating weaker bonding and thence susceptibility to flexibility; however, in the solid-state molecular structure of the *nido* ten-vertex subcluster of the metallaborane species it is 'stretched' to 2.089 Å, outside reasonable interboron 'bonding distance' [24]. In the $[B_{22}H_{21}(OEt)]^{2-}$ dianion, which also has a *nido* ten-vertex subcluster, this 'gunwale' distance is stretched to 2.23 Å (schematic **11**) [25]. Any computational approach to this sort of phenomenon in macropolyhedral boron-containing cluster chemistry is not yet established.

10

11

Nevertheless, at the basic level, an extension of the Wade-Williams type of approach to accommodate fused cluster species, devised by Jemmis as his '*mno* rule' [26], can be used to reconcile the electron counts of macropolyhedral species with the electron counts of the individual cluster types when taken together with specific electron-count adjustments that depend on how many atoms the two clusters hold in common. Aspects of this approach are dealt with in more detail in Chap. 4 of this Volume. As generally applied, this *mno* rule is, however, a post-rationalization, because, for a given macropolyhedral species, an adjustment of the electron count can either engender changes of constituent subclusters among the formalisms of Figs. 6.1 or 6.2, or engender changes in the intercluster fusion mode, specifically by changing the number of vertices held in common, and so is of limited predictive use. Inconvenient transition-element behavior [10] may complicate matters further.

Fused cluster compounds, or their modes of fusion, can be categorized in terms of the number of vertices held in common, one, two, three, and so on (see also Chap. 4 in this volume). The simplest is to have one atom in common. These are sometimes called '*commo*' species. A common boron atom occurs in the pure borane species $B_{15}H_{23}$, (schematic **12**) [27], but this is rare for a boron atom. More common is to have an atom of a metallic element in common, either main-group or a transition-element. Similar bonding principles to those of boron apply to other main-group elements, *e.g.* the cadmium dimer $[(Et_2O)_2Cd(B_{10}H_{12})_2Cd(OEt_2)_2]$ (schematic **13**) [28], which starts to illustrate the interesting structural diversities that may occur in macropolyhedral boron chemistry. A more elaborate manifestation of this type of structural motif is in $[(Et_2C_2B_4H_4)Co(Et_2C_2B_3H_3Bu)Ru]_4$, a tetrameric species [29]. Compounds worth mentioning here in the context of this volume are the species $[B_5H_{10}BeB_5H_{10}]$ (schematic **14**) and $[BH_4BeB_5H_{10}]$ [30, 31], in which the common atom of fusion is of beryllium. Beryllaborane chemistry is sparsely represented [30–33], principally because of the toxicity problems associated with working with the element. The propensity for cluster formation from both of the neighbors of beryllium, boron and lithium, is of course well recognized. In view of this, beryllaborane, and indeed beryllium cluster chemistry itself, are areas very suitable for predictive structural and behavioral calculational work [33].

12

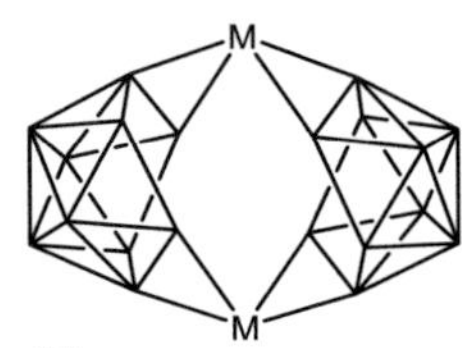

13

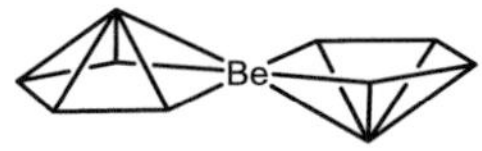

14

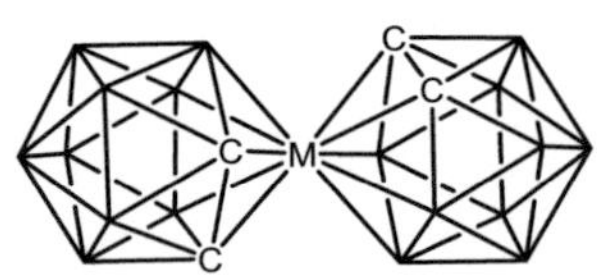

15

Examples with a commonly held transition-element atom are much more common, and long-recognized [34–36]. Well examined examples here are the 'bis(dicarbollide)' complexes (schematic **15**), such as the $[Ni(C_2B_9H_{11})_2]^{2-}$ dianion, and the series of 'cosane' molecules based on the $[Co(C_2B_9H_{11})_2]^-$ monoanion. In most one-atom-in-common *commo* species, however, the commonly held atom, whether of boron, other main-group element, or transition element, displays normal sp^3 dsp^2 or d^2sp^3 hybridization behavior, and so it can be argued that cluster multicenter bonding characteristics are not carried through the point of conjunction, and therefore do not concur with the macropolyhedral definition, although of course in most cases there will be bonding contributions from OMOs that extend through the conjoining atom into both subclusters.

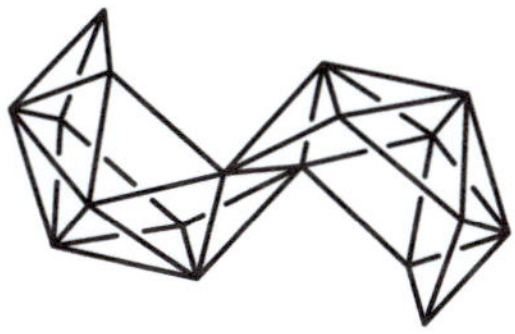

16

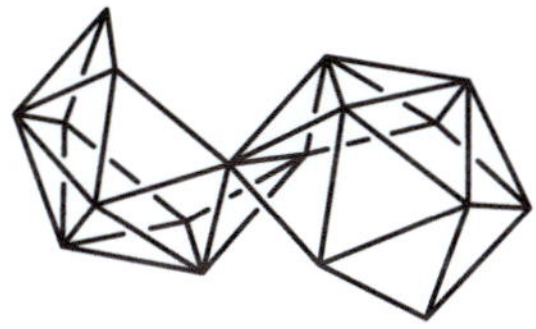

17

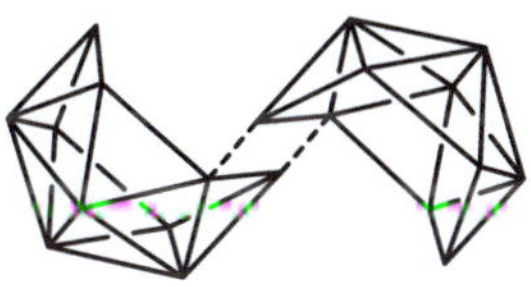

16 A

17 A

With two atoms held in common, then multicenter bonding aspects certainly can and do form part of the fusion bonding. For isolated compounds, such two-atom conjunctions are by far the most ubiquitous in the literature at the time of writing. The *anti* and *syn* isomers of $B_{18}H_{22}$ (schematic skeletal structures **16** and **17**) are the most readily accessible and are *nido:nido* species [37]. $B_{16}H_{20}$ and $B_{14}H_{18}$ are also in this *nido:nido* category [38–44], but the difficulty of their preparation means that they are little examined to date [45–47]. The $B_{18}H_{22}$ isomers are illustrative. They can be regarded in terms of two *nido*-decaboranyl units fused with two boron atoms held in common (Schematics **16 A** and **17 A**). The bonding at the conjunction can be considered in terms of the bonding of the *nido* ten-vertex single-cluster model $B_{10}H_{14}$. Using a simplistic Lipscomb-type 'semi-localized' bonding scheme for $B_{10}H_{14}$ results in electronic structures **18 A** and **18 B** that do not mimic the two-fold symmetry of the molecule, and so a partial three-center bonding as in **18 C** is invoked to account for this, being regarded as the hybrid of the two canonical forms **A** and **B**,

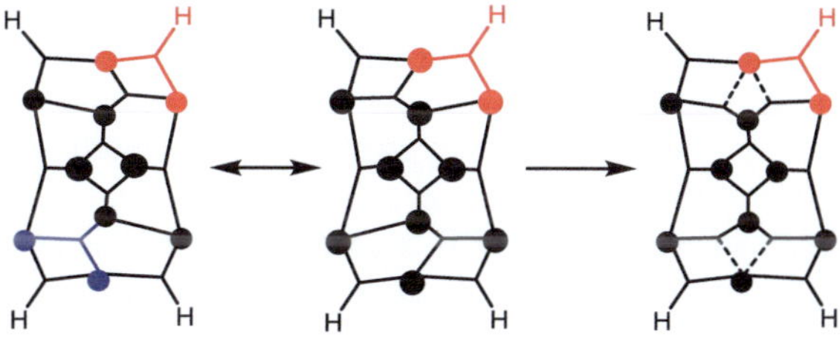

18 A **18 B** **18 C**

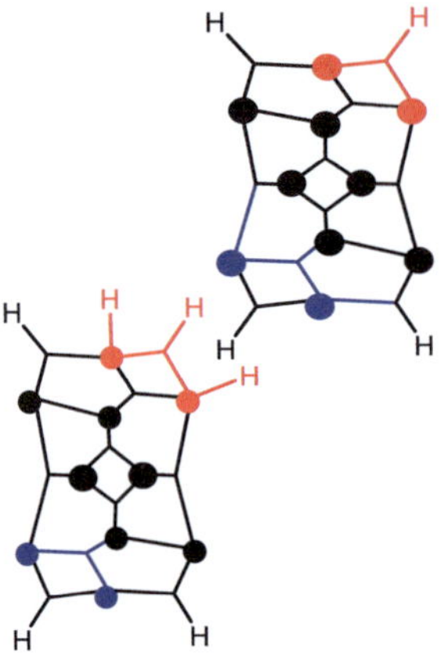

19 A

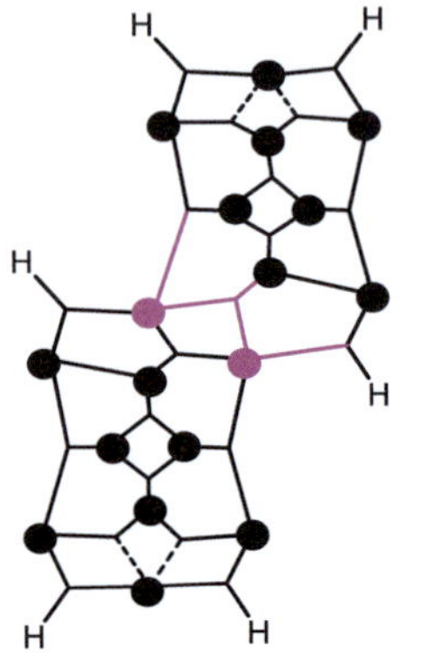

19 B

It can be seen from schematic **19 A** that the three-center bonding in $B_{10}H_{14}$ to a bridging hydrogen atom (red), together with the two two-electron bonds to *exo* H atoms (also red), mimics the internal bonding to the boron atom in a nominal B(2) position (blue). Consequently, when two *nido*-decaboranyl units nominally fuse together to give *anti*-$B_{18}H_{22}$ (and an equivalent fusion can be invoked for the *syn* isomer), they do so with minor perturbation to the internal bonding and thence much of the *nido*-decaboranyl character of the two resulting subclusters is retained (schematic **19 B**). Although there is little perturbation, nevertheless it can be seen that the effective two-fold symmetry of the 'partial' three-center bonds near the point of conjunction is destroyed to give one three-center and one two-center bond. This and other factors do produce variations on the chemistry over that encountered for $B_{10}H_{14}$. One factor not apparent from these simplistic electronic considerations is that for the fused species there is an inhibition of a cluster rearrangement due to the fixing of the two boron atoms involved in the conjunction. In this regard a number of reactions undergone by $B_{10}H_{14}$ cannot be mimicked by the *nido*-decaboranyl subclusters of $B_{18}H_{22}$ because they involve a cluster rearrangement and a concomitant interchanging and/or scrambling of the boron atom positons in the cluster. For example, the very ready reaction of $B_{10}H_{14}$ with SMe_2, to give *arachno*-$B_{10}H_{12}(SMe_2)_2$ and dihydrogen, which is believed to involve a cluster rearrangement [48], is not duplicated by *anti*-$B_{18}H_{22}$; reaction of SMe_2 with $B_{18}H_{22}$ in fact requires an oxidizing agent, and the product $B_{18}H_{20}(SMe_2)$ still retains the *nido* : *nido* characteristics of $B_{18}H_{22}$ itself [49], with no *arachno* subcluster formation. Another aspect here is that the $B_{18}H_{22}$ isomers are held to be highly acidic [50, 51], with a Brønsted acidity comparable to those of the strong mineral acids; by contrast, the pK_a value 2.70 of the $B_{10}H_{14}$ single-cluster model [52, 53] is comparable to that of $CH_2ClCOOH$. It would be interesting to see a computationally backed rigorous theoretical interpretation of this phenomenon. Conversely, of course, some reactions of macropolyhedral species, for example those that involve more than one subcluster, are not available to their single-cluster analogues.

As mentioned above, the two-borons-in-common, or, more generally, the two-atoms-in-common, mode of conjunction is by far the most common. Often, of known compounds, they are *nido-nido* conjunctions and can be rationalized, at least simplistically, in a similar way to that rehearsed for $B_{18}H_{22}$ above. The method also readily applies to conjunctions between other cluster types, for example *nido* : *arachno* species $[B_{18}H_{20}(NC_5H_5)_2]$ (Schematics **20 A** and **20 B**) [54–56]. Here the *nido* subcluster will have similar perturbations to those discussed above for *anti*-$B_{18}H_{22}$, but it can be seen that the bonding in the *arachno* subcluster appears to be little changed compared to its single-cluster analogue $[B_{10}H_{12}(NC_5H_5)_2]$. Although little research has been carried out on $[B_{18}H_{20}(NC_5H_5)_2]$, it appears that that its *arachno* subcluster may indeed have very similar properties to $[B_{10}H_{12}(NC_5H_5)_2]$; for example their very similar fluorescence properties suggest very similar electronic structures, and the nuclear magnetic shielding properties of $[B_{10}H_{12}(NC_5H_5)_2]$ and the *arachno* subcluster of $[B_{18}H_{20}(NC_5H_5)_2]$ are also very similar [56].

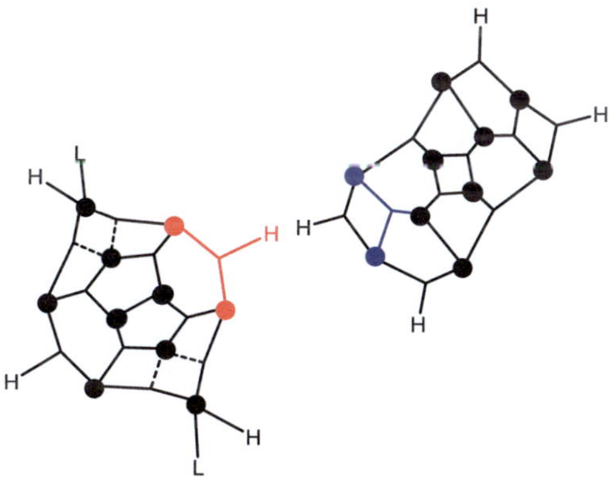

20 A

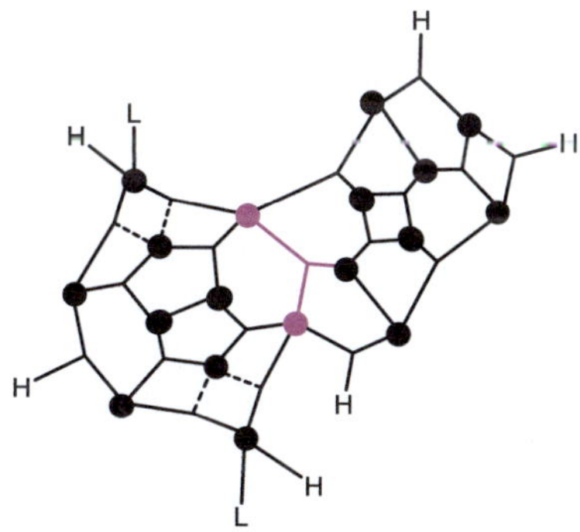

20 B

There are instances when this type of simple analysis does not hold, and individual subcluster character is uncertain. This is illustrated by the eight-vertex : eight-vertex two-borons-in-common species $B_{14}H_{20}$ (schematic structure **21**) [57, 58]. Overall, this has a relatively open C_2v 14-vertex structure of formal *arachno* B_nH_{n+6} formulation, but, because of its very open aspect, to regard its architecture as based on a 16-vertex closed deltahedron by the removal of two vertices is far-fetched. In terms of a two-borons-in-common cluster fusion the formulation implies the fusion of *arachno* and *nido* eight-vertex cluster types. This in turn implies an asymmetric structure due to the two different eight-vertex subcluster types, in contrast to the observed hydrogen-atom disposition and overall two-fold symmetry with two identical subclusters. It seems therefore that the $B_{14}H_{20}$ structure may be at present *sui generis* and thence a possible model for new *genera* of extended macropolyhedral structural classes, and it will be interesting to see the results of any calculational work to investigate potential viability of symmetrically extending the structure to incorporate more vertices. A related problem occurs in attempts to rationalize the observed structure of $B_{12}H_{16}$ and its conjugate anion $[B_{12}H_{15}]^-$ in terms of specific *nido* or *arachno* individual subclusters (schematic structure **22**) [59].

21

22

The ubiquity of the two-atoms-in-common mode is well illustrated by the variety of structures that have been found for the sulfur-containing macropolyhedral thiaboranes. To date these include *(a)* the 18-vertex neutral *n*-$[S_2B_{16}H_{16}]$ (schematic **24**) [60, 63] and *iso*-$[S_2B_{16}H_{16}]$ (schematic **25**) [61] isomers, neutral $[S_2B_{16}H_{14}(PPh_3)]$ (schematic **26**) [62] and the *n*-$[S_2B_{16}H_{15}]^-$ (schematic **27**) [61], $[S_2B_{16}H_{17}]^-$ (schematic **28**) [63] and $[SB_{17}H_{19}]^-$ (schematic **29**) [64] anions, *(b)* 19-vertex neutral $[S_2B_{17}H_{17}(SMe_2)]$ (schematic **30**) [65], neutral $[S_2B_{17}H_{17}]$ (schematic **31**) [66], and the $[S_2B_{17}H_{18}]^-$ (schematic **32**) [67] and $[S_2B_{17}H_{16}]^-$ (schematic **33**) [66] anions, and *(c)* 20-vertex neutral $[S_2B_{18}H_{20}]$ (schematic **34**) [68] and the *n*-$[S_2B_{18}H_{19}]^-$ (schematic **35**) [69] and *iso*-$[S_2B_{18}H_{19}]^-$ (schematic **36**) [68] anions. Some of these show an intercluster bridging *sigma* linkage in addition to the two-atoms-in-common mode, for example the *iso*-$S_2B_{16}H_{16}$ species represented by **25** compared to its *n*-$S_2B_{16}H_{16}$ isomer **24**. A similar bridging feature is also seen in some macropolyhedral metallaboranes and is consequently a potentially common motif. These thiaboranes generally seem much more flexible and adaptable than pure binary boranes, and several of the structures revealed by their pursuance, and by the pursuance of their metal derivatives (see below), have led, and can in future lead, to quite unexpected formulations.

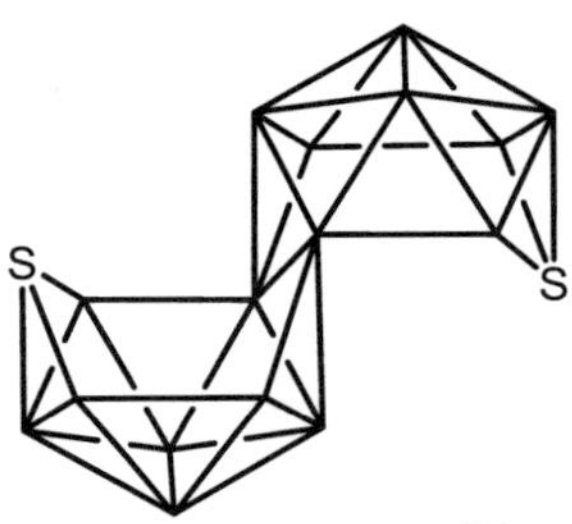

24

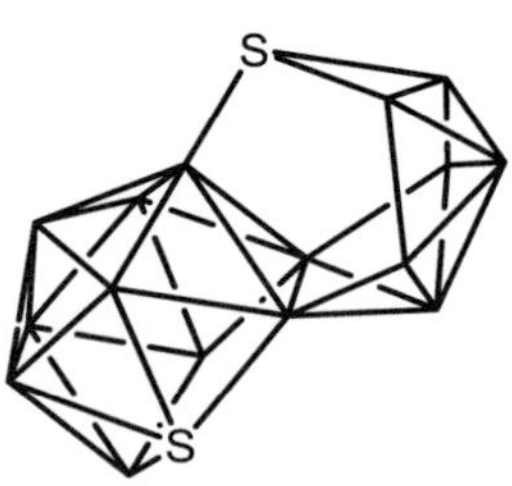

25

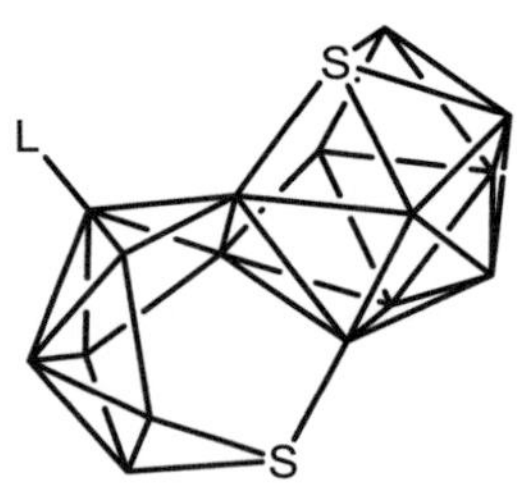

26

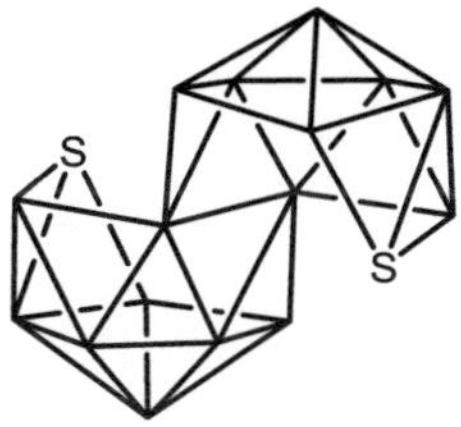

27

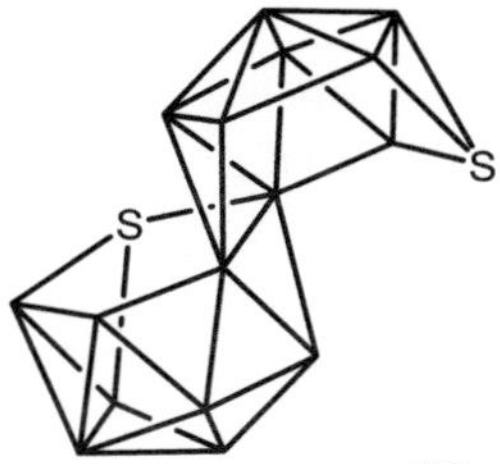

28

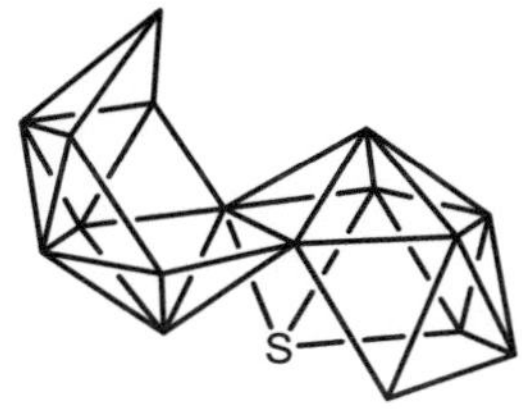

29

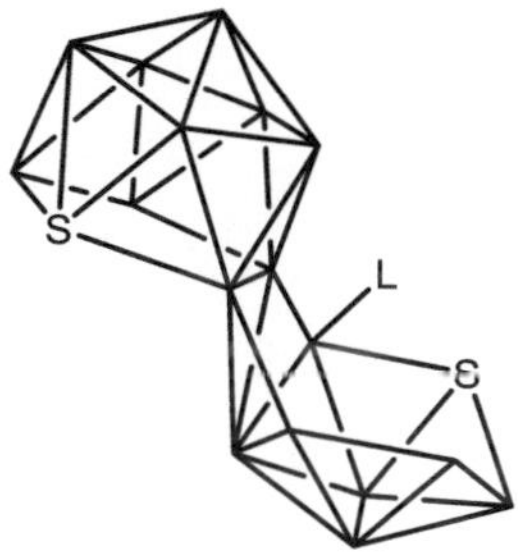

30

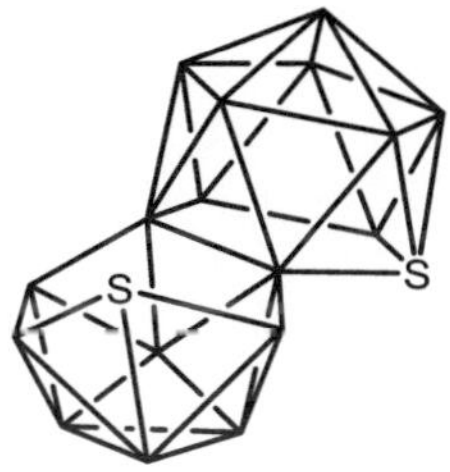

31

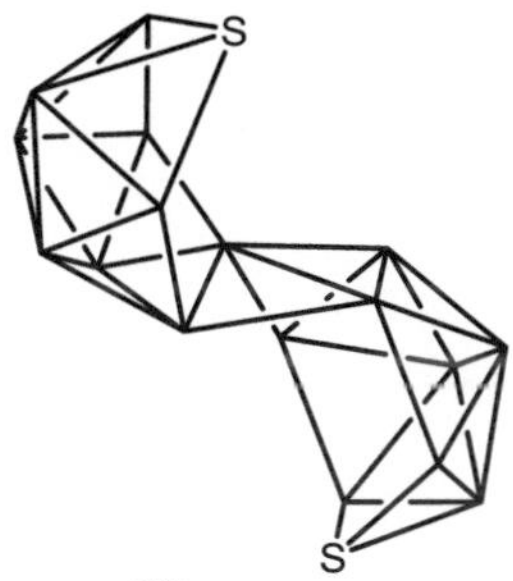

32

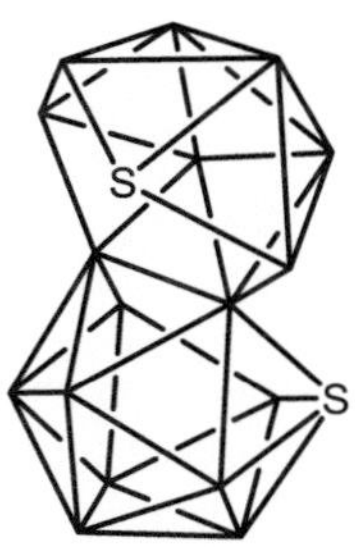

33

34

35

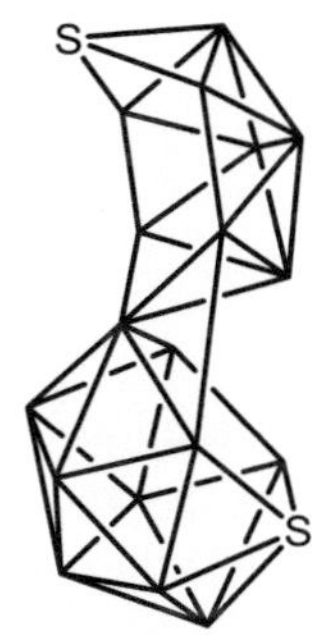

36

Reactions that generate macropolyhedral compounds from single-cluster species are discussed below. There are no general transferable generic methods for cluster fusion, and so an alternative method of exploring what macropolyhedral chemistry has to offer is to conduct reactions on pre-formed macropolyhedral species. Many of the thiaboranes just mentioned derive from this approach, *viz.* reactions of elemental sulfur with the macropolyhedral binary boranes *syn* and *anti*-$B_{18}H_{22}$ or their anions. Other predominant work leading to new structural types has been associated with reactions with various transition-element compounds to generate macropolyhedral metallaboranes. Reactions of $B_{18}H_{22}$, particularly the more accessible *anti* isomer, have figured predominantly here, and the smaller compounds $B_{14}H_{18}$ and $B_{16}H_{20}$ are not well-examined. Nevertheless, with [$PtMe_2(PMe_2Ph)_2$], $B_{14}H_{18}$ has yielded [$(PMe_2Ph)_2PtB_{14}H_{16}$] (schematic **37 A**) and $B_{16}H_{20}$ has yielded [$(PMe_2Ph)_2PtB_{16}H_{17}Me$] (schematic **37 B**) [45]. $B_{16}H_{20}$ with [$RuCl_2(PPh_3)_3$] has yielded [$(PPh_3)_2RuB_{16}H_{20}$] (schematic **37 C**) with a novel ten-vertex '*neo-nido*' subcluster [46] and with [$RhCl_2(C_5Me_5)$]$_2$ the bimetallic compound [$(C_5Me_5)_2Rh_2B_{16}H_{19}Cl$] (schematic **38 A**) is formed as well as the interesting [$(C_5Me_5)_2Rh_2B_{11}H_{15}$] species (schematic **38 B**) *via* the loss of five boron atoms from the macropolyhedral cluster assembly [47]. The last compound has an unprecedented 13-vertex macropolyhedral cluster core based on a *nido* 10-vertex {MB_9} subcluster and a *nido* 5-vertex {MB_4} subcluster fused with their open-face {B_2} edges in common.

37 A

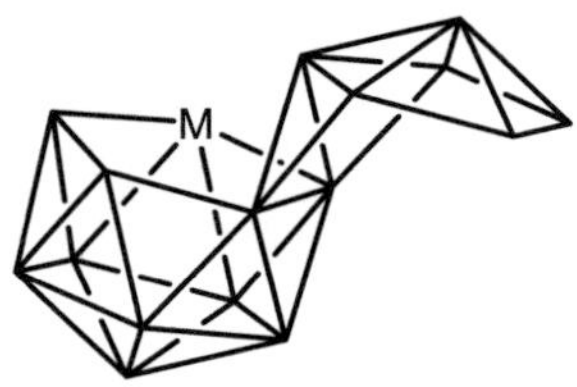

37 B

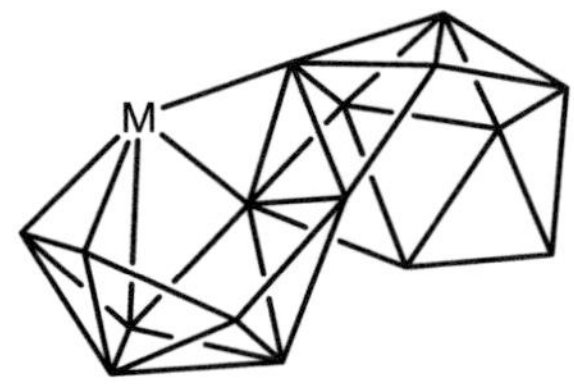

37 C

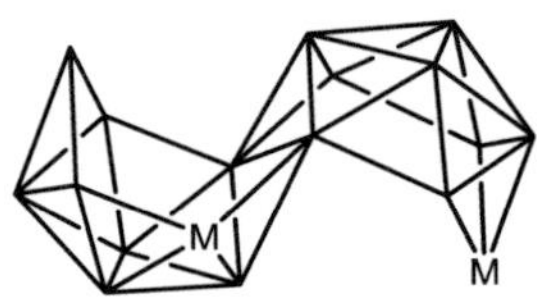

38 A

38 B

Reactions of the $B_{18}H_{22}$ isomers, pioneered by Sneath, Todd and co-workers over 40 years ago [54], have exhibited the incorporation of one, two and three metal centers into the macropolyhedral matrix. Two common reactions are the mimicking of *nido*-$B_{10}H_{14}$, which is known to give *nido*-type 11-vertex $MB_{10}H_{12}$ species and also *nido*-type 10-vertex MB_9H_{13} species *via* boron vertex loss [70], and a number of compounds of general formulation $M_{18}H_{20}$, with an 11-vertex *nido*-type {MB_{10}} subcluster [71], and $MB_{17}H_{19}$, with a 10-vertex *nido*-type {MB_9} subcluster [72] (schematic skeletal structures **39** and **40**) are known. Here and throughout M represents a transition-element center and its associated *exo*-polyhedral ligands. A related structural type, as represented by an isomer of [$(PMe_2Ph)_2PtB_{18}H_{20}$], has the metal center bridging between the two subclusters (schematic skeletal structure **41**) [71]. The second *nido*-decaboranyl subcluster within the $MB_{18}H_{22}$ and $MB_{17}H_{19}$ structural types can similarly add a metal center, so that overall two metal centers can add to the cluster, as exemplified by [$(PMe_2Ph)_4Pt_2B_{18}H_{17}Cl$] of schematic skeletal structure **42** [24].

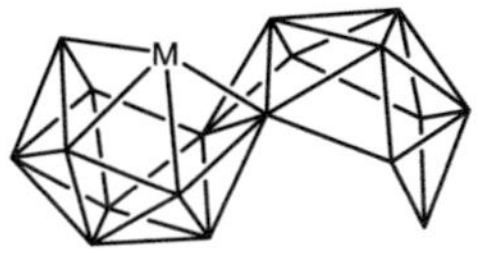

39

40

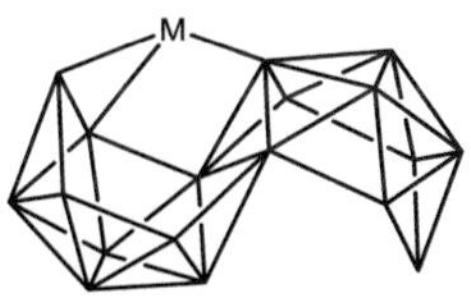

41

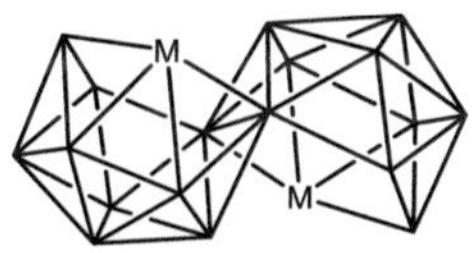

42

With transition-element centers such {Ru(C_6H_6)}, {Rh(C_5Me_5)} and {Ir(C_5Me_5)}, the metal center can become bound to both subclusters, as in [(C_5Me_5)Ir$B_{18}H_{20}$] (schematic skeletal structure **43**) [73]. Here the interaction with the second subcluster is more intimate than that in the [$(PMe_2Ph)_2PtB_{18}H_{20}$] isomer **39** above, so that the compound now exhibits a three-atoms-in-common mode of intercluster fusion. Factorization into the two subclusters (schematics **43 A** and **43 B**) reveals a 12 vertex *nido*-type {MB_{11}} unit **A** fused to a *nido*-decaboranyl {B_{10}} unit **B**, with three boron atoms held in common. In non-metallated compounds, at the time of writing, the three-atoms-in-common fusion mode is rare, and limited to the anions $[B_{20}H_{18}]^{2-}$ and $[B_{21}H_{18}]^-$ and their derivatives [74, 75], and the long-known bis(ligand) adducts of $B_{20}H_{16}$ such as [$(MeCN)_2B_{20}H_{16}$] [76].

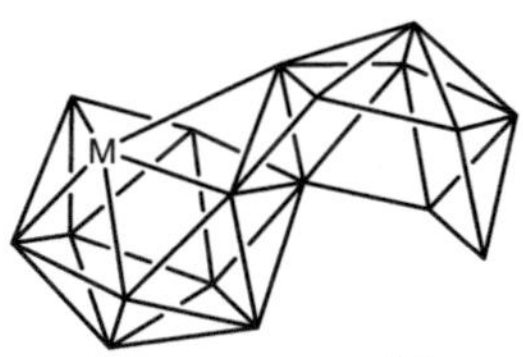

43

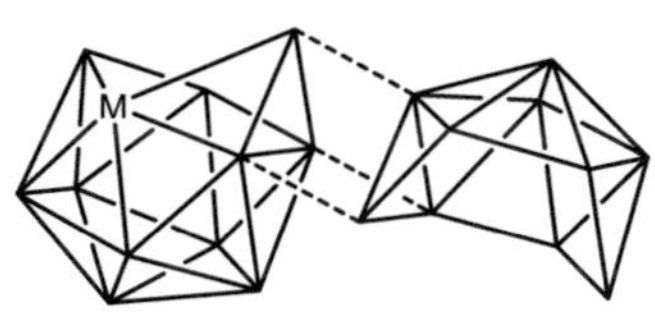

43 A 43 B

A second such center inserted into the other subcluster, as in [{(C_6Me_6)Ru}$_2B_{18}H_{18}$] [73] generates a matrix based on two 12-vertex *nido*-type {MB_{11}} units, now showing increased intercluster intimacy as they are fused with four boron atoms held in common (skeletal structures **44 A** and **44 B**).

44 A

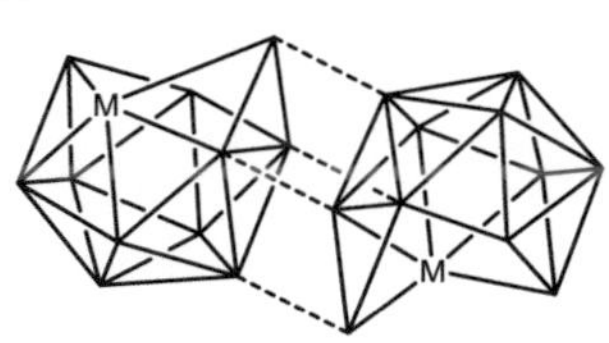

44 B

Triple addition of a metal center, as in [{(C_5Me_5)Ir_3}$B_{16}H_{15}$(OH)] (skeletal structure **45 A**) [77], however, now gives a very condensed species that does not similarly factorize so readily into two recognisable subclusters: the individual subcluster architectures are essentially lost, and motifs for a more condensed general structural type start to emerge. In particular, there is now an incidence of a centrally positioned tetrahedral-based core structure to the cluster. Thus, an alternative factorization as in schematic **45 B** has been adopted, in which two 12-vertex clusters (one *closo* {MB_{11}} unit and one with a partially open-faced '*isonido*' {M_2B_{10}} configuration), both fuse *via* a 5-vertex {MB_4} linking unit which includes a tetrahedral {MB_3} core. It has thence been held that consideration of this very condensed structure introduces an interesting structural concept with potentially important general connotations. Specifically, it has been held that the central non-hydridic tetrahedral 'core' leads to a 'megaloborane' concept for globular boron-containing structures that is based on central borons-only cores surrounded by units based on boron hydrides which themselves may or may not be mutually fused [78–82].

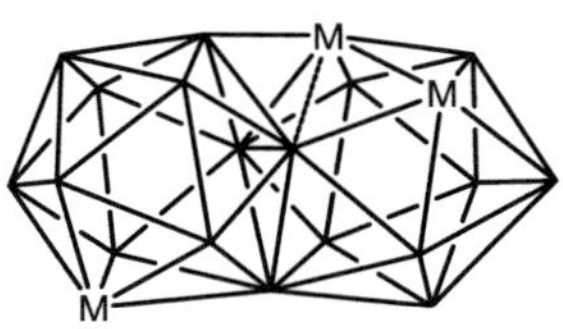

45 A

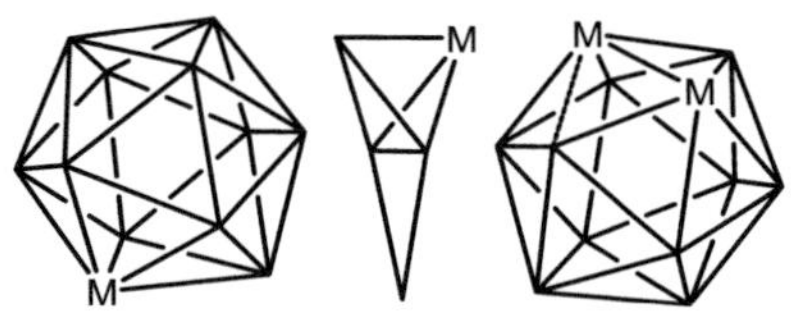

45 B

This {Ir_3B_{18}} assembly is formed by multiple-atom *Aufbau* on the *syn*-$B_{18}H_{22}$ isomer [77]. In view of the implications of the structure, it is of interest to seek other means of generation of such species. In a second approach to this type of condensed compound, therefore, [$(PMe_3)_4$(CO)$Ir_2B_{26}H_{24}$] (schematic **46**) has been found to result from fusion of [$(PMe_3)_2$(CO)HIrB_8H_{12}] with $B_{10}H_{14}$, using the Bould concept of *molten boranes as solvents* [83], *viz.*, in this particular case, using molten $B_{10}H_{14}$ itself as the solvent [83]. A third type of approach, the conjunction of two anionic $[B_{10}H_{10}]^{2-}$ residues in a reaction with [$PdCl_2(PPh_3)_2$], results in [$(PPh_3)_3(PPh_2)_2Pd_4B_{20}H_{16}$] (schematic **47 A**) [81]. These last two macropolyhedrals also have a globular boron-core cluster architecture (*e.g.* **47 B**), and have several other interesting features other than the {$IrB_{18}H_{14}$} and {$PdB_{20}H_{16}$} globules themselves [81, 83, 84]. With increasing size of the globular species, the increasingly larger surface of the globule can start to exhibit topographical features that engender particular chemistries. For example, the {(PPh_3)Pd(PPh_2)Pd(PPh_2)Pd(PPh_3)} unit of the palladium compound snakes very freely back and forth over a topographically 'smoother' extended cluster surface [81, 84], whereas, by contrast, the iridium compound shows a unique very restricted rotation within its {Ir$(PMe_3)_2$} unit because the unit is intimately held and constrained in a crevice on the macropolyhedral surface [82, 83]. These interesting fluxional phenomena have not yet received theoretical attention, although recently, in single-cluster work, DFT techniques have been successfully applied to clarify the rotationally fluxional behaviour in the ‘slipped *closo*’ platinadicarbaborane series [$(PR_3)_2PtC_2B_9H_{11}$] [22]; this set of compounds has long been a subject of theoretical examination [8, 85, 86].

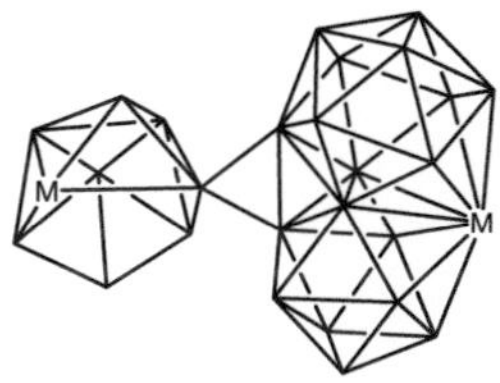

46

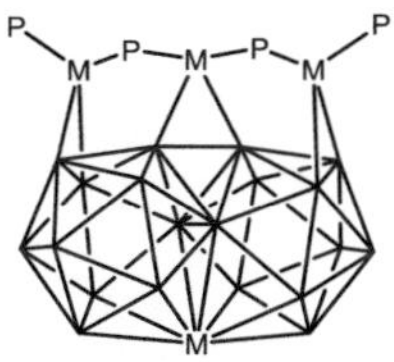

47 A

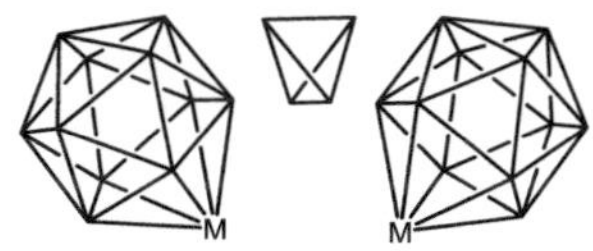

47 B

Use of the molten borane technique in another system, the fusion of [$(PMe_2Ph)_2PtB_8H_{12}$] with $B_{10}H_{14}$, yields compounds that show an elegant progression of intimacy in intercluster fusion (schematics **48** → **50**) [84]. and extension of the technique to the use of pre-formed macropolyhedrals themselves gives contiguous triple-cluster [$(PMe_2Ph)PtB_{26}H_{24}(PMe_2Ph)$] (schematic **51**) from *anti*-$B_{18}H_{22}$ and [$(PMe_2Ph)_2PtB_8H_{12}$] [87]. This is held to be the first molecular boron cluster compound so intimately to fuse more than two clusters [82–84]. Also, its structure very nicely links the **48** → **49** → **50** progression of increasing intimacy of intercluster fusion to the culminating very condensed species such as **45**, **46** and **47** [84]. An important future pointer here is that there is clearly high potential for the use of reactive boranes as solvents for the generation of new, unpredictable, macropolyhedral types. This synthetic approach is limited by the lack of suitable boranes, which need to melt in order to give a reactive molten solvent before suitable reaction temperatures are reached, but here the greater scope of known carbaboranes augurs well for macropolyhedral carbaborane and macropolyhedral metallacarbaborane discovery synthesis. An interesting possibility for good temperature control in this type of process is the use of ionic liquids that contain potentially reactive borane species as one of the ionic components. In an initial approach to this usage, the ionic liquid [C_5H_5N-*n*-C_5H_{11}][6-Ph-*nido*-6-CB_9H_{11}] has been made, and reaction with [$(PMe_2Ph)_2PtB_8H_{12}$] gives [$(PMe_2Ph)_2PtB_8H_{11}CPh$] and [4-(*n*-$C_5H_{11}NC_5H_4$)-6-Ph-nido-6-$CB_9H_{10}$] [88]; albeit not macropolyhedral species, the latter products nevertheless demonstrate appropriate and very interesting reactivity and thence show potential for macropolyhedral synthesis.

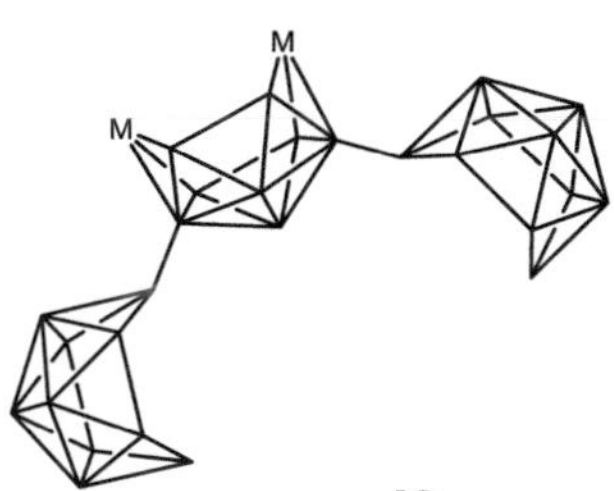

48

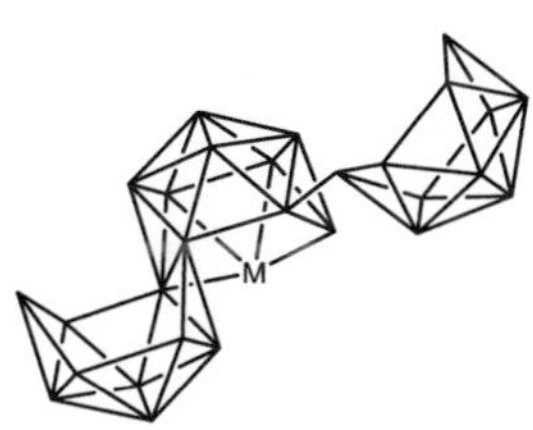

49

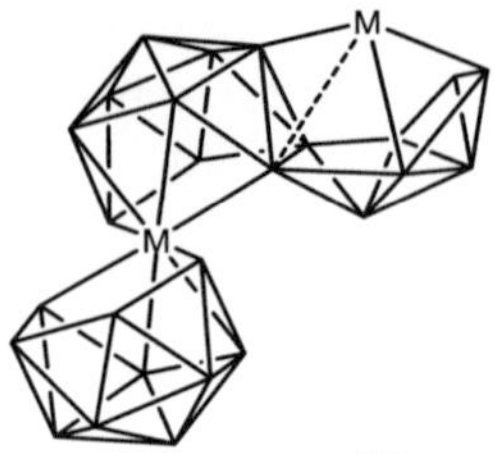

50

51

In contrast to large hollow-shell perceptions that are based on fullerene chemistry [89], or that are based on extrapolations of the classical *closo* $[B_nH_n]^{2-}$ structures to high values of *n* [90], structures such as **45**, **46** and **47** therefore suggest an emerging new borane-based structural principle based on filled big shells, rather than hollow ones. The implicit new chemistry here in principle extends beyond the simple four-atom core unit, and is an area in which predictive computational chemistry may be valuable (see also Chap. 4 in this Volume). A structural premise here is that many well-recognised cluster species can be related to fragments of solid-state extended matrices, with the peripheries of the fragments bounded by covalently bound terminators, for example metal carbonyl cluster compounds can thus be regarded as based on fragments of crystalline metal lattices, and polyoxometallates can be regarded as based on fragments of ceramic lattices. In accord with this concept, models have been proposed for globular megaloborane species based on fragments of crystalline elemental boron lattices with the peripheries bounded by hydrogen atoms. Here STO-3G *ab initio* calculations (held at the time to be at a reasonably high level for preliminary surveys of such big molecules) [82] on model systems that are based on domains of elemental boron matrices are found to stabilise in particular regions, *e.g.* neutral $B_{27}H_{21}$ and $B_{84}H_{54}$ based on sections of the solid-state matrix of β-rhombohedral boron (Fig. 6.4) [83, 84]. Of these, $B_{84}H_{54}$ as calculated has a surface mostly of *endo* and bridging hydrogen atoms, and, based on conventional small single-cluster behaviour, may be expected to be reactive, although of course it would be interesting to see how the properties of such *endo* and bridging hydrogen atoms may be modified when associated with such an extended cluster surface. In this regard $B_{27}H_{21}$ and $B_{84}H_{54}$ approach two extremes of surface character, because $B_{27}H_{21}$ has a surface of predominantly *exo* hydrogen atoms, and may therefore prove to be somewhat less reactive. Overall it has been suggested that since species such as $B_{27}H_{21}$ and $B_{84}H_{54}$ are at energetic minima on a multidimensional structure-energy continuum, they may therefore be inherently isolatable [91]. The concept was subsequently assimilated [92] and the idea developed in terms of borons-only clusters such as $\{B_{84}\}$ [93], but this introduces the very speculative theoretical area, involving, for example, swarm structure-searching calculations, as well as very esoteric experimental areas of borons-only cluster chemistry and chemical physics [94, 95], which are outside the scope of this chapter (But see Chap. 1 in this Volume). Certainly at present most such speculatively proposed and/or transiently identified borons-only species transcend Sir Ronald

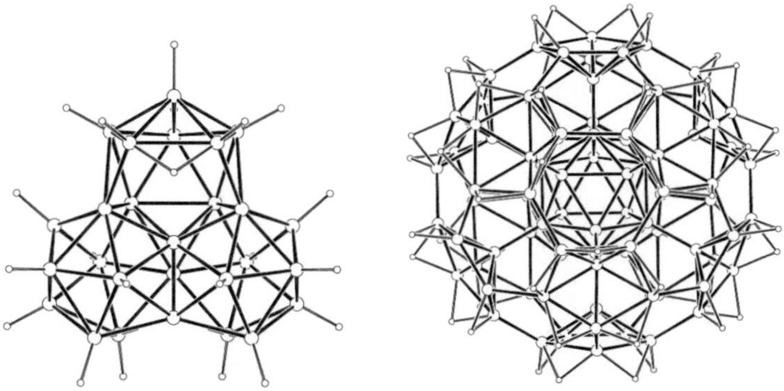

Fig. 6.4 *Left*: a representation of the STO-3G *ab initio* calculated structure of neutral $[B_{27}H_{21}]$, in which the $\{B_{27}\}$ unit has many similarities with the $\{B_{27}\}$ unit in the β-rhombohedral allotrope of elemental boron. There is a seven-atom 'borons-only' core, with a boron hydride skin consisting of 20 peripheral boron atoms with 20 *exo*-hydrogen atoms and one bridging hydrogen atom. *Right*: a representation of the AM-1 semi-empirical energy-minimized structure of neutral $[B_{84}H_{54}]$, in which the $\{B_{84}\}$ unit mimics that in β-rhombohedral boron. There are 24 atoms in the borons-only core, and a $\{B_{60}H_{54}\}$ boron hydride skin. Its architecture may be visualized as an icosahedral $\{B_{12}\}$ unit, like that in the [*closo*-$B_{12}H_{12}]^{2-}$ anion, fused to 12 *nido*-hexaboranyl units that are themselves fused together (see schematic **52**) [81]

Nyholm's criterion of chemical realism – "But can you put it in a bottle?" – although they do attract publicity [96].

The filled-ball architectural principles of such globular 'megaloborane' species will be quite different to those of the hollow fullerenes, and their novel properties will consequently complement those of the hollow carbon cages. The construction of such new large molecular nanoglobules, and their tailoring by substituent chemistry to modify and enhance interactions important for host-guest interlocking, for crevice and anchor generation for molecular recognition, and to tune domains of hydrophilic versus lipophilic character, and so on, is of high potential usage. Implicit in this is the development of clean *Aufbau* techniques from simple boron-containing starting substrates. However, generic routes for the required intimate cluster condensations are as yet lacking (for other aspects of macropolyhedral synthesis, see below, near schematic structures **58**). Similar problems in principle arise in carbon hydride chemistry, but, in practice, syntheses of polycondensed aromatics from smaller hydrocarbons such as benzene are not required. This is because condensed hydrocarbon feedstock compounds are readily available, albeit, when their origins in primordial forests in the carboniferous era are considered, in very low yield and from very inefficient processes [91].

A possible route for syntheses of the larger globular filled-ball megaloboranes is suggested by the structure of $B_{84}H_{54}$ mentioned above (Fig. 6.4, right). This globular structure can be regarded as constructed of a central *closo* $[B_{12}H_{12}]^{2-}$ unit, of which the 12 *exo*-hydrogen atoms are replaced by 12 apically bound *nido* B_6H_{10}-type units, of which the pentagonal bases are themselves mutually fused *via* dihydrogen loss to create the outer $\{B_{60}H_{54}\}$ skin. This leads to the supposition that a synthetic route may be constructed similarly, *viz.* by the stepwise assembly of boron hydride units about a central cluster followed by their ultimate cross-linking (*e.g.* schematic **52**) [91]. Purported initial feasibility steps along this pathway have been established by the isolation of *(a)* [6,9-$(SMe_2)_2$-*arachno*-$B_{10}H_{10}$-1,5-(6′-*nido*-$B_{10}H_{13})_2$] (schematic cluster structure **53**) [97], in which two *nido*-decaboranyl units are assembled on adjacent sites around an *arachno*-decaboranyl core (see also the platinaborane structure **48** mentioned above), *(b)* [*nido*-$B_{10}H_{12}$-(*nido*-$B_5H_8)_2$] (compound **54**), with two smaller *nido* clusters (now based on *nido*-B_5H_9) around the central boron core, rather than larger $\{B_{10}H_{14}\}$-derived clusters, thus approximating more closely to the {*nido*-B_6H_{10}}-based construction of $B_{84}H_{54}$ [91], and *(c)* [$(CH_2CH_2C_5H_4N)$-*arachno*-$B_{10}H_{10}(NC_5H_4$-*closo*-$C_2B_{10}H_{10})$] (schematic skeletal structure **55**) [91], which illustrates the principle of linking the peripheral encapsulating units, and also introduces the ultimate concept of the assembly of borane units to encapsulate carbon-based cores, rather than boron-based cores. For example, with a benzenoid core, a discoidal aspect would result, and it would be interesting to see speculative calculational work in that direction.

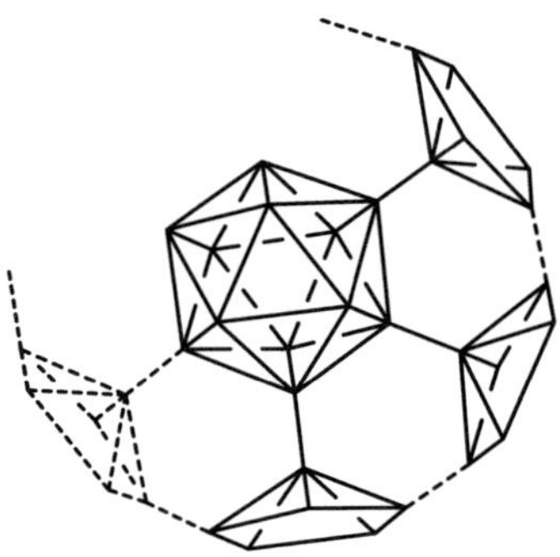

52

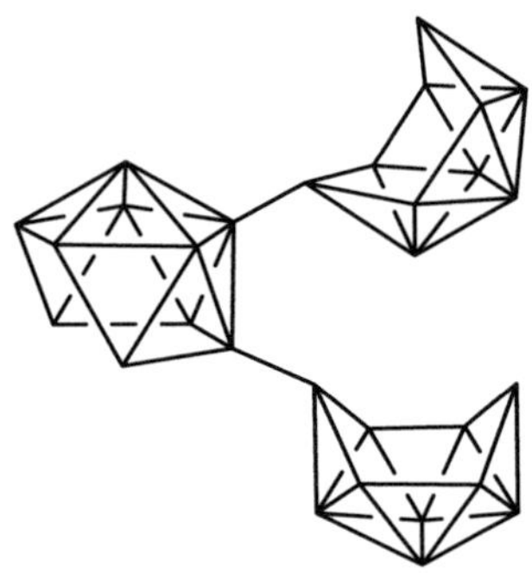

53

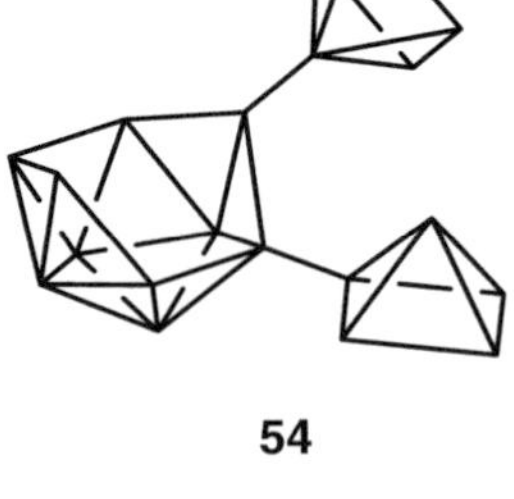

54

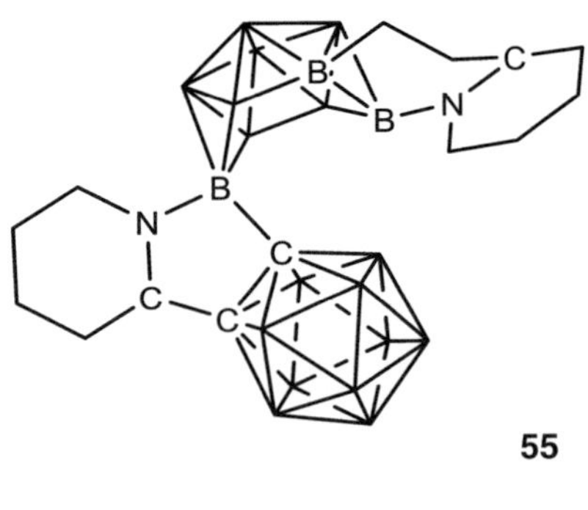

55

The preceding paragraph highlights the overall problem of macropolyhedral synthesis. After the pioneering systematic work of Riley Schaeffer and co-workers, in which $B_{13}H_{19}$ [41, 98], $B_{14}H_{20}$ [57, 58], $B_{14}H_{22}$ [42], $B_{15}H_{23}$ [27], and $B_{16}H_{20}$ [42, 43], were made by appropriate elegant individual strategies, no transferable generic route for cluster fusion has emerged. Metal-induced and metal-catalyzed fusion has been used in isolated cases, but these approaches are not generally applicable. An interesting process is that of Grimes and co-workers which has been used to make 12-vertex $B_{12}H_{16}$ (discussed above, schematic structure **22**) from the [*nido*-B_6H_9]$^-$ anion by treatment with $FeCl_2$ [59, 99], and a similar metal-induced route gives high-carbon-content 12-vertex $R_4C_4B_8H_8$ carbaboranes from $R_2C_2B_4H_6$ 6-vertex cluster species [100]. There are hints in the literature that attempts have been made to apply the method to the {$C_2B_9H_{11}$} 'carbollide' entity, but with no success, and so it would appear that this otherwise promising-looking approach is not generally applicable. The iridium and palladium species [$(PMe_3)_4(CO)Ir_2B_{26}H_{24}$] (schematic skeleton **46**) and [$(PPh_3)_3(PPh_2)_2Pd_4B_{20}H_{16}$] (schematic skeleton **47 A**) mentioned above clearly result from fusion about a metal-atom center, as does the 19-vertex molybdenum-containing [$(CO)_2MoB_{16}H_{15}C_2Ph_2$]$^-$ anion [101], from a fusion of two nine-vertex [1-Ph-*nido*-1-CB_8H_{11}] *nido*-monocarbaborane units with [$Mo(CH_3CN)_3(CO)_3$], to give a rare example of a macropolyhedral metallacarbaborane species. Here it can be noted *en passant* that, in spite of the tremendous variety of known single-cluster carbaborane chemistry [102], only two macropolyhedral monocarbaboranes, *viz.* 18-vertex [$(Me_3CNH_2)CB_{17}H_{19}(CN)$] [103], and 19-vertex [9-($^{ter}BuNH_2$)-(*anti*)-{9-$CB_{18}H_{20}$}] [104] are known (schematic skeletons **56** and **57**). One factor inhibiting their synthesis has been that protocols for carbon-atom insertion that work well with single-cluster species have not so far been generally successful when applied to macropolyhedral boranes (the latter predominantly the $B_{18}H_{22}$ isomers in reports so far, for reasons of availability).

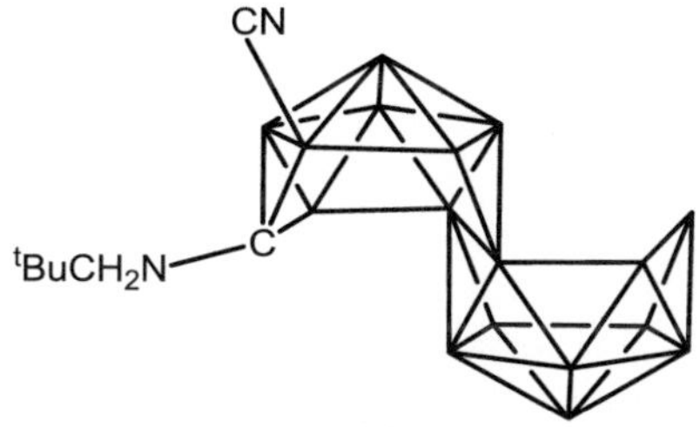

56

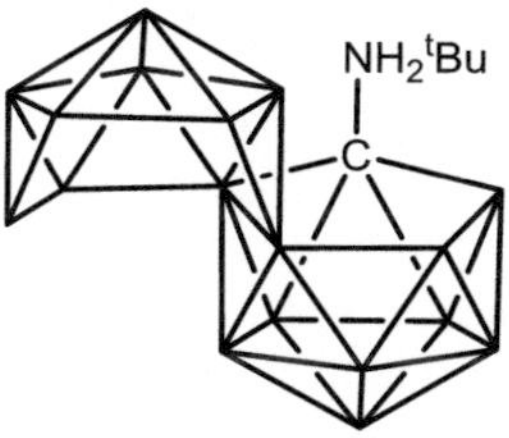

57

One route to the synthesis of the $B_{18}H_{22}$ isomers involves auto-fusion from nine-vertex *arachno* species such as [$B_9H_{13}(SMe_2)$] [40, 105, 106] and there is also a route starting from the *nido* nine-vertex binary borane [B_9H_{12}]$^-$ anion [107]. Certain other nine-vertex *arachno* species are susceptible to inter-cluster autofusion under conditions of mild heating. Thus SB_8H_{12} gives a mixture of two isomers of formulation $S_2B_{16}H_{16}$ (schematic structures **25** and **26** above) [60, 61], and [$SB_8H_{12}(SMe_2)$] gives [$S_2B_{17}H_{17}(SMe_2)$] (schematic structures **58 A** and **58 B**) which has a near-unique six-connected boron atom in a novel ten-vertex *arachno*-type subcluster [65]; the only other experimentally isolated example that the author is aware of is also in a macropolyhedral thiaborane, the *n*-$[S_2B_{18}H_{19}]^-$ anion (schematic **35** above, see also Fig. 6.5 below) [68, 69]. In view of the usefulness of these two *arachno* nine-vertex thiaboranes as entries into macropolyhedral chemistry, and also the usefulness of *arachno*-type nine-vertex metallaboranes [1, 2, 108, 109] as exemplified below, it is interesting but somewhat disappointing that cluster fusion has so far not been found to occur when corresponding *arachno* nine-vertex carbaborane compounds are treated similarly.

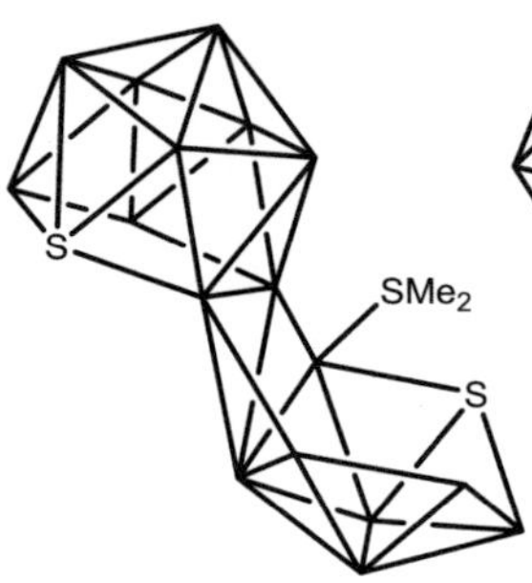

58 A

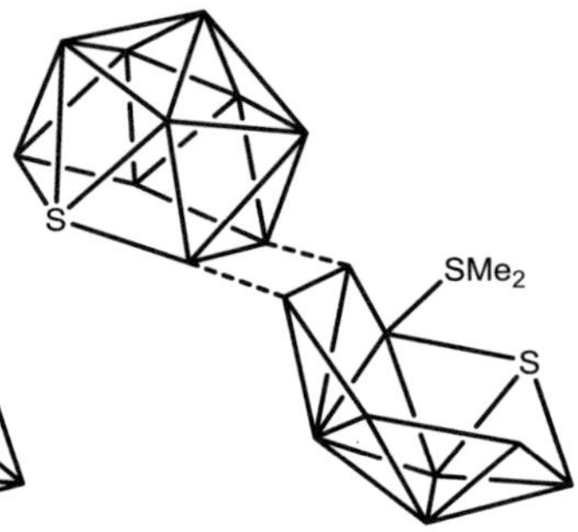

58 B

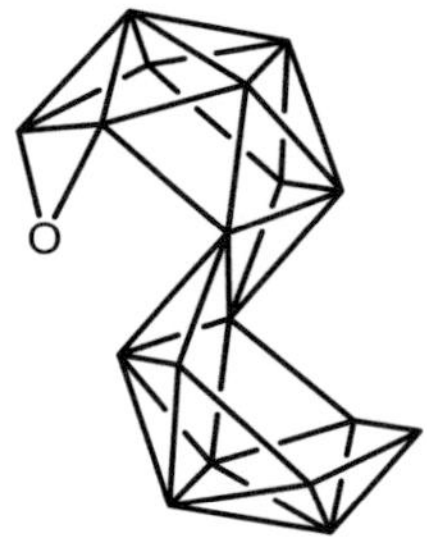

59

Alternative routes to macropolyhedral thiaboranes involve direct incorporation of elemental sulfur into the $B_{18}H_{22}$ isomers or their anions, as mentioned above, and the first reported macropolyhedral thiaborane, the $[S_2B_{17}H_{18}]^-$ anion (schematic **32** above) [67] and the species $S_2B_{17}H_{17}$ (schematic **31** above) [66] are both formed in this manner. Mechanistically related to the first stages of the sulphur incorporation reactions may be the reaction of the [*anti*-$B_{18}H_{19}]^-$ anion with H_2O to give the 19-vertex $[OB_{18}H_{21}]^-$ oxaborane anion (schematic **59**) [110] in which the heteroatom is not as intimately bound as in the two thiaboranes, and therefore may model an intermediate step. Boron-containing clusters that also contain oxygen as the heteroatom center are in fact rare, and, as far as we are aware, are limited to the *nido-type* 12-vertex $[OB_{11}H_{12}]^-$ anion [111–115] and the 'macropolyhedral' 19-vertex $[OB_{18}H_{21}]^-$ anion just mentioned [111–115], together with five neutral oxametallaboranes, 11-vertex *nido*-type $[(\eta^5\text{-}C_5Me_5)RhOB_{10}H_9Cl(PMe_2Ph)]$ [116] and $[(\eta^5\text{-}C_5Me_5)RhOB_{10}H_{10}(NEt_3)]$ [117], both closely related to the $[OB_{11}H_{12}]^-$ anion [114], *nido* 10-vertex $[(\eta^6\text{-}C_6H_3Me_3)FeOB_8H_{10}]$ [118], *arachno*-type 10-vertex $[(PMe_2Ph)_2PtOB_8H_{10}]$ [119] and *arachno*-type 11-vertex $[(\eta^6\text{-}C_6Me_6)RuOB_9H_{13}]$ **7** [120]. These have mostly been recognized serendipitously, and in this context it is perhaps surprising that stoichiometric use of potential oxygen-donor reagents for directed synthetic attempts seems not to have been carried out. There is certainly scope here for a calculational probe into the viability of isolating other stable oxaborane species, most particularly single-cluster species. Based initially on the many known stable heteroborane configurations that are known, in the first instance calculational entries could be made by simple oxygen-for-heteroatom replacement [121]. As with the reactions of transition-element centers with the binary borane macropolyhedrals discussed above, similar reactions of metal centers with the macropolyhedral thiaboranes yield a large variety of interesting macropolyhedral architectures. These are exemplified by *syn* and *anti* $[(C_5Me_5)_2Rh_2S_2B_{15}H_{14}(OH)]$ (schematics **60 A** and **60 B**) [122], $[(PMe_2Ph)_2PtS_2B_{15}H_{14}(NHCOMe)]$ (schematic **61**) [123], $[(PMe_2Ph)_2PtS_2B_{16}H_{16}]$ (schematic **62**) [124], $[(MeC_6H_4{}^{iso}Pr)RuS_2B_{15}H_{15}]$ (schematic **63 A**) [125], $[(MeC_6H_4{}^{iso}Pr)RuS_2B_{16}H_{16}]$ (schematic **63 B**) [125] and $[(PPh_3)NiS_2B_{16}H_{12}(PPh_3)]$ (schematic **64**) [62]. Syntheses may also be effected by the addition of elemental sulphur to macropolyhedral metallaboranes [126]. As well as the metal-center additions and the occasional boron-vertex loss seen with the binary borane reactions, these thiaborane systems that incorporate metal centers seem more prone to cluster rearrangements and also more prone to the loss of boron vertices during the course of the reactions.

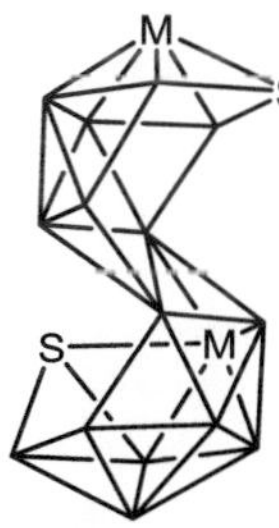

60 A

60 B

61

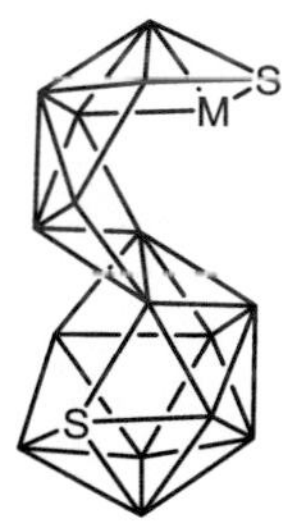

62

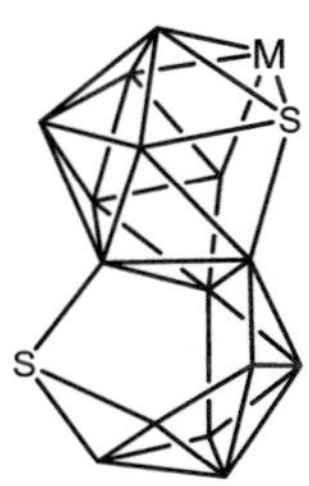

63 A

63 B

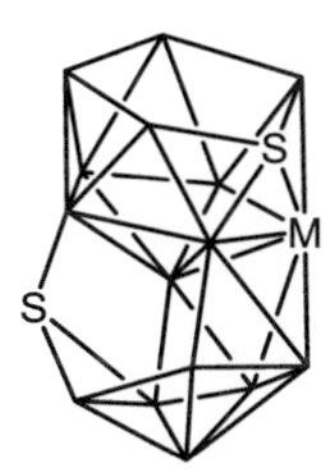

64

As mentioned above, thermolyses of nine-vertex *arachno* boranes and thiaboranes yield macropolyhedral boranes and thiaboranes. These cluster fusion processes suggest that *arachno* nine-vertex metallaboranes may similarly be prone to autofusion to give macropolyhedral metallaboranes. In accord with this idea, it is found that simple thermolyses of the *arachno* nine-vertex metallaboranes of palladium and platinum of general formulation $[(PR_3)_2PdB_8H_{12}]$ and $[(PR_3)_2PtB_8H_{12}]$, and of closely related species such as $[(CO)(PMe_3)_2IrB_8H_{12}]$, also yield a variety of macropolyhedral metallaborane types. Many of the products are of individually unique constitution, and constitute a clear augury for a huge variety of macropolyhedral architectures and behaviour to be encountered as the area develops, which in turn will engender new challenging horizons for computational approaches. Examples include [$(PMe_2Ph)PtB_{16}H_{18}(PMe_2Ph)$] (schematic **65**) [108], [$(PMe_2Ph)_4Pt_3B_{14}H_{16}$] (schematic **66**) [108], [$(PMe_2Ph)_2Pt_2B_{16}H_{15}(C_6H_4Me)(PMe_2Ph)$] (schematic **67**) [109], [$(PPh_3)_4ClPd_4B_{16}H_{17}(PPh_3)_2$] (schematic **68**) [127], [$(CO)(PMe_3)_2IrB_{16}H_{14}Ir(CO)(PMe_3)_2$] (schematic **69**) [128], [$(PMe_2Ph)_2Pd_2B_{16}H_{22}(PMe_2Ph)_2$] (schematic **70**) [129], and [$(PMe_2Ph)_3Pt_2B_{16}H_{20}(PMe_2Ph)$] (schematic **71**) [129]. Some of these, for example the last two species, of less-condensed architecture, clearly model intermediates on the way to the more condensed species [129, 130], and, as calculational techniques and the ability to exploit them evolve with more rapid processing, it will be interesting to see theoretical mechanistic insight into macropolyhedral borane assembly processes. In the formation of the two isomers of $S_2B_{16}H_{16}$ from relatively simple SB_8H_{12}, for example, reasonable intermediate steps have been proposed [61], and this reaction could therefore constitute a good initial prospect for elucidation in this regard.

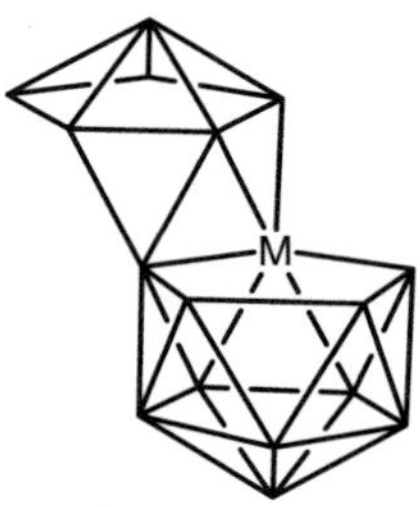

65

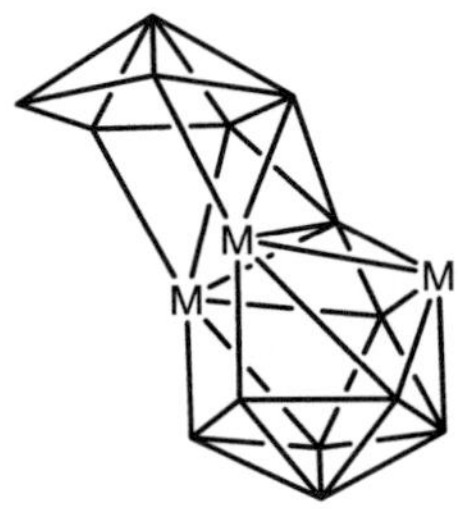

66

67

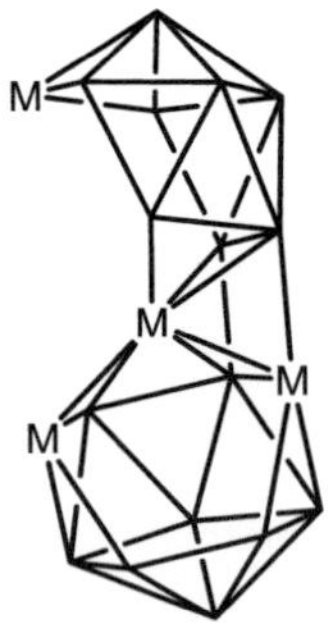

68

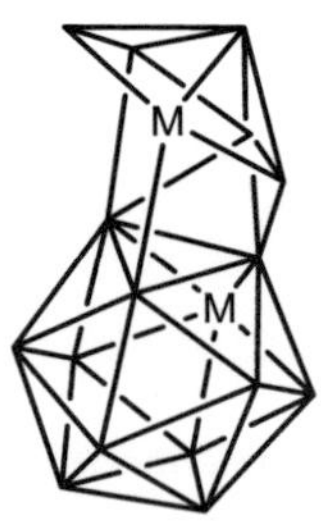

69

70

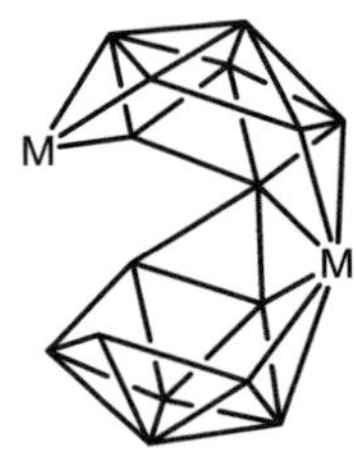

71

In accord with the general philosophy of this volume, computational calculations and potential opportunities for work in new directions have been pointed out as they arise above. In general, the bulk of the reported work in boron-containing cluster chemistry has been concerned with DFT structure minimizations and GIAO-based nuclear magnetic shielding computations to confirm suppositions, and to show consistency with experimental NMR chemical shifts. This approach is especially useful when no direct molecular structure evidence is available, *e.g.* when results from single-crystal X-ray work are not available. Certainly such calculational work has been effectively applied with advantage in the development of macropolyhedral chemistry, with structural and nuclear shielding results that match experimentally obtained values, even with heavy transition elements in the matrix, as with the confirmation of 15-vertex $[(PMe_2Ph)_2PtB_{14}H_{16}]$ (schematic **37 A** above) [45]. Commonly, both in single-cluster and macropolyhedral work, this usage has become quite routine and pedestrian unless it is directed at truly novel work, although it is generally of course gratifying when calculated molecular structure and nuclear shieldings, experimentally determined molecular structure, and experimentally determined NMR shieldings all fit nicely together. The necessity of experimental checkpoints is emphasised [131], and there are additional caveats in the use of this combined approach.

One such instance is bond rotamer effects on calculated structures, energies, nuclear magnetic shieldings and geometries. This is exemplified by work on phenyl-group rotation in the 11-vertex *nido* compound [7-Ph-*nido*-7,8,10-$PC_2B_8H_{10}$], which resulted in two related conclusions [132]. Firstly, although many boron-containing clusters may appear ostensibly rigid, their inherent flexibility (see discussion near Fig. 6.4 above) means that there can be a significant differential effect on the cluster geometry and cluster nuclear shielding arising from different rotamers about bonds to *exo*-polyhedral substituents. This is a general phenomenon, and likely to be most marked for *pi*-interactive substituents, for example unsaturated groups or groups with lone pairs *alpha* to the cluster. It is important to consider contributions from all bond rotamers in predictions of nuclear shielding, particularly when the rotamers are of similar energy. Thus lower-level calculations may not reveal all rotamer minima, and so, if used to pinpoint approximate minima, which are then subsequently refined at a higher calculational level, then other important rotational minima may be overlooked. Secondly, and implicit in the nuclear shielding and geometrical changes within the cluster, is that the differences between rotamers in their bonding *exo* to the cluster are manifested in changes in the cluster molecular orbitals. In the case of [7-Ph-*nido*-7,8,10-$PC_2B_8H_{10}$], the bonding involving the cluster atoms adjacent to a phenyl-substituted site is most affected. In the general case, however, there is the possibility that more extended cluster molecular orbitals may be involved, so that noticeable effects distal from the substituent, *e.g.* in antipodal positions, might be observable; a related possibility is that, in extreme cases, much more significant distortions of cluster geometry may occur – a cluster fragility phenomenon (Fig. 6.3 above). Again, these effects are likely to be most marked for *pi*-interactive substituents [133, 134]. Related effects will obtain in transition-element/borane systems that exhibit rotational fluxionality [22, 135].

An additional danger is that focus on calculated gas-phase minima in isolation may not recognise an adjacent deep alternative energetic well on the structure-energy surface. Tunnel vision here also may obscure a strong thermodynamic tendency to fragment into smaller very stable entities, which may preclude any prospect of realistic isolation of a calculated molecule if the barrier leading from the minimum along to the reaction coordinate of the fragmentation is low. In single cluster work this was recently emphasised in a DFT-based analysis, backed up by experimental checkpoints from single-crystal X-ray analyses and from NMR spectroscopy, of the keynote nine-vertex closed-cluster system based on $[(PMe_3)_2IrB_8H_8]$ and $[(PMe_3)_2IrB_8H_7Cl]$ [20]. It was found that the effects of the chlorine substituent versus hydrogen subtly dictated whether *isocloso* or *closo* forms (Figs. 6.1 and 6.2 above) were thermodynamically the more stable. It was determined that there was a difference in energy of only a few kJ mol^{-1} between the *isocloso* and *closo* structural forms, but that a high energetic barrier precluded isomerisation between the two forms. Ostensibly a nice rounded story, but the work strongly emphasised that these energetics are relative to each other – experimentally it was found that both compounds, of which only the *isocloso* forms are in fact known, are very delicate, and unstable with respect to a general decomposition. This represents a general danger when calculational predictions and speculations slip beyond the bounds of the intellectual control engineered by reference to experimental checkpoints and common sense: they perhaps (or perhaps not?) should never transgress the bounds of what may be considered chemically realistic.

Some applications, however, particularly when applied to chemical change, and involving transition-state determinations supported by frequency analyses, are helpful in distinguishing mechanism. This type of application is somewhat more progressive and challenging than simpler single-structure work, and has so far been of limited use in macropolyhedral chemistry. An example that illustrates the potential is the analysis of the remarkable sequence of behaviour exhibited by the 20-vertex $[S_2B_{18}H_{19}]^-$ anion and its neutral conjugate acid $[S_2B_{18}H_{20}]$ (schematic skeletal structure **34** above) [68]. The system is very unusual because it also involves a hexagonal pyramidal $\{B_7\}$ motif [65] within one of the subclusters [69]. Upon deprotonation of $[S_2B_{18}H_{20}]$, a markedly different species is observed, implying instant cluster isomerism; upon re-protonating, the starting compound is not regenerated, but a fluxional species is formed which does however slowly isomerise on standing to regenerate the starting $[S_2B_{18}H_{20}]$ species. Definitive multinuclear NMR spectra having been obtained for each stage, along with NMR-determined estimations of activation energies, and along with a checkpoint molecular structure of the starting compound from single-crystal X-ray work, DFT B3LYP/6-31G* calculations of ground-state structures and transition states, and thence GIAO calculations of magnetic nuclear shieldings, readily defined the entire sequence of processes as summarized in Fig. 6.5.

Another interesting rearrangement, albeit not yet tackled computationally, occurs upon the treatment of the 19-vertex anion $[S_2B_{17}H_{18}]^-$ (schematic skeletal structure **32** above) with oxidising acids. This procedure quantitatively generates neutral $[S_2B_{17}H_{17}]$ (schematic **31** above), characterised as its conjugate base, the $[S_2B_{17}H_{16}]^-$ anion (schematic **33** above) [66]. This is a dehydrogenative oxidation, and can be considered in simple mechanistic terms as the closing of a sulfur-to-boron and a boron-to-boron vector at the point of fusion of the two subclusters (arrows in

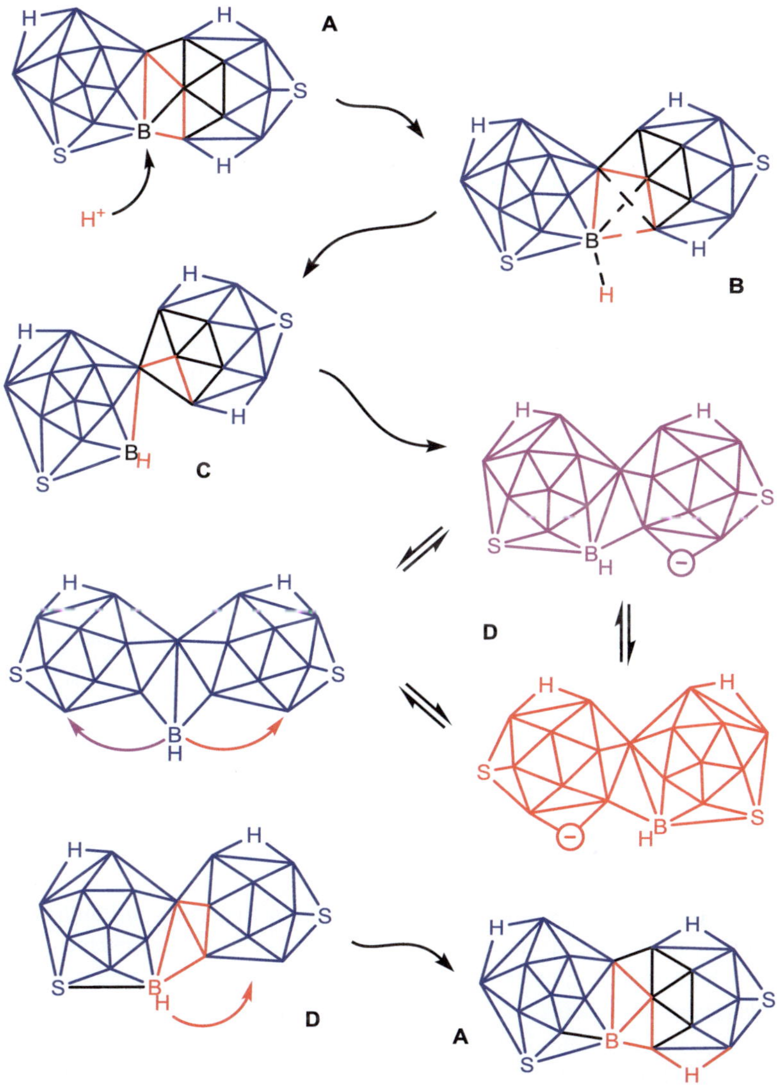

Fig. 6.5 The $[S_2B_{18}H_{19}]^-$ anion **A** has an unusual *arachno*-type 11-vertex $\{SB_{10}\}$ subcluster that has an open hexagonal pyramidal $\{B_7\}$ structural feature. This is conjoined, with two boron atoms in common, to a second $\{SB_{10}\}$ subcluster of conventional *nido* 11-vertex geometry. Protonation of $[S_2B_{18}H_{19}]^-$ forms neutral $[S_2B_{18}H_{20}]$ **C** *via* a diamond-square-diamond process **B** that dismantles the $\{B_7\}$ pyramidal feature. Deprotonation of **C** yields the fluxional $[S_2B_{18}H_{19}]^-$ anion **D**, which is isomeric with anion **A**. Anion **D**, as well as neutral **C**, does not have the $\{B_7\}$ hexagonal pyramidal feature. Neutral **C** consists of conventional *nido* 11-vertex $\{SB_{10}\}$ and *arachno* ten-vertex $\{SB_9\}$ subclusters conjoined with a single *spiro* boron atom in common. Anion **D** is closely related to neutral **C**, but with an additional inter-boron intercluster link. Anion **D** spontaneously reverts to anion **A** over a few hours at room temperature, remarkable in that the open $\{B_7\}$ hexagonal pyramid is regenerated. The fluxionality of **D** involves the inter-subcluster transfer of a $\{BH\}$ unit. The reassembly of **A** from **D** involves two successive hydrogen-atom hops and a reverse of the initial **A** → **C** diamond-square-diamond rearrangement (After Ref. [68])

schematic **72**). This closure engenders a substantive geometric change of some 33° in the relative orientation of the two subclusters (compare hatched lines in schematics **72** and **73**). This large nanomechanical change in mutual orientation induced by a facile two-electron oxidation is held to have interesting implications for molecular device development [66], for example if the switching subclusters can be associated with borane-based rod-like molecules. Here it can be noted that redox-flexible transition-element centres can be positioned at points of intercluster conjunction, for example as in the structurally related platinaborane $[(PMe_2Ph)PtB_{16}H_{18}(PMe_2Ph)]$ (structure **65** above) [108], and that related transition-element centres in borane clusters can be tailored to undergo reversible geometric changes associated with very small free-energy differences between metal oxidation states, for example as in the geometrically *closo*-type $\{PtC_2B_8\}$ system [135]. In this general context, the converse phenomenon of a reductive decrease in the intimacy of macropolyhedral intersubcluster bonding upon the conversion of $[(C_5Me_5)Ir$-*syn*-$B_{18}H_{20}]$ to give the anion $[(C_5Me_5)IrB_{18}H_{19}S]^-$ is also noteworthy [126], as are the successive one-electron *closo* → *isonido* and *isonido* → *closo* switch-on/switch-off processes in the $\{MoC_2B_9\}$12-vertex system: $2[(SPh)_2MoC_2B_9H_{11}]^{2-} \rightarrow [H_{11}B_9C_2Mo(SPh)_4MoC_2B_9H_{11}]^{2-} \rightarrow [H_{11}B_9C_2Mo(SPh)_4MoC_2B_9H_{11}]^-$ [136].

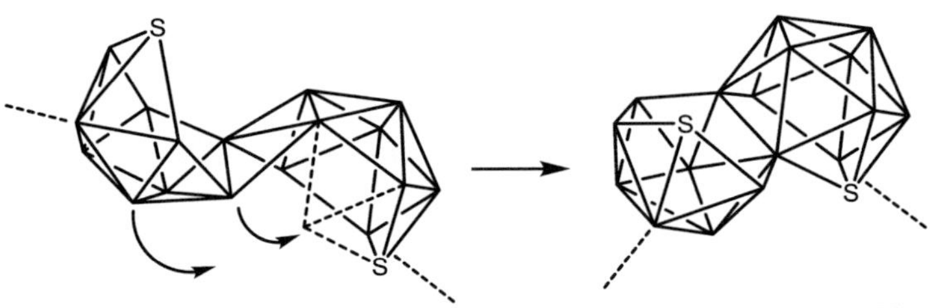

72 **73**

Another important aspect of rearrangement chemistry in boron-containing cluster chemistry is that of rearrangements within single clusters themselves, a general topic in which several cluster types have attracted theoretical analyses since the fascinating rearrangements among the $C_2B_{10}H_{12}$ isomers were first examined some 50 years ago by Lipscomb [137]. in a process that itself has continually attracted theoretical attention over the intervening years [138]. However, few if any theoretical rationalizations of intra-subcluster rearrangement behaviour in macropolyhedral chemistry have so far been reported, although it is a clear when comparing reaction products with starting materials – and there are many instances given in this chapter – that considerable intra-subcluster rearrangement can occur, and so there is good future potential for theoretical and calculational mechanistic elucidation. More challenging is the examination of bimolecular reaction chemistry (see Chap. 9 in this Volume), and there is clear future scope here also for useful calculational approaches in the macropolyhedral area, particularly allied with any cluster rearrangements which must often occur as a consequence of site of attack. If megaloborane chemistry develops, and interactions of smaller molecules with crevices or extended surfaces becomes important, there will be clear scope for the theoretical examination of the interactions of the smaller molecules with the various topographical features of the surfaces. Perhaps more subtle are calculational attempts to define what factors may engender changes associated with the concepts of cluster flexibility and fragility discussed above (near Fig. 6.3), but at the time of writing these have had limited application even in single-cluster chemistry [21, 22], and such potential usage in big-borane chemistry is yet to be realised.

An important currently increasing use of the techniques of DFT in boron-containing cluster chemistry is in a predictive capacity, and in a capacity to analyse behavior, rather than in the simpler basic calculations of ground-state molecular structure. Calculation of viability of reaction chemistry, or of potential product stability, and thence calculations on structural and substituent variants to determine relative stabilities, can be a vast help in saving tedious experimental bench time in that a most likely outcome can then be specifically targeted in synthetic work without wasting time on unsuitable or unviable syntheses. An example here, albeit in single-cluster work, deals with the sequestration of small molecules by the metal-metal vector of a 12-vertex $\{M_2B_{10}H_{10}\}$ matrix [139]. Here, the modifications required to engender reversible binding of SO_2 on the metal-metal vector after non-reversible binding was discovered [140] were assessed by DFT work, and a resulting suitable candidate $[(PMe_2Ph)_2PtPd(phen)B_{10}H_{10}]$ was duly synthesized and duly exhibited reversible SO_2 binding [141]. This type of approach is not yet applied to macropolyhedral work, though is of potential usage when, for example, the megaloborane area becomes sufficiently advanced to build and exploit cluster topographical features such as crevices and extended surfaces which may usefully bind or otherwise interact with smaller reactive molecules or biologically significant entities.

More sophisticated analytical and predictive applications to macropolyhedral boron-containing cluster chemistry are emerging at the time of writing. This is exemplified by the analysis of the detailed ground- and excited-state characteristics of the *syn* and *anti* $B_{18}H_{22}$ isomers [142], and thence the elucidation of the factors that dictate that the *anti* isomer exhibits fluorescence whereas the *syn* does not. Calculational elucidation of the very rapid and therefore experimentally unmeasurable geometrical changes within the exceptionally short excited-state lifetimes thence give clues as to how future syntheses may be tailored, for example by substituent chemistry [143], to enhance or modify the photophysical properties in advantageous directions. This is of relevance to the fluorescence properties of $B_{18}H_{22}$ which has elegantly been shown to have lasing ability in the blue region [144]. This work is dealt with more thoroughly in Chap. 3 of this Volume. Extension of the approach to the brick-red fluorescence of $B_{18}H_{20}(NC_5H_5)_2$ (schematics **20 A** and **20 B** above) suggests appropriate advantageous modifications by the positing and tailoring of the substituents on the pyridine moieties, and also suggests a significant influence of rotamer orientation about the B-N linkages [56]. The point about criticality of rotamer behaviour [132] has been made above.

Above near Fig. 6.4 the possibility of future structures based on extensions or adaptations of known structural features is discussed, specifically the postulation of stable filled globular boron hydrides based on fragments of elemental boron matrices [81, 82, 84]. An alternative approach is to postulate structures based on the extension of known polyhedral borane cluster characteristics for the prediction of new architectural constructs. In this regard, examples of seven types of possible polymeric structures were discussed by Lipscomb some 35 years ago [145]. More recently, a consideration of the two-borons-in-common fusion mode of the two known $B_{18}H_{22}$ isomers leads to the recognition that other isomers are in principle possible [146]; two of these are designated as *cisoid-anti* and *cisoid-syn* (schematics **74** and **75**), as distinct from a *transoid-anti* and *transoid-syn* nomenclature for the two known isomers (schematics **16** and **17** above). Although calculated to be of some 200 or more kJ mol^{-1} to higher energy than the two known isomers, nevertheless it is held that they may well be kinetically stable and thence inherently isolatable. An extension of the two-*nido*-decaboranes two-atoms-in-common fusion mode would lead to more extended oligomers of general formulation $B_{10+8n}H_{14+8n}$, but as far as the author is aware these have not specifically been examined computationally. The next in the series would be $B_{28}H_{30}$; the only compound experimentally established that approximates to this type of triple-cluster fusion is the platinaborane [$(PMe_2Ph)PtB_{26}H_{24}(PMe_2Ph)$] mentioned above (schematic **51**) [84, 87]. On the other hand, an alternative factorisation of the $B_{18}H_{22}$ skeletal structures, and indeed the basic $B_{10}H_{14}$ structure, into fused pentagonal pyramids based on *nido*-B_6H_{10} ultimately leads to the postulation of a macrocyclic $B_{40}H_{40}$ species of D_{5d} symmetry (schematic **76**) in which the axial apices of the pentagonal pyramids point alternately above and below the toroidal plane. Extension of this idea by capping the open faces leads to a $B_{50}H_{30}$ unit of D_{10h} symmetry, based on a ring of ten *closo-type* pentagonal bipyramidal {B_7H_7} units [146].

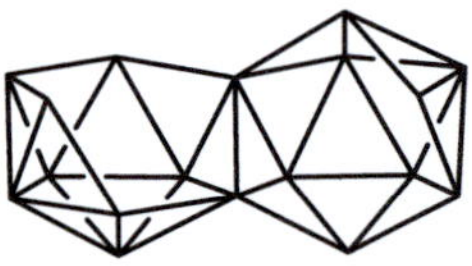

74

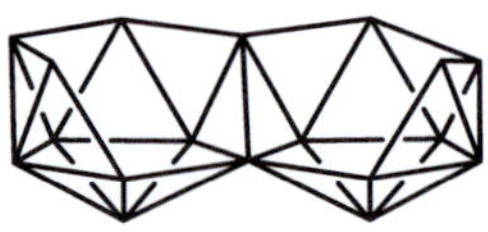

75

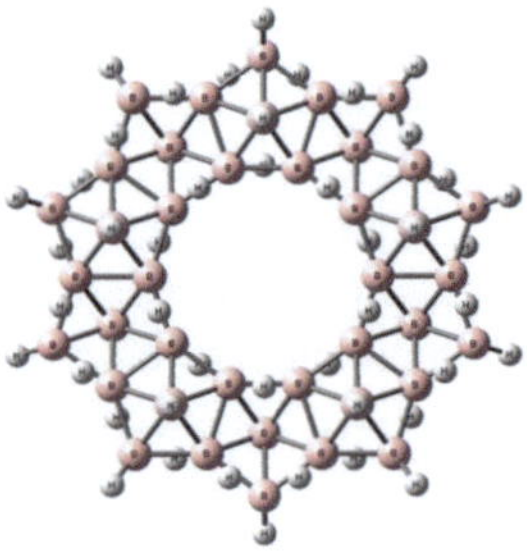

76

A related extension of the four-borons-in-common fusion mode exhibited by the globular borane $B_{20}H_{16}$ (see Chap. 8 of this Volume) leads to the speculation of an extended series of oligomers of general formulation {$B_{(8n+4)}H_{(4n+8)}$} (*e.g.* diagrams **77–80**) [147, 148]. The first member of the series, *i.e.* for $n=0$, would be of atomic constitution {$B_{12}H_{12}$}, but the known incidence and high stability of the [*closo*-$B_{12}H_{12}$]$^{2-}$ dianion together with the neutrality of known stable $B_{20}H_{16}$ leads to the idea that the most stable series would have a progressive change in charge, *i.e.* of general formulation $[B_{(8n+4)}H_{(4n+8)}]^{(2n-2)+}$. This series is subject to ongoing investigation [147]. Due to the difficulty of acquisition, either by preparative routes [149–151], or by isolation from its trace incidence [147] in crude $B_{10}H_{14}$ [151, 152], little experimental chemistry has been carried out on four-borons-in-common $B_{20}H_{16}$ [76, 151, 153, 154]. Other aspects of this type of conjunction, and aspects of three-borons-in-common fusion, are considered in Chap. 4 of this Volume. Three-borons-in-common fusion is manifested experimentally in some of the metallaboranes mentioned above in this chapter, in the long-recognized bis(ligand) adducts of $B_{20}H_{16}$ [76, 151], and also in the pair of interesting $[B_{20}H_{18}]^{2-}$ and $[B_{21}H_{18}]^{-}$ anions; [74, 75] these latter represent and presage yet another extended *generis* of basic binary borane structure, *i.e.* the three-atoms-in-common fusion mode which is another potentially repeating motif for speculative calculation of oligomers and polymers.

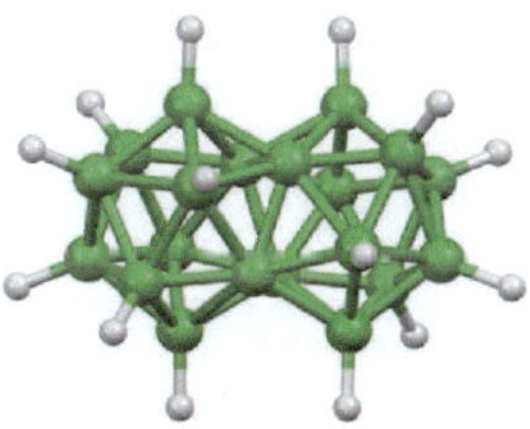

77

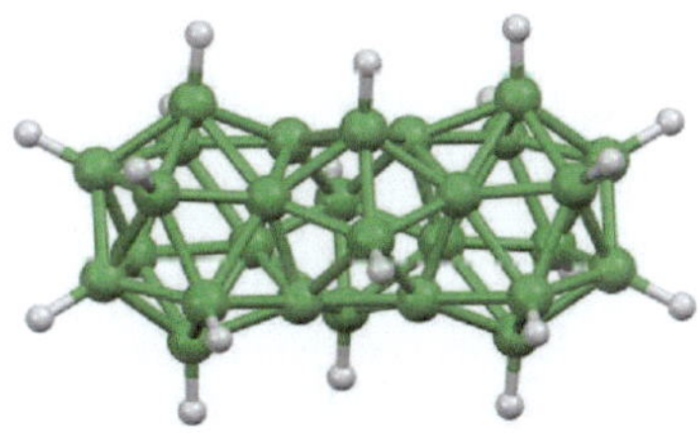

78

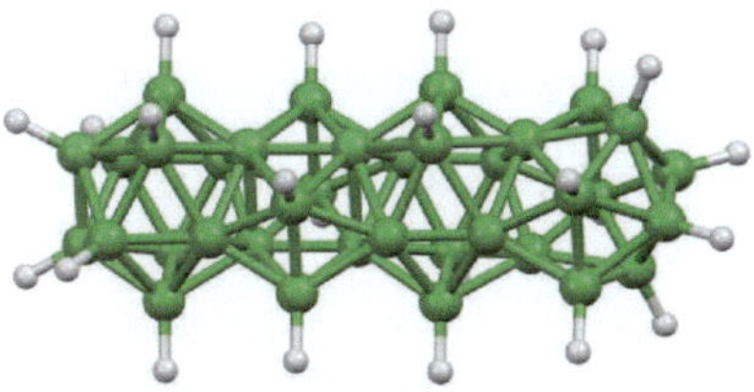

79

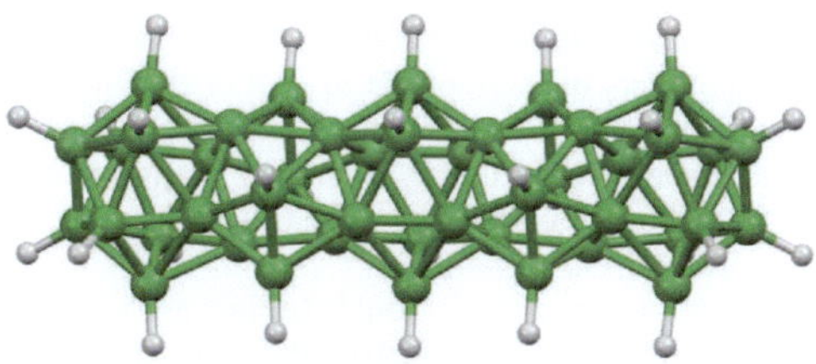

80

References

1. Beckett MA, Crook JE, Greenwood NN, Kennedy JD, McDonald WS (1982) Molecular structure of the seventeen-vertex conjuncto platinaheptadecaborane $[(PMe_2Ph)PtB_{16}H_{18}(PMe_2Ph)]$. J Chem Soc Chem Commun 1982:552–553
2. Beckett MA, Crook JE, Greenwood NN, Kennedy JD (1983) A macropolyhedral polymetallaborane cluster: molecular structure of the seventeen-vertex triplatinaheptadecaborane $[(PPhMe_2)_4Pt_3B_{14}H_{16}]$. J Chem Soc Chem Commun 1983:1228–1230
3. Williams RE (1971) Carboranes and boranes: polyhedra and polyhedral fragments. Inorg Chem 18:210–214
4. Williams RE (1976) Coordination number pattern recognition theory of carborane structures. Adv Inorg Chem Radiochem 18:67–142

5. Wade K (1971) The structural significance of the number of skeletal bonding electron pairs in carboranes, the higher boranes and borane anions, and various transition-metal carbonyl cluster compounds. Chem Commun 1971:792–793
6. Wade K (1976) Structural and bonding patterns in cluster chemistry. Adv Inorg Chem Radiochem 10:1–66
7. Mingos DMP (1972) A general theory for cluster and ring compounds of the main group and transition elements. Nat Phys Sci 236:99–102
8. O'Neill ME, Wade K (1982) Chapter 1: Structural and bonding features in metallaboranes and metallacarboranes. In: Grimes RN (ed) Metal interactions with boron clusters. Plenum, New York, pp 1–41. ISBN 0-306-40933-X
9. Mingos DMP (1984) Polyhedral skeletal electron-pair approach. Acc Chem Res 17:311–319
10. Kennedy JD (1998) Chapter 3: Disobedient Skeletons. In: Casanova J (ed) The borane carborane carbocation continuum. Wiley, New York/Chichester, pp 85–116. ISBN 0-471-18075-0
11. Enrione RE, Boer FB, Lipscomb WN (1964) Preparation, isolation, and structure of B_8H_{12}. J Am Chem Soc 86:1451–1452
12. Enrione RE, Boer FB, Lipscomb WN (1964) Octaborane(12). Inorg Chem 3:1659–1666
13. See, for example: Guggenberger LJ (1969) Crystal structure of tetraamminezinc octahydrooctaborate(−2). $Zn(NH_3)_4B_8H_8$. Inorg Chem 8: 2771–2774; and references therein
14. Jelínek T, Štíbr B, Kennedy JD, Thornton-Pett M (1995) Eight-vertex polyhedral monocarbaborane chemistry. Three closo anions, $[CB_7H_8]^-$, $[CB_7H_7I]^-$, $[CB_7H_6I_2]^-$. Preparation and structural studies. J Chem Soc Dalton Trans 1995:431–437
15. Wiersema RE, Hawthorne MF (1973) Electrochemistry and boron-11 nuclear magnetic resonance spectra of monocarbon carboranes. Inorg Chem 12:785–788
16. Tolpin EI, Lipscomb WN (1973) Fluxional behavior of undecahydroundecaborate(2-) ($B_{11}H_{11}^{2-}$). J Am Chem Soc 95:2384–2386
17. Volkov O, Paetzold P (2003) The chemistry of the undecaborates. J Organomet Chem 680:301–311
18. Štíbr B, Kennedy JD, Drdáková E, Thornton-Pett M (1994) Nine-vertex polyhedral iridamonocarbaborane chemistry. Single-crystal X-ray diffraction analysis and NMR studies of the products of thermolysis of $[(CO)(PPh_3)_2IrCB_7H_8]$. Emerging alternative systematic cluster-geometry patterns. J Chem Soc Dalton Trans 1994:229–236
19. Bould J, Kennedy JD, Thornton-Pett M (1992) Ten-vertex metallaborane chemistry. Aspects of the iridadecaborane closo-isonido-isocloso structural continuum. J Chem Soc Dalton Trans 1992:563–576
20. See, for example: Bould J, Harrington RW, Clegg W, Kennedy JD (2012) Nine-vertex metallaborane chemistry. Preparation and characterisation of $[1,1,1\text{-}(PMe_3)_2H\text{-isocloso-}IrB_8H_7\text{-}8\text{-}X]$, where X=Cl or H. J Organomet Chem (TP Fehlner 75th Birthday Edition) 721/722:155–163
21. Bould J, Kennedy JD (2014) An Assessment of the use of DFT calculations to investigate the intercarbon stretching phenomenon in C-substituted 'pseudocloso'-3,1,2-MC_2B_9 metalladicarbaboranes. J Organomet Chem 749:163–173
22. Kennedy RD, Kennedy JD (2015) The contrarotational fluxionality of $[3,3\text{-}(PMe_2Ph)_2\text{-closo-}3,1,2\text{-}PtC_2B_9H_{11}]$ and related species. Dalton Trans 44:9620–9629
23. Fehlner TP (1999) Connections Between ^{11}B NMR Chemical Shifts and Electronic Structure in Metallaboranes. A Précis. Collect Czech Chem Commun 64:767–782
24. Londesborough MGS (2002) New developments in macropolyhedral metallaborane chemistry. Thesis, University of Leeds, Leeds
25. Volkov O, Rath NP, Barton L (2003) Derivatization of the macropolyhedral cluster $[B_{22}H_{22}]^{2-}$: Isolation and characterization of the 5′-OEt- and 4'-OH derivatives. J Organomet Chem 608:212–217

26. See, for example: Jemmis ED, Balakrishnarajan MM, Pancharatna PD (2002) Electronic requirements for macropolyhedral boranes. Chem Rev 102:93–144
27. Rathke J, Schaeffer R (1977) New boron hydride, pentadecaborane(23). J Am Chem Soc 95:3402–3402
28. Greenwood NN, McGinnety JA, Owen JD (1972) Crystal structure of bis [bis(diethyl ether)-μ-(dodecahydro-*nido*-decaborato)-cadmium], $[(Et_2O)_2Cd(B_{10}H_{12})]_2$. J Chem Soc Dalton Trans 1972:989–992
29. Parker KG, Russell JM, Sabat M, Grimes RN (1999) A novel heterooctametallic metallacarborane tetramer. Collect Czech Chem Commun 64:819–828
30. Gaines DF, Walsh JL (1978) Chemistry of 2-berylla-nido-hexaborane(11) compounds. Insertion of beryllium into a borane cage. Inorg Chem 17:1238–1241
31. Gaines DF, Walsh JL, Calabrese JC (1978) Low-temperature crystal and molecular structures of 2-tetrahydroborato-2-berylla-nido-hexaborane(11) and 2,2′-commo-bis [2-berylla-nido-hexaborane(11)]. Inorg Chem 17:1242–1248
32. Gaines DF (1980) Recent advances in the chemistry of pentaborane(9). In: Parry RW, Kodama G (eds) Boron Chemistry-4. Pergamon, Oxford/New York, pp 73–79. ISBN 0-08-025256-7
33. Bicerano J, Lipscomb WN (1979) Molecular orbital studies of nido-beryllaboranes, $B_5H_{10}BeX$, where X is borane(4), pentaborane(10), methyl, or cyclopentadiene. Inorg Chem 18:1565–1571
34. Hawthorne MF, Young DC, Wegner PA (1965) Carbametallic boron hydride derivatives. I. Apparent analogs of ferrocene and ferricinium Ion. J Am Chem Soc 87:1818–1819
35. Hawthorne MF, Pilling PL (1965) Carbametallic boron hydride derivatives. III. The π-$C_5H_5Fe(\pi$-$B_9C_2H_{11})$ system. J Am Chem Soc 87:3987–3988
36. Hawthorne MF, Andrews TD (1965) Carborane analogues of cobalticinium ion. Chem Commun 1965:443–444
37. Pitochelli AR, Hawthorne MF (1962) The Preparation of a New Boron Hydride $B_{18}H_{22}$. J Am Chem Soc 82:3218–3218
38. Heřmánek S, Fetter K, Plešek J (1972) A new stable borane $B_{14}H_{18}$. Chem Ind (London) 1972:606–606
39. Heřmánek S, Fetter K, Plešek J, Todd LJ, Garber AR (1975) Tetradecaborane(18). Preparation and structure. Inorg Chem 14:2250–2253
40. Plešek J, Heřmánek S, Hanousek F (1968) Chemistry of Boranes X. Novel stable borane $B_{16}H_{20}$. Collect Czech Chem Commun 33:699–705
41. Rathke J, Moody DC, Schaeffer R (1974) Boranes. XLI. New boron hydride, tridecaborane(19). Inorg Chem 13:3040–3042
42. Rathke J, Schaeffer R (1974) Boranes XXXVIII. Reactions of hexaborane(10) with boron hydride Lewis acids. Inorg Chem 13:3008–3011
43. Dolan PJ, Moody DC, Schaeffer R (1981) Studies of boranes. 48. Reactions of hexaborane(10) with Lewis acids to yield acid–base complexes and synthesis of halogenated hexaborane(10) derivatives *via* halogen-transfer reactions. Inorg Chem 20:745–748
44. Dixon DA, Kleier DA, Halgren TA, Lipscomb WN (1974) Localized orbitals in large boron hydrides. Hexadecaborane and related molecules. J Am Chem Soc 96:2293–2295
45. Bould J, Jelínek T, Barrett SA, Coles SJ, Hursthouse MB, Thornton-Pett M, Štíbr B, Kennedy JD (2005) Macropolyhedral boron-containing cluster chemistry. The reaction of $B_{16}H_{20}$ and $B_{14}H_{18}$ with $[PtMe_2(PMe_2Ph)_2]$ to give $[(PMe_2Ph)_2PtB_{16}H_{17}Me]$ and $[(PMe_2Ph)_2PtB_{14}H_{16}]$. Dalton Trans 2005:1499–1503
46. Carr MJ, Perera SD, Jelínek T, Kilner CA, Štíbr B, Kennedy JD (2005) Macropolyhedral boron-containing cluster chemistry. An unusual 'neo-nido' ten-vertex subcluster configuration in a $[(PPh_3)_2RuB_{16}H_{20}]$ species. J Organomet Chem (EUROBORON 3 Special Edition) 690:2857–2859
47. Carr MJ, Perera SD, Jelínek T, Štíbr B, Clegg W, Kilner CA, Kennedy JD (2007) Macropolyhedral boron-containing cluster chemistry. The unique nido-five-vertex-$\langle B_2\rangle$-

nido-ten-vertex conjuncto structure of [(η^5-C_5Me_5)$_2Rh_2B_{11}H_{15}$] *via* an unexpected cluster-dismantling. Chem Commun 2007:3559–3561
48. Bould J, Dörfler U, Thornton-Pett M, Kennedy JD (2001) A rearrangement of the ten-boron nido/arachno decaboranyl cluster. Inorg Chem Commun 4:544–546
49. Jelínek T, Grüner B, Císařová I, Štíbr B, Kennedy JD (2007) Macropolyhedral boron-containing cluster chemistry. The reaction of syn-$B_{18}H_{22}$ with SMe_2 and I_2 in monoglyme. Structure of [7-(SMe_2)-syn-$B_{18}H_{20}$]. Inorg Chem Commun 10:125–128
50. Olsen FP, Vasavada RC, Hawthorne MF (1968) The chemistry of n-$B_{18}H_{22}$ and i-$B_{18}H_{22}$. J Am Chem Soc 90:3946–3951
51. Heřmánek S, Plotová H (1970) Chemistry of Boranes. XXII. The acidity of boranes. Collect Czech Chem Commun 36:1639–1643
52. Heřmánek S, Plotová H, Plešek J (1975) On the acidity characteristics of decaborane(14) and its benzyl derivatives in organic solvent-water systems. Collect Czech Chem Commun 40:3593–3601
53. Stanko VI, Chapovskii YA, Brattsev VA, Zakharkin LI (1965) The chemistry of decaborane and its derivatives. Russ Chem Rev 34:424–439
54. Sneath RL, Todd LJ (1973) Transition metal and carborane derivatives obtained from octadecaborane(22). Inorg Chem 12:44–48
55. Carr MJ, Perera SD, Hamilton Mcleod AR, Londesborough MGS, Jelínek T, Štíbr B, Kilner CA, Thornton-Pett M, Kennedy JD (2010) Aspects of macropolyhedral boron-containing cluster chemistry – August 2010. Abstracts Fifth European Meeting on Boron Chemistry (EUROBORON 5). Edinburgh, Scotland, 29 August – 02 September 2010, Abstract no. O25, p 46
56. Londesborough MGS, Dolanský J, Cerdán L, Lang K, Jelínek T, Garcia-Moreno I, Oliva J, Hnyk D, Roca-Sanjuán D, Nikl M, Kennedy JD (2015) Products of the reaction between anti-$B_{18}H_{22}$ and pyridine. Work in preparation for publication
57. Huffman JC, Moody DC, Schaeffer R (1975) New boron hydride. Tetradecaborane(20). J Am Chem Soc 97:1621–1622
58. Huffman JC, Moody DC, Schaeffer R (1981) Studies of boranes. 47. Synthesis and x-ray crystallographic study of tetradecaborane(20). Inorg Chem 20:741–745
59. Brewer CT, Grimes RN (1984) Metal-induced oxidative fusion of boranes. Synthesis of dodecaborane(16), the first neutral dodecaborane. J Am Chem Soc 106:2722–2723
60. Jelínek T, Kennedy JD, Štíbr B (1994) Macropolyhedral boron-containing cluster chemistry. Cluster fusion to give the novel nido-nido eighteen-vertex dithiaoctadecaborane (anti)-[9,9′-$S_2B_{16}H_{16}$]. J Chem Soc Chem Commun 1994:1415–1416
61. Dosangh PK, Bould J, Londesborough MGS, Jelínek T, Thornton-Pett M, Štíbr B, Kennedy JD (2003) Macropolyhedral boron-containing cluster chemistry. Aspects of the $S_2B_{16}H_{16}$ system. Preparation, structure, NMR spectroscopy and isomerism. J Organomet Chem 680:312–322
62. Kaur P, Thornton-Pett M, Clegg W, Kennedy JD (1996) Macropolyhedral boron-containing cluster chemistry. An interesting nineteen-vertex dithianickelaborane, [(PPh_3)$NiS_2B_{16}H_{12}$(PPh_3)]. J. Chem Soc Dalton Trans 1996:4155–4157
63. Carr MJ, Clegg W, Kilner CA, Londesborough MGS, Kennedy JD (2010) Macropolyhedral boron-containing cluster chemistry. [$S_2B_{16}H_{17}$]$^-$ - a new eighteen-vertex thiaborane anion. Collect Czech Chem Commun (B Štíbr 70th Birthday Edition) 75:807–812
64. Jelínek T, Kilner C, Thornton-Pett M, Kennedy JD (1999) Macropolyhedral boron-containing cluster chemistry. The [$SB_{17}H_{19}$]$^-$ anion: a nido-ten-vertex : arachno-ten-vertex cluster architecture and the first single-sulphur macropolyhedral thiaborane. Chem Commun 1999:1905–1906
65. Kaur P, Holub J, Rath NP, Bould J, Barton L, Štíbr B, Kennedy JD (1996) Macropolyhedral boron-containing cluster chemistry. Nineteen-vertex $S_2B_{17}H_{17}$(SMe_2). An unusual apical

boron atom of cluster connectivity six that introduces a new polyhedral borane building block. Chem Commun 1996:273–275
66. Jelínek T, Kennedy JD, Štíbr B, Thornton-Pett M (1998) Macropolyhedral boron-containing cluster chemistry. An interesting angular change in mutual subcluster orientation in the oxidative protonation of $[S_2B_{17}H_{18}]^-$ to give $[S_2B_{17}H_{16}]^-$. Inorg Chem Commun 1:179–181
67. Jelínek T, Kennedy JD, Štíbr B, Thornton-Pett M (1994) Macropolyhedral boron-containing cluster chemistry. The isolation and characterisation of the first macropolyhedral thiaborane $[9,9'\text{-}S_2B_{17}H_{18}]^-$. Angew Chem Int Edn 33:1599–1601
68. Ormsby DL, Greatrex R, Kennedy JD (2008) Macropolyhedral boron-containing cluster chemistry. The $\{S_2B_{18}\}$ system. The reversible disassembly and reassembly of the hexagonal pyramidal $\{B_7\}$ feature in the $[S_2B_{18}H_{19}]^-$ anion. An establishment of molecular structures, intermediates and transition states by the DFT-structure/GIAO-NMR method. Dalton Trans (K Wade Special Edition) ('Hot Paper') 2008:1625–1634
69. Jelínek T, Cisařová I, Štíbr B, Kennedy JD, Thornton-Pett M (1998) Macropolyhedral boron-containing cluster chemistry. The $[S_2B_{18}H_{19}]^-$ anion, and the reversible generation of an apical boron cluster site with cluster connectivity six. J Chem Soc Dalton Trans (Dalton Commun) 1998:2965–2967
70. Kennedy JD (1986) The polyhedral metallaboranes, part II. Prog Inorg Chem 34:211–434. ISBN 0-471-81948-4
71. Cheek YM, Greenwood NN, Kennedy JD, McDonald WS (1982) New modes of bonding in some platinum derivatives of $B_{18}H_{22}$: X-ray structures of $[(PMe_2Ph)_4Pt_2\text{-}\eta^1,\eta^2\text{-}B_{18}H_{16}]$ and of three structural isomers of $[(PMe_2Ph)_2PtB_{18}H_{20}]$. J Chem Soc Chem Commun 1982:80–81
72. Fontaine XLR, Greenwood NN, Kennedy JD, MacKinnon PI, Thornton-Pett M (1986) Preparation of $[(C_5Me_5)_2Rh_2B_{17}H_{18}]$ *via* a degradative insertion from anti-$B_{18}H_{22}$, and a mechanism for anti→syn macropolyhedral interconversion. J Chem Soc. Chem Commun 1986:1111–1113
73. Shea SL, Bould J, Londesborough MGS, Perera SD, Franken A, Ormsby DL, Jelínek T, Štíbr B, Holub J, Kilner CA, Thornton-Pett M, Kennedy JD (2003) Polyhedral boron-containing cluster chemistry. Aspects of architecture beyond the icosahedron: some recent supermolecular and supramolecular developments. Pure Appl Chem 75:1239–1248
74. Bernhardt E, Brauer DJ, Finze M, Willner H (2007) *closo*-$[B_{21}H_{18}]^-$: a face-fused diicosahedral borate ion. Angew Chem Int Ed 46:2927–2930
75. Schlüter F, Bernhardt E (2012) Fluorierung von *closo,closo*-$[B_{21}H_{18}]^-$ mit *a*HF und F_2 zu *closo,closo*-$[B_{21}H_{18-x}F_x]^-$ ($x=1$–3) und *closo,closo*-$[B_{21}F_{18}]^-$. Z Anorg Allg Chem 638:594–561
76. Enemark JH, Friedman LB, Lipscomb WN (1966) The molecular and crystal structure of $B_{20}H_{16}(NCCH_3)_2 \cdot CH_3CN$. Inorg Chem 5:2165–2172
77. Shea SL, Jelínek T, Štíbr B, Thornton-Pett M, Kennedy JD (2000) Macropolyhedral boron-containing cluster chemistry. Isolation and structure of the twenty-one-vertex globular cluster compound $[(\eta^5\text{-}C_5Me_5)_3Ir_3B_{18}H_{15}(OH)]$. Inorg Chem Commun 3:169–172
78. Kennedy JD (1997) Macropolyhedral boron-containing cluster chemistry. Overview and recent developments. July 1996. In: Siebert W (ed) Advances in Boron Chemistry. Royal Society of Chemistry, Cambridge, pp 451–462. ISBN 0-08504-722-0
79. McGrath TD, Kennedy JD, McInnes YM, Thornton-Pett M (1997) Macropolyhedral platinaboranes. In: Siebert W (ed) Advances in Boron Chemistry. Royal Society of Chemistry, Cambridge, pp 480–483. ISBN 0-08504-722-0
80. Jelínek T, Štíbr B, Kennedy JD, Thornton-Pett M (1997) Excursions into the country of fused-cluster heteroboranes. In: Siebert W (ed) Advances in Boron Chemistry. Royal Society of Chemistry, Cambridge, pp 426–429
81. Yao H-J, Hu C-H, Sun J, Jin R-S, Zheng P-J, Bould J, Greatrex R, Kennedy JD, Ormsby DL, Thornton-Pett M (1999) Isolation and structure of $[(PPh_3)_3(PPh_2)_2Pd_4B_{20}H_{16}]$. A possible prognostic for new globular borane-based cluster architectures. Collect Czech Chem Commun 64:927–937

82. Bould J, Ormsby DL, Yao H-J, Hu C-H, Sun J, Jin R-S, Shea SL, Clegg W, Jelínek T, Rath NP, Thornton-Pett M, Greatrex R, Zheng P-J, Barton L, Štíbr B, Kennedy JD (2000) Macropolyhedral boron-containing cluster chemistry. Further Progress beyond the Icosahedron. July 1999. In: Davidson M, Hughes AK, Marder TB, Wade K (eds) Contemporary boron chemistry. Royal Society of Chemistry, Cambridge, UK, pp 171–174. ISBN 0-85404-835-09
83. Bould J, Kennedy JD, Barton L, Rath NP (1997) Macropolyhedral boron-containing cluster chemistry. Triple cluster fusion and the molecular structure of $[(PMe_3)_2IrB_{26}H_{24}Ir(CO)(PMe_3)_2]$. A 28-vertex metallaborane cluster with a polyboron core. J Chem Soc Chem Commun 1997:2405–2407
84. Bould J, Clegg W, Teat SJ, Barton L, Rath NP, Thornton-Pett M, Kennedy JD (1999) An approach to megaloboranes. Mixed and multiple cluster fusions involving iridaborane and platinaborane cluster compounds. Crystal structure determinations by conventional and synchrotron methods. In: Boron Chemistry at the Millennium, Special Edition of Inorg Chim Acta 289:95–124
85. Mingos DMP, Forsyth ML, Welch AJ (1977) *X*-Ray crystallographic and theoretical studies on 'slipped' metallacarboranes. J Chem Soc Chem Commun 1977:605–607
86. Mingos DMP, Forsyth ML, Welch AJ (1978) Molecular and crystal structure of 3,3-bis(triethylphosphine)-1,2-di-carba-3-platinadodecaborane(11), and molecular-orbital analysis of the 'slip' distortion in carbametallaboranes. J Chem Soc Dalton Trans 1978:1363–1374
87. Bould J, Barrett SA, Barton L, Rath NP, Kennedy JD (1998) Macropolyhedral boron-containing cluster chemistry. Isolation and characterisation of the 27-vertex contiguous triple-cluster species $[(PMe_2Ph)_2PtB_{26}H_{26}(PMe_2Ph)]$. Inorg Chem Commun 1:365–367
88. Bullen NJ (2009) New aspects of monocarbaborane chemistry. Thesis, University of Leeds, Leeds
89. Lipscomb WN, Massa L (1992) Examples of large closo boron hydride analogs of carbon fullerenes. Inorg Chem 31:2297–2299
90. See, for example: Boustani I, Rubio A, Alonso JA (2000) *Ab initio* study of boron hydride spheres. In: Davidson M, Hughes AK, Marder TB, Wade K (eds) Contemporary boron chemistry, Royal Society of Chemistry, Cambridge, pp 493–496. ISBN 0-85404-835-09
91. Bould J, Londesborough MGS, Ormsby DL, MacBride JAH, Wade K, Kilner CA, Clegg W, Teat SJ, Thornton-Pett M, Greatrex R, Kennedy JD (2002) Macropolyhedral boron-containing cluster chemistry. Models for intermediates en route to globular and discoidal megaloborane assemblies. Structures of $[nido\text{-}B_{10}H_{12}(nido\text{-}B_5H_8)_2]$ and $[(CH_2CH_2C_5H_4N)\text{-}arachno\text{-}B_{10}H_{10}(NC_5H_4\text{-}closo\text{-}C_2B_{10}H_{11})]$ as determined by synchrotron X-ray diffraction analysis. J Organomet Chem 657:256–261
92. Jemmis ED, Balakrishnarajam MM (2001) Polyhedral boranes and elemntal boron: direct structural relations and diverse electronic requirements. J Am Chem Soc 123:4324–4330
93. Prasad DLVK, Jemmis ED (2008) Stuffing improves the stability of fullerenelike boron clusters. Phys Rev Lett 100:1665504–1665504
94. See, for example: Lv J, Wang Y, Zhu L, Ma Y (2014) B_{38}: an all-boron fullerene analogue. Nanoscale 6:11692–11696
95. Huang W, Sergeeva AP, Zhai H-J, Averkiev BB, Wang L-S, Boldyrev AI (2010) A concentric planar doubly π-aromatic B_{19}^- cluster. Nat Chem 2:202–206
96. See, for example: Buckyball boron, in Chemistry World, 2014 (August), page 6
97. Bould J, Dörfler U, Clegg W, Teat SJ, Thornton-Pett M, Kennedy JD (2001) Triple linking of the decaboranyl cluster. Structure of $[(SMe_2)_2B_{10}H_{10}(B_{10}H_{13})_2]$ as determined by synchrotron X-ray diffraction analysis. J Chem Soc Chem Commun 2001:1788–1789
98. Huffman JC, Moody DC, Schaeffer R (1976) Boranes. XLV. Crystal and molecular structure, improved synthesis, and reactions of tridecaborane(19). Inorg Chem 15:227–232
99. Brewer CT, Swisher RG, Sinn E, Grimes RN (1985) Metal-promoted fusion of $B_6H_9^-$. Directed synthesis and structural characterization of dodecaborane(16), $B_{12}H_{16}$. J Am Chem Soc 107:3558–3564

100. Maynard RB, Grimes RN (1982) Oxidative fusion of carborane ligands in iron and cobalt complexes: a systematic study. J Am Chem Soc 104:5983–5986
101. Carr MJ, Franken A, Kilner CA, Kennedy JD (2004) Macropolyhedral boron-containing cluster chemistry. Cluster assembly about a molybdenum centre. Formation of the nineteen-vertex $[(CO)_2MoB_{16}H_{15}C_2Ph]^-$ anion. Dalton Trans (Dalton Commun, Hot Paper) 2004:2612–2613
102. Grimes RN (2011) Carborane chemistry, 2nd edn. Academic/Elsevier, Amsterdam. ISBN 978-0-12-374170-7
103. Jelínek T, Kennedy JD, Štíbr B, Thornton-Pett M (1995) Macropolyhedral boron-containing cluster chemistry. The eighteen-vertex monocarbaborane $[(Me_3CNH_2)CB_{17}H_{19}(CN)]$. J Chem Soc Chem Commun 1995:2407–2408
104. Jelínek T, Císařová I, Štíbr B, Kennedy JD (2007) Macropolyhedral boron-containing cluster chemistry. Synthesis of the nineteen vertex monocarbaborane [9-($^{ter}BuNH_2$)-(anti)-{9-$CB_{18}H_{20}$}] by direct carbon-atom Aufbau. Dalton Trans (Dalton Commun) 2007:4766–4768
105. Plešek J, Heřmánek S, Štíbr B, Hanousek F (1967) Chemistry of Boranes. VII. A new synthesis of borane $B_{18}H_{22}$; an application of three-center bond theory on the interpretation of reaction mechanisms. Collect Czech Chem Commun 32:1095–1103
106. Dobson J, Keller PC, Schaeffer R (1968) Boranes. XXIII. Isononaborane-15. Inorg Chem 7:399–402
107. Li Y, Sneddon LG (2006) Improved synthetic route to *n*-$B_{18}H_{22}$. Inorg Chem 45:470–471
108. Beckett MA, Crook JE, Greenwood NN, Kennedy JD (1986) Synthesis, molecular structures, and n.m.r. properties of $[(PMe_2Ph)PtB_{16}H_{18}(PMe_2Ph)]$ and $[(PMe_2Ph)_4Pt_3B_{14}H_{16}]$, and a discussion of the bonding at platinum in these macropolyhedral platinaborane clusters. J Chem Soc Dalton Trans 1986:1879–1893
109. Beckett MA, Greenwood NN, Kennedy JD, Salter PA, Thornton-Pett M (1986) Identification and molecular structure of the eighteen-vertex macropolyhedral diplatinaoctadecaborane $[(PMe_2Ph)_2Pt_2B_{16}H_{15}(C_6H_4Me)(PMe_2Ph)]$. J Chem Soc Chem Commun 1986:556–557
110. Jelínek T, Kennedy JD, Štíbr B, Thornton-Pett M (1995) Macropolyhedral boron-containing cluster chemistry. An interesting nineteen-vertex oxaborane anion, $[OB_{18}H_{21}]^-$. J Chem Soc Chem Commun 1995:1665–1666
111. Ouassas A, R'Kha C, Mongeot H, Frange B (1991) Reaction of $B_{11}H_{14}NR_4$ (R = Et, Bu) with M_2O_3 (M = As, Sb, Bi) in biphasic systems. Inorg Chim Acta 180:257–261
112. Ouassas A, Fenet B, Mongeot H, Frange B (1994) The first oxaborane species: 12-vertex-28e-nido-$B_{11}H_{11}O^-$. In: Kabalka G (ed) Current topics in the chemistry of boron. The Royal Society of Chemistry, Cambridge, pp 363–366. ISBN 0-85186-535-6
113. Ouassas A, Fenet B, Mongeot H, Frange B, Gautheron B, Barday E (1995) Oxygen in an electron-deficient borane skeleton: the oxa-*nido*-dodecaborate anion $[OB_{11}H_{12}]^-$. J Chem Soc Chem Commun 1995:1663–1664
114. Frange B, Kennedy JD (1996) Polyhedral oxaborane chemistry. Some comparative ^{11}B NMR shielding patterns within the twelve-vertex nido-type geometry. Main Group Met Chem 19:175–181
115. Serra C, Ouassas A, Boutalib A, Barday E, Gautheron B, Hanquet B, Frange B (1997) New results about the oxaborane anion $OB_{11}H_{12}^-$. Main Group Met Chem 20:247–254
116. Fontaine XLR, Fowkes H, Greenwood NN, Kennedy JD, Thornton-Pett M (1985) An unusual open twelve-vertex oxametallaborane cluster compound: synthesis and X-ray structural characterisation of $[(C_5Me_5)RhB_{10}OH_9Cl(PMe_2Ph)]$. J Chem Soc Chem Commun 1985:1722–1723
117. Ditzel EJ, Fontaine XLR, Fowkes H, Greenwood NN, Kennedy JD, MacKinnon P, Zhu S, Thornton-Pett M (1990) Oxarhodaborane chemistry. The formation of $[(C_5Me_5)RhOB_{10}H_{10}(NEt_3)]$ and $[(C_5Me_5)RhB_9H_{12}]_2O$ from $[(C_5Me_5)RhB_{10}H_{13}Cl]$. J Chem Soc Chem Commun 1990:1692–1694

118. Micciche RP, Brigugliio JJ, Sneddon LG (1984) Metal atom synthesis of metallaboron clusters. 5. Synthesis of the first (η^6-arene)metallaborane and (η^6-arene)metallaoxaborane clusters. Structural characterizations of 5-[η^6-$C_6(CH_3)_3H_3$]FeB_9H_{13} and 2-[η^6-$C_6(CH_3)_3H_3$]Fe-6-OB_8H_{10}. Inorg Chem 23:3992–3999
119. Kim Y-H, Brownless A, Cooke PA, Greatrex R, Kennedy JD, Thornton-Pett M (1998) Polyhedral oxaplatinaborane chemistry. Characterisation of [9,9-$(PMe_2Ph)_2$-arachno-9,6-$PtOB_8H_{10}$] and its metallaborane non-oxa cognate [6,6-$(PMe_2Ph)_2$-arachno-6-PtB_9H_{11}-9-(PMe_2Ph)]. Inorg Chem Commun 1:19–22
120. Bould J, Bown M, Kennedy JD (2005) Polyhedral oxaruthenaborane chemistry. Characterisation of a [(η^6-C_6Me_6)$RuOB_9H_{13}$] species of arachno eleven-vertex cluster character. Collect Czech Chem Commun 70:410–429
121. Yang X, Jiao H, Schleyer PR (1997) Structures of the 12-vertex oxa- and thia-*nido*-dodecaborates and $B_{13}H_{13}^{2-}$: a theoretical DFT/GIAO/NMR investigation. Inorg Chem 36:4897–4899
122. Kaur P, Kennedy JD, Thornton-Pett M, Jelínek T, Štíbr B (1996) Macropolyhedral boron-containing cluster chemistry. Syn and anti [$(C_5Me_5)_2Rh_2S_2B_{15}H_{14}(OH)$]; an interesting nineteen-vertex isomeric pair. J Chem Soc Dalton Trans 1996:1775–1777
123. Kaur P, Brownless A, Perera SD, Cooke PA, Jelínek T, Kennedy JD, Thornton-Pett M, Štíbr B (1998) Macropolyhedral boron-containing cluster chemistry. Reaction of [$PtMe_2(PMe_2Ph)_2$] with [9,9′-$S_2B_{16}H_{16}$] to give the eighteen-vertex dithiaplatinaborane [$(PMe_2Ph)_2PtS_2B_{15}H_{14}(NHCOMe)$]: an entry into macropolyhedral metallaheteroborane chemistry. J Organomet Chem 557:181–185
124. Carr MJ, Londesborough MGS, Bould J, Císařová I, Kennedy JD (2005) Macropolyhedral boron-containing cluster chemistry. A metallathiaborane from $S_2B_{17}H_{17}$. Isolation and characterisation of [$(PMe_2Ph)_2PtS_2B_{16}H_{16}$]. A neo-arachno ten-vertex cluster shape, and the constitution of the [arachno-$B_{10}H_{15}$]$^-$ anion. Collect Czech Chem Commun 70:430–440
125. Carr MJ, Londesborough MGS, Hamilton McLeod AR, Kennedy JD (2006) Macropolyhedral boron-containing cluster chemistry. Metallathiaboranes from $S_2B_{17}H_{17}$: isolation and characterisation of [(η^6-$MeC_6H_4{}^{iso}Pr$)$RuS_2B_{16}H_{16}$] and [(η^6-$MeC_6H_4{}^{iso}Pr$)$RuS_2B_{15}H_{15}$]. Dalton Trans 2006:3624–3626
126. Shea SL, McGrath T, Jelínek T, Štíbr B, Thornton-Pett M, Kennedy JD (1998) Metallaborane reaction chemistry. Macropolyhedral metallaheteroborane synthesis by direct hetero-atom insertion. Formation of the twenty-vertex [(η^5-C_5Me_5)$IrB_{18}H_{19}S$]$^-$ anion from nineteen-vertex syn-[(η^5-C_5Me_5)$IrB_{18}H_{20}$]. Inorg Chem Comm 1:97–100
127. Londesborough MGS, Kilner CA, Thornton-Pett M, Kennedy JD (2002) Macropolyhedral boron-containing cluster chemistry. A novel triple-cluster structural motif. Isolation and characterization of contiguous twenty-vertex [$(PPh_3)_4ClPd_4B_{16}H_{17}(PPh_3)_2$]. J Organomet Chem 657:262–266
128. Barton L, Bould J, Kennedy JD, Rath NP (1996) Macropolyhedral boron-containing cluster chemistry. The isolation and characterisation of the eighteen-vertex nido-5′-iridaoctaborano-(3′,8′:1,2)-closo-4-iridadodecaborane [$(CO)(PMe_3)_2IrB_{16}H_{14}Ir(CO)(PMe_3)_2$]. J Chem Soc Dalton Trans 1996:3145–3149
129. Londesborough MGS, MacLean EJ, Teat SJ, Thornton-Pett M, Kennedy JD (2005) Macropolyhedral boron-containing cluster chemistry. Synchrotron X-ray structural analysis of [$(PMe_2Ph)_2Pd_2B_{16}H_{22}(PMe_2Ph)_2$] and [$(PMe_2Ph)_3Pt_2B_{16}H_{20}(PMe_2Ph)$]: Models of intermediates to more condensed metallaboranes from the [$(PMe_2Ph)_2PtB_8H_{12}$] thermolysis system. Chem Commun (‘Hot Paper’) 2005:1584–1586
130. Londesborough MGS, MacLean EJ, Teat SJ Bould J, Kilner CA, Thornton-Pett M, Kennedy JD (2003) Macropolyhedral boron-containing cluster chemistry. Pointers towards the mechanism of homofusions of nine-vertex arachno metallaboranes. In: Bubnov Y (ed) Boron chemistry at the beginning of the 21st century. Editorial URSS, Moscow, pp 248–254. ISBN 5-9519-0019-0
131. Bould J, Oro LA, Macías R, Kennedy JD, Londesborough MGS (2011) A DFT and crystallographic reinvestigation of the [$L_2RuC_2B_7H_9$] and [$L_3RuC_2B_7H_9$] ‘hypercloso’ and closo systems. Polyhedron 30:2140–2145

132. Ormsby DL, Greatrex R, Štíbr B, Kennedy JD (2000) Bond rotamers and calculated ^{11}B NMR chemical shifts in boron-containing cluster chemistry. Some effects in the nido-{7,8,10-PC_2B_8} system. J Organomet Chem 614/615:61–65; with erratum (2002) J Organomet Chem 657:279
133. Ditzel EJ, Fontaine XLR, Greenwood NN, Kennedy JD, Thornton-Pett M (1989) Direct evidence for $N \rightarrow B$ π-donation into a polyhedral borane cluster. J Chem Soc Chem Commun 1989:1115–1116
134. Roth M, Meyer F, Paetzold P (1997) Opening of closed azadecaboranes RNB_9H_9 by Amines. Collect Czech Chem Commun 62:1299–1309
135. Kennedy JD, Štíbr B, Jelínek T, Fontaine XLR, Thornton-Pett M (1993) Eleven-vertex polyhedral dicarbaplatinaborane chemistry. Aspects of the chemistry of some closo structured {1,2,3-PtC_2B_8} species and some related compounds. Collect Czech Chem Commun 58:2090–2120
136. Kim J-H, Lamrani M, Hwang J-W, Do Y-K (1997) First tetrathiolate-bridged dinuclear molybdacarbaboranes: scission and formation of the C–C cluster bond during oxidation processes. J Chem Soc Chem Commun 1997:1761
137. Lipscomb WN (1966) Framework rearrangement in boranes and carboranes. Science 153:373–378
138. See, for example: Brown CA, McKee ML (2006) Rearrangements in icosahedral boranes and carboranes revisited. J Mol Model 12. doi:10.1007/s00894-006-0111-5
139. Bould J, McInnes YM, Carr MJ, Kennedy JD (2004) Metallaborane reaction chemistry. Part 9. A facile and reversible dioxygen capture by a B-frame-supported bimetallic: structure of $[(PMe_2Ph)_4(O_2)Pt_2B_{10}H_{10}]$. Chem Commun 2004:2380–2381
140. Bould J, Kilner CA, Kennedy JD (2005) Metallaborane Reaction Chemistry. Part 11. The capture of dioxygen, carbon monoxide and sulphur dioxide by $[(PMe_2Ph)_4Pt_2B_{10}H_{10}]$. Dalton Trans 2005:1574–1582
141. Bould J, Kennedy JD (2008) Metallaborane reaction chemistry. Part 13. A predicted and found tailored facile and reversible capture of SO_2 by a B-frame-supported bimetallic: structures of $[(PMe_2Ph)_2PtPd(phen)B_{10}H_{10}]$ and $[(PMe_2Ph)_2Pt(SO_2)Pd(phen)B_{10}H_{10}]$. Chem Commun 2008:2447–2449
142. Londesborough MGS, Hnyk D, Bould J, Serrano-Andrés L, Sauri V, Oliva JM, Kubát P, Polívka T, Lang K (2012) Distinct photophysics of the isomers of $B_{18}H_{22}$ explained. Inorg Chem 51:1471–1479
143. Sauri V, Oliva JM, Hnyk D, Bould J, Braborec J, Merchán M, Kubát P, Císařová I, Lang K, Londesborough MGS (2013) Tuning the photophysical properties of anti-$B_{18}H_{22}$: efficient intersystem crossing between excited singlet and triplet states in new 4,4′-$(HS)_2$-anti-$B_{18}H_{20}$. Inorg Chem 52:9266–9274
144. Cerdán L, Braborec J, Garcia-Moreno I, Costela A, Londesborough MGS (2015) A borane laser. Nature Commun 6:5958–5964
145. Lipscomb WN (1980) Examples of possible polymeric borane structures. Inorg Chem 19:1415–1416
146. Oliva JM, Rué J, Hnyk D, Kennedy JD, Rosenfeld VR (2013) Borane polyhedra as building blocks for unknown but potentially isolatable new molecules – Extensions based on computations of the known $B_{18}H_{22}$ isomers. Croat Chem Acta (D J Klein 70th Birthday Edition) 86:485–494
147. Hnyk D, Holub J, Jelinek T, Macháček J, Londesborough MGS (2010) Revisiting $B_{20}H_{16}$ by means of a joint computational/experimental NMR approach. Collect Czech Chem Commun (B Štíbr 70th Birthday Edition) 75:1115–1123
148. Bould J, Hnyk D, Kennedy JD, Macháček J, Oliva JM (2015) Could linear megaloboranes $B_{(8n+4)}H_{(4n+8)}$ exist? Ongoing work
149. Friedman LB, Dobrott RD, Lipscomb WN (1963) Preparation and structure of a new boron hydride, $B_{20}H_{16}$. J Am Chem Soc 85:3505–3505
150. Miller NE, Muetterties EL (1963) A new boron hydride, $B_{20}H_{16}$. J Am Chem Soc 85:3506–3506

151. Miller NE, Forstner JA, Muetterties EL (1964) Chemistry of boranes. XXI. Icosaborane-16. Inorg Chem 3:1690–1694
152. Greenwood NN, Kennedy JD, Taylorson D (1978) Mass-spectrometric evidence for icosaborane(26). J Phys Chem 82:623–625
153. Dobrott RD, Friedman LB, Lipscomb WN (1964) Molecular and crystal structure of $B_{20}H_{16}$. J Chem Phys 40:866–872
154. Kaur P, Perera SD, Jelínek T, Štíbr B, Kennedy JD, Clegg W, Thornton-Pett M (1997) Macropolyhedral boron-containing cluster chemistry. Isolation and characterization of the twenty-one-vertex rhenaborane $[(PMe_2Ph)_3HReB_{20}H_{15}Ph(PHMe_3)]$. Chem Commun 1997:217–218

Chapter 7
Electronic Requirements and Structural Preferences for Large Polyhedral Boranes

Musiri M. Balakrishnarajan and Pattath D. Pancharatna

Abstract The structural and electronic preferences of boron-based polyhedral systems with 12 or fewer vertices is well understood by the combined use of localized and delocalized bonding paradigms. Current research in polyhedral boranes predominantly involves scaling the size within the single polyhedron or by having multiple polyhedral units. Though large and multiple polyhedra are experimentally known for long in molecules as well as solids, the beginning of the current century witnessed comprehensive understanding of bonding in these extended systems. Here, we address the various bonding features exhibited by boron-based large polyhedral systems with the scaling of the system size, their electronic and structural preferences.

7.1 Introduction

The intriguing position of boron in the periodic table makes its bonding unique, compared to any of its adjacent elements. Unlike its neighbors beryllium and aluminium, boron seldom exhibits purely ionic bonding. The irony is that despite being referred as a metalloid or semi-metal, the first ionization energy of boron is less than that of beryllium due to its loosely bound 2p electron. However, removal of two or more electrons is much more difficult as they are more strongly bound in its 2s orbital. This is presumably due to the increased positive charge which makes the existence of B^{3+} ion impossible whether free or solvated. Compared to aluminium with a similar valence electronic configuration, boron valence orbitals have fewer radial nodes and also experience reduced screening from core electrons, which makes these orbitals less diffuse and their electrons tightly bound, hampering ionic bonding. This is clearly reflected in the first three ionization energies of boron which are ~50 % more than aluminium consistently.

Exclusive ionic bonding in isolation being ruled out, boron as a main group element is forced to exhibit covalent bonding. However, with only three valence elec-

M.M. Balakrishnarajan (✉) • P.D. Pancharatna
Department of Chemistry, Pondicherry University, Pondicherry 605014, India
e-mail: mmbkraja@gmail.com; pdpancha@gmail.com

D. Hnyk, M.L. McKee (eds.), *Boron*, Challenges and Advances in Computational Chemistry and Physics 20,
DOI 10.1007/978-3-319-22282-0_7

trons, classical electron pair bonding leaves a reactive empty orbital and necessitates either a negative charge or a coordinate bond for attaining Lewis octet configuration. When bonded to highly electronegative elements, boron indeed forms either Lewis acids or hosts a formal negative charge in its compounds. But while bonding with itself or with elements of comparable electronegativity such as carbon, boron radically changes this classical behavior. Usually covalent bonding between non-metals comprises of low lying σ-bonding MOs, with non-bonding and/or π-MOs in the frontier. These are successfully described classically by localized σ-bonds, π-bonds and lone pairs with some degree of resonance between them. However, the non-classical bonding of boron clearly steers away from this familiar pattern. This is mainly originating from the comparable diffuseness and energetic proximity of 2s and 2p orbitals of boron, which favors relatively strong sp mixing even compared to carbon. Hence, when boron interacts with p block elements of comparable electronegativity to form single bonds, the entailing 'avoided-crossing' keeps its σ-bonding levels close to the frontier, significantly above its π-MOs. In the case of unsaturated boron compounds, the occupied π-bonding MOs are extremely reactive, in contrast to the moderately reactive C≡C bonds and extremely stable N≡N. The few compounds that are known to have multiple bonding between boron atoms are stabilized by the presence of strong π-acceptors like nitrogen substituted heterocyclic carbenes (NHC) which are adequately substituted with bulky groups to offer steric protection. [1] These compounds, though described by localized π-bonds between boron atoms in their representative Lewis structure, are stabilized by the substantial π-delocalization arising from back-bonding. The allene like near D_{2d} geometry of the recently reported compound [2] with a formal B≡B bond serves as an ideal illustration. Though the bonding between boron atoms is claimed to be triple [3], its doubly degenerate HOMO clearly shows the two interpenetrating allyl like linear B-B-C units in orthogonal directions, each contributing only half π-bond order. Hence π-bonds between boron atoms are stable only when substituted with strong π-acceptor ligands. In the absence of such a special environment, boron, in an attempt to engage all its valence orbitals for stability, interacts with many neighbors, much more than what is classically expected based on Lewis theory. This leads to cluster motifs that are seen in much of the boron-rich compounds and solids in which even other elements are forced into non-classical bonding.

The dogma of electron pair bond arising from the Lewis concept – typically represented as a line connecting the two bonded atoms – is deeply entrenched in classical chemical structure theory. Hence understanding and appreciating the non-classical nature of bonding in boron required overcoming daunting mental blockades. The structure and the bonding description of even the simplest diborane molecule created much debate for several decades [4]. Due to the stoichiometric similarity and experimentally observed stronger B-B stretching force constants, the classical ethane-like structure was initially supported by Pauling. He justified the electron deficiency of diborane by means of resonance structures involving one-electron bonds. But when the position of bridging hydrogen atoms symmetrically above and below the B_2H_4 plane was eventually proved, it was clear that the rules of classical valence theory need to be broken to account for its bonding. The typical chemical penchant for associating an electron pair with a localized bond gave birth to the concept of the three-center two-electron (3c-2e) bond, a distinct bond that involves three atoms represented by two or more lines between different atoms. Thus the convention of a line representing an electron-pair bond was broken. However, violating the monovalence of hydrogen atoms by bridging gained much easier acceptance among chemists than depicting boron atom with more than four bonded neighbors. There is still a considerable resistance in drawing a line between the two boron atoms in B_2H_6. This is despite the fact that the distance between the two boron atoms (1.78 Å) lie well within the B-B multicentered bonding range and most chemical probes show bonding interaction between them. It is interesting to note that similar bridged structures among organic systems like vinyl cation [5], ethane dication [6] etc., the latter being isostructural/isoelectronic to diborane, are usually represented with a line connecting the two carbon atoms. In fact, the ethane dication is unstable against dissociation into two methyl cations unlike diborane. Longuet-Higgins seminal theoretical arguments [7] that firmly supported the hydrogen bridged structure without the direct B-B bond is presumably intended to delineate its bonding from ethane, as advocated by Pauling, and this proposed structure is widely accepted even today.

Extending the molecular orbital study of diborane to polyhedral structures, Longuet Higgins successfully predicted the dianionic requirement of octahedral [8] and icosahedral [9] borane clusters. This prediction is based on the observation that closed-shell molecules are more stable than those open-shell systems in which the highly degenerate highest occupied MOs are partially occupied, similar to Hückel's derivation of 4n + 2 rule in annulenes. He also introduced the localized two-center two-electron (2c-2e) description for *exo* polyhedral bonds [8], conceptually separating them from the rest of the skeletal MOs that needed a delocalized description. Subsequently, attempts were made to employ Pauling resonance theory with the newly included 3c-2e bonds to explain the bonding in other known polyhedral boranes [10]. The 'styx' formalism classified cluster bonding into varieties of localized bonds, often in resonance [11]. The main aim was to retain the octet configuration around boron by subtle classification of bonds into 2c-2e and 3c-2e bond categories. Though this approach was successful in several open systems, for highly symmetric *closo* boranes these categorization led to uncertainties even in the assessment of electronic requirements [12]. Notwithstanding the failure of localized bonding theory involving 3c-2e/2c-2e bonds, a general understanding of polyhedral bonding of boranes remained elusive despite the experimental characterization of several borane/carborane systems. With the increased availability of structural information for several open polyhedral boranes, William's realization [13] that they are not always derived from icosahedral $B_{12}H_{12}^{2-}$ led to the major breakthrough in polyhedral bonding. The experimental characterization of $B_{12}H_{12}^{2-}$ and other *closo* borane $B_nH_n^{2-}$ dianions (n = 6–12) along with their carborane analogs finally allowed the formulation of a general rule for polyhedral bonding. Wade's rule [14] came as a major breakthrough in polyhedral bonding, as it is derived prudently from several experimental facts and theoretical insights using statistical reasoning to fit the majority of experimentally known systems in a general rule, casting away the known exceptions.

Wade's theory assumed the boron atoms in boranes to be sp hybridized, with the inward pointing sp hybrids and the two tangential p orbitals involved in skeletal bonding. The outward pointing sp hybrid was presumed to form a classical 2c-2e bond with the hydrogen atom. Further Wade asserted that the n inward pointing sp hybrids of the *closo* B_nH_n system leads to just one bonding molecular orbital (MO) while the two tangential surface p orbitals of every boron atom together form n bonding and n antibonding MOs. This explains the requirement of n+1 electron pairs for *closo* boranes. This explanation was also extended to partially open systems by further emphasizing that the required number of electron pairs increase by one with increase in cluster faults due to the removal of a vertex from the idealized *closo* system. This necessitates an increase in negative charge or bridging hydrogen atoms to provide the additional electrons. The simplicity of Wade's formalism proved to be enormously helpful in rationalizing the intricate structures of open polyhedral boranes, allowing eloquent categorization.

It is to be noted that earlier MO theoretical studies that totally avoided the mixing of radial and tangential MOs led to erroneous electron count even for the icosahedral cluster [15], hindering generalization. Hiding behind the simplicity of Wade's rule is the chemical intuition that allowed its derivation despite the known lack of MO theoretical rigor. It is well known that not all known *closo* boranes have a clear demarcation between bonding and antibonding canonical MOs. Subsequent studies based on tensor harmonic theory [16] showed that highly symmetric *closo* boranes ($B_nH_n^{2-}$, where n=5,6,7,10 and 12) in which the major symmetry axis passes through a pair of vertices capping one or more rings have degenerate MOs in the frontier that helps in unambiguously identifying predominantly bonding MOs. Fortuitously, these are the most stable frameworks, with rich chemistry. A detailed MO analysis on these systems indeed reveals that the magic behind the Wade's n+1 rule is the appropriate mixing of radial and tangential MOs enforced by the underlying symmetry [17]. Since boron typically have large number of neighbors in the cluster, the fact that the *closo* electron count remains unaltered by the removal of one or more vertices is quite justified, as the resulting perturbation is not significant enough to alter the total number of bonding MOs. Hence, Wade's cluster bonding formulation that drew heavily from earlier results [14] is one of the seminal theory that successfully employs the abstract blending of conceptually separated localized bonds (B-H) and delocalized MOs (skeletal) in a single theory. This effectively filled the void left by resonance theory of localized bonds, and remains popular even today as a single unique theory to understand the non-classical bonding of simple polyhedral boranes.

The concept of separating the chemically passive regions with localized bonds and active regions with delocalized fragment MOs (that are subtly distinct from canonical MOs) proved to be enormously successful in understanding polyhedral bonding. This approach is effectively extended to transition metal clusters by using the isolobal analogy [18] and in explaining the stability and structural preferences of polyhedral boranes using the ring-cap matching principle [19]. The latter approach also assisted in reaffirming [13] the bonding connection between polyhedral boranes and organic aromatic systems by the six interstitial electron rule [20] justifying the term three-dimensional aromaticity [21] for polyhedral boranes.

Though bonding in polyhedral systems with twelve or fewer vertices is understood comprehensively by the abstract conceptual separation into localized and delocalized bonding, the structural and electronic preferences for larger polyhedral boranes remained unexplored till the beginning of this century despite being known for a long time. Here, we categorize the different possible ways by which boron typically extends its cluster bonding to grow in size, their preferred structural motifs and variations in their electronic requirements along with the mechanisms by which it attempts to reduce the building up of charges.

7.2 Sharing of Edges Between Polyhedra

Shortly after the experimental characterization of icosahedral $B_{12}H_{12}^{2-}$, many large polyhedral boranes formed by sharing one or more edges between the individual open/*closo* cages are also reported. The *closo* $B_{20}H_{16}$ was experimentally characterized [22, 23] shortly after the isolation of icosahedral [24] $B_{12}H_{12}^{2-}$and serves as a unique example that has two icosahedral units sharing four edges spread across four vertices. In the same year, the *n*-$B_{18}H_{22}$ (C_i) that has two *nido* B_{10} units sharing an edge was reported experimentally, which is probably the first structurally characterized macropolyhedral borane with an open face. Subsequently, its topological isomer *iso*-$B_{18}H_{22}$ as well as a *closo-nido* combination that shares a deltahedral face between a *closo* B_{12} and *nido* B_{11} units are reported. These are some of the earliest known examples (Fig. 7.1) for macropolyhedral boranes. There is renewed interest in the two $B_{18}H_{22}$ isomers for their distinct photophysical properties [25]; the anti-$B_{18}H_{22}$ is shown to exhibit florescence and capability for producing singlet oxygen which is also found to be a promising alternative laser material with remarkable photo-stability [26].

The presence of multiple polyhedral cages is easy to perceive as the shared atoms have no axial hydrogen atoms, similar to naphthalene. The maximum number of edges that possibly can be shared between two deltahedra is limited to four, since

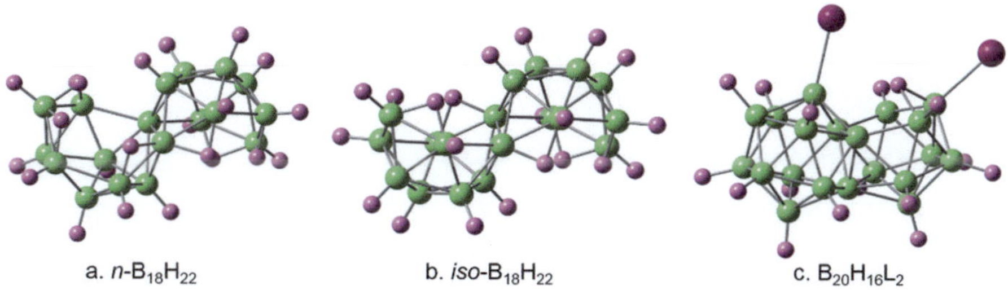

Fig. 7.1 Some of the earliest known macropolyhedral boranes with open cages

five-vertex sharing lacks the negative curvature that is required for individual cages even for the largest known monopolyhedral *closo* borane $B_{12}H_{12}^{2-}$. Other than the unique $B_{20}H_{16}$ molecule, the sharing of either one edge or three edges that simultaneously forms a triangular face are commonly observed in various experimentally reported systems [27]. Polyhedral sharing of two edges is precluded by the requirement of the deltahedral geometry to exhibit convex surfaces for the individual cages.

To assess the bonding in these systems, the first and foremost information is the number of electron pairs required for its stable existence. It is interesting to note that polyhedral skeletal electron-pair theory that uses isolobal analogy to assess the electronic requirements of transition metal clusters was successfully extended to cover edge-shared metal clusters [28]. It was found that the required number of electron pairs is a function of the number of edges that are shared between the polyhedra. However, extrapolating these results to polyhedral boranes proved futile, as the isolobal analogy is not applicable for the shared vertices.

The systematic inquiry using extended Hückel (eH) calculations on idealized geometry of various edge-shared *closo* systems with two or three mono polyhedral boranes (Fig. 7.2) clearly revealed the total number of electrons required for bonding [29]. This is assessed by examining the energy levels of these systems and looking for an ideal electron count that results in a larger HOMO-LUMO gap. The optimized geometry of these systems with the charges obtained from eH calculations indeed were shown to be minima on their potential energy surface with the standard density functional calculations (B3LYP/6-31G*), confirming the validity of eH predictions.

These results show that sharing of a single edge retains the dianionic requirement of the monopolyhedral *closo* boranes, irrespective of the number of cages in the system. However, sharing of three edges (face-sharing) between polyhedra show that charge requirement changes with the number of cages. Contrary to the usual expectation of increased charges for these multicage systems, calculations show

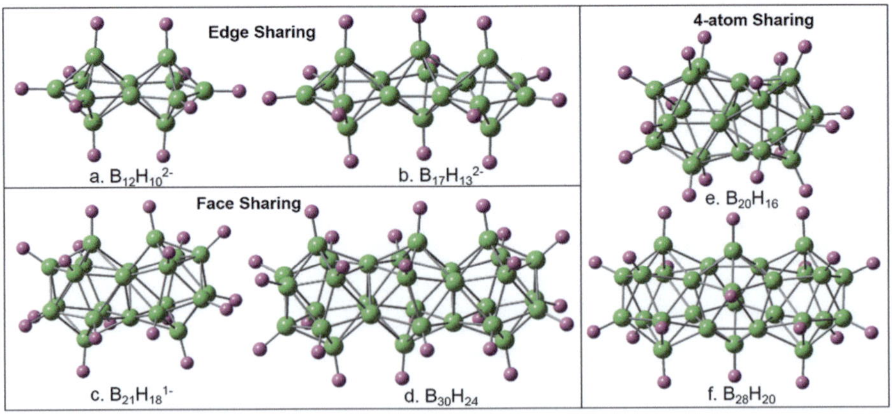

Fig. 7.2 Idealized examples of macropolyhedral boranes formed by sharing one, three and four edges with their charges

that the dianionic charge of the monopolyhedral *closo* systems reduces by one electron per sharing. As illustrated in Fig. 7.2, face sharing of two icosahedral B_{12} units leads to the $B_{21}H_{18}$ system that shows a large HOMO-LUMO gap for −1 charge. For the model system with three B_{12} units sharing a face between adjacent units, a neutral species has a large HOMO-LUMO gap. Extending further, it was clear that the charge of the system is reduced by one for every face-sharing junction, indicating the build-up of positive charge for further extensions.

Similar calculations on model systems sharing four edges as in $B_{20}H_{16}$, show that each junction reduces the charge requirement by two. $B_{20}H_{16}$ has a large HOMO-LUMO gap for the neutral species as reported experimentally. Extending it to another cage leads to D_{2h} symmetric $B_{28}H_{20}$ borane that requires a 2+ charge for a large HOMO-LUMO gap.

Though the charges vary systematically with respect to the number of edges shared, the total number of electron pairs required for skeletal bonding remains independent of the shared edges. The overall requirement of skeletal electron pairs for a general macropolyhedral system that share one or more edges can be succinctly represented by a general formula n + m, where n and m represent the total number of vertices and number of mono polyhedral cages respectively [29]. The n + m formula holds good for all macropolyhedral systems that exhibit sharing of edges, irrespective of the number and nature of individual polyhedral units involved. Representation of the electron counting scheme in terms of electron pairs involved in polyhedral bonding as in Wade's n + 1 rule formalism, instead of Mingo's total electron pairs, has its benefits. It allows the slight modification of Wade's empirical MO explanation, i.e., the number of tangential bonding MOs remains n but the radial bonding MOs become m, so that the more generalized n + m rule reduces to n + 1 for monopolyhedral systems.

The n + m rule of condensed macropolyhedral boranes can also be extended to open systems with one or more missing vertices, similar to monopolyhedral cases. The number of bridging hydrogen atoms observed in a variety of experimentally reported open systems was successfully explained by the rule. It also helped in correcting the erroneous assignment of the number of bridging hydrogens in the edge-shared 19-vertex macropolyhedral borane [30]. The validity of the n + m rule for condensed systems was extensively tested for a variety of model systems using ab-initio calculations and the vibrational calculations unambiguously showed the accuracy of assigned charges using the n + m rule by characterizing them as minima on the PES. Several isomeric frameworks that are possible with open edge-sharing architectures are also explored computationally using this rule [31].

The geometry of all these condensed polyhedral systems in general show slight shortening of the shared B-B distances and elongation of bonds emanating from the shared boron to other boron atoms, presumably due to steric repulsion between the proximate B-H groups of the adjoining polyhedra. This factor also primarily determines the overall stability of the condensed polyhedral boranes and clearly explains the computed preference of the large polyhedra for a smaller polyhedra in face sharing [32]. These steric factors are also responsible for the marked absence of *closo* boranes that share a single edge but are characterized in open systems. Here one or more vertices inflicting steric repulsion is absent, and host bridging hydrogen atoms for compensating the electron count. Expectedly, the edge-sharing *nido-macropolyhedral* systems B_nH_{n+4} in which these destabilizing steric repulsions are absent show remarkable additivity in their total energy and these macropolyhedral frameworks are preferred over single *nido* polyhedra for systems having more than twelve vertices (n > 12) [33]. In $B_{20}H_{16}$, this steric repulsion is relieved by the elongation of the B-B bonds across the diagonal of the shared B_4 ring that result in a high D_{2d} symmetry with substantially deformed deltahedral geometry of the icosahedral cage. The LUMO of the $B_{20}H_{16}$ is found to be concentrated on these shared boron atoms and determines the reactivity of this molecule [34] as expected.

7.3 Sharing of Single Vertex Between Polyhedra

According to the ring-cap matching principle, boron favors capping a pentagonal B_5 ring and to some extent a square B_4 ring. When a single boron atom is shared between two polyhedra, the B-H groups around the shared atom come too close (Fig. 7.3a), leading to substantial steric repulsion. Hence, sharing a single vertex between two polyhedra seldom occurs in pure polyhedral boranes. These destabilizing HB....BH interactions can be reduced if the shared vertex is occupied by a larger hetero atom. Several such systems are known [27] and Fig. 7.3b, c illustrates some of the typical examples. In these cases, it is fairly understood that the two

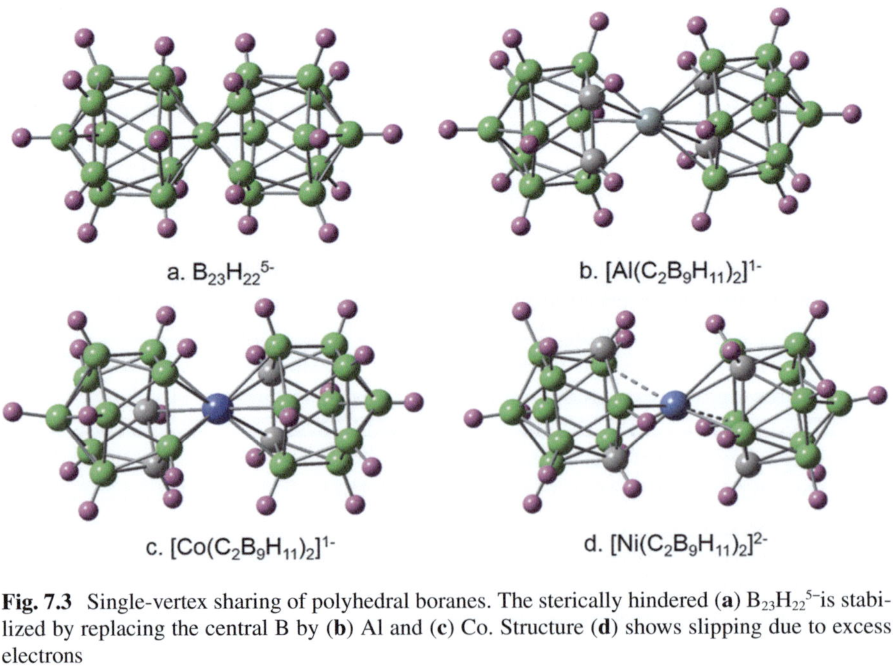

Fig. 7.3 Single-vertex sharing of polyhedral boranes. The sterically hindered (**a**) $B_{23}H_{22}^{5-}$is stabilized by replacing the central B by (**b**) Al and (**c**) Co. Structure (**d**) shows slipping due to excess electrons

cages behave as if they are totally independent. In other words, the central atom shares its orbitals between the clusters but not its electronic contribution. Hence, unlike the various edge-sharing systems discussed above, the electron count for single vertex sharing polyhedra does not fit in the n + m formula. It requires one more electron pair to allow both the polyhedra to fill their own unshared skeletal bonding MOs.

There are numerous single-vertex-sharing polyhedra reported experimentally where the shared vertex is occupied by large elements from the second and third row, most of them consistently differing from n + m electron count by one additional electron pair, as expected. Instead of treating all these systems as an exception to the n + m rule, Jemmis proposed an additional parameter to include these systems within a single generalized electron counting formula. Hence, shortly after the framing of the n + m rule, it was amended to include one more parameter 'o', which counts the number of such single vertex sharing in the macropolyhedral system. Referred to as Jemmis' mno rule [35], this electron-counting scheme is generalized enough to cover most of the known macropolyhedral boranes, including those with transition metals, by careful usage of the isolobal analogy. This can also be extended to cover hybrid systems containing both organic aromatic rings and polyhedral boranes. In fact, the compounds formed with late transition metals at the sharing site (Fig. 7.3d) are found to exhibit the characteristic slipping [36] due to the presence of an extra electron pair than that is required by mno rule. This further establish the strength of the generalized mno rule in rationalizing the bonding in these systems. The mno rule also has exceptions, most notably in metallaboranes, especially when multiple metals present are directly connected. A detailed review [27] illustrating the application of the mno rule to many of the known macropolyhedral boranes, including those that do not involve sharing, showed that the exceptions to the rule constitute only a small component. These can be identified just by visual inspection of the skeletal geometry as they often involve multiple transition metals with metal-metal bond(s).

7.4 Exo-linking of Polyhedra

Instead of sharing vertices, polyhedral boranes also prefer connecting through a formal 2c-2e bond to make larger boranes. This motive is omnipresent in all boron rich solids, but often appears under various disguise in molecular systems. The B_{10} dimers illustrate the various possibilities succinctly (Fig. 7.4) [27]. Unlike the HOMO of monopolyhedral boranes which are dominated by tangential MOs, the HOMO of 2c-2e linked systems is dominated by this exo-B-B σ-bonding. Hence these 2c-2e bonds often appear with bridging hydrogens that stabilize the HOMO thereby increasing the HOMO-LUMO gap. Irrespective of the presence of these bridging hydrogen atoms, the original formula of n tangential orbitals and m radial orbitals holds good, and the general mno rule can be applied effectively.

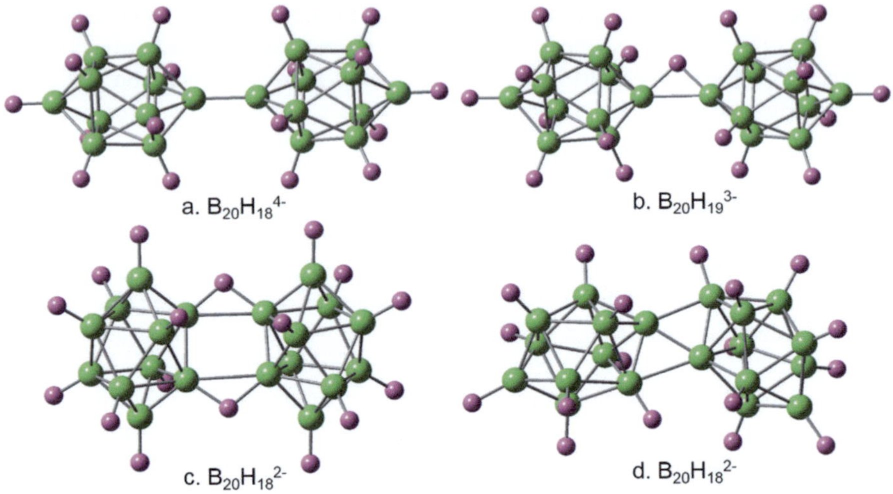

Fig. 7.4 Various types of exo-linkages known with B_{10} dimer

In addition to the directly linked 2c-2e bonds and hydrogen bridged 3c-2e bonds, polyhedra like B_{10} also exhibit a rhombic linkage that involves four boron atoms – two adjacent boron atoms from each polyhedron. Since the four boron atoms of the rhombus lie in the same plane, it cannot be considered as a fragment of a *closo* polyhedron, and this raises questions about the nature of bonding in the rhombus unit. This rhombic B_4 skeleton also is characterized in isolation, and shows enormous flexibility to accommodate varying number of exo-substituents [37, 38]. A detailed theoretical investigation on the bonding in these systems shows that the rhombic framework requires only two electron pairs and can be treated as having two 3c-2e bonds for simplicity [39]. For electron counting purposes, this rhombic skeleton can be conveniently ignored while counting the number of individual cages m, to fit into the generalized mno rule. The rhombic structural motif is also found in sodium boride Na_3B_{20} in which two pentagonal bipyramidal B_7 units exhibit such linkage between two adjacent boron atoms of the five-membered rings [40].

Hence the central B-B σ-bond in exo-linked polyhedral boranes is often stabilized by bridging hydrogen atoms, which also reduce the charge requirement. The rhombic mode of linking polyhedra further imparts stability and reduces the buildup of charges by two.

7.5 Bonding in Supraicosahedral Systems

Another natural way of thinking about large polyhedral boranes is to increase the overall size of the cage beyond the icosahedral limits. Despite the intense efforts to characterize such systems, the mono cage *closo* boranes ($B_nH_n^{2-}$ where $n > 12$)

remain elusive, though there are several computational studies predicting their possible structures [41, 42]. The difficulty is mainly attributed to the remarkable stability associated with the icosahedral cage, though there are also inherent geometrical reasons that discourage boron from forming these supraicosahedral clusters.

The icosahedron being the largest possible regular convex deltahedron, any attempts to increase the cluster size beyond 12 has to compromise its regularity by having one or more vertices capping a hexagonal ring. In an ideal deltahedron with equal B-B bond lengths, a B-H group bonding to six adjacent neighbors brings the cap down into the middle of the hexagonal ring. This proves to be disastrous for polyhedral bonding, since the two tangential p orbitals of the B-H cap cease to interact with the doubly degenerate π-frontier MOs of the ring, due to symmetry mismatch. To retain the pyramidalization for capping so as to allow overlap between ring and cap, either the cap to ring distance has to elongate or the hexagonal ring has to compress. This brings in bond-alternation leading to instability. However, if the B-H cap is replaced by an isolobal fragment with more diffuse orbitals, such structures are readily formed and several of them are known experimentally [43–45]. The fact that polyhedral boranes with B-H capping a hexagonal ring are known experimentally in a few cases [46, 47] gives hope towards the experimental realization of supraicosahedral boranes.

Assuming B-H caps on hexagonal rings as strain points, Lipscomb theoretically evaluated various geometrical possibilities for supraicosahedral boranes [41]. He also predicted the relative stabilities of their isomers based on uniformity of distribution of hexagonal B-H caps on the cluster surface to reduce the overall strain. However, later computational studies [42] identified a cluster, $B_{16}H_{16}^{2-}$ (D_{4d}) with square faces instead of the usual deltahedra to obey the *closo* electron count. Though all-boron supraicosahedral boranes are yet to be characterized, there are several carboranes synthesized recently with more than 12 vertices [47, 48]. Some of these systems also have open square faces, but still follow the *closo* electron count. The recent systematic study on various supraicosahedral boranes [49] showed that the hexagonal B-H caps prefer to be adjacent to each other but with a broken bond between these strained B-H groups. This results in the formation of nearly square faces and eliminates the unstable sixth coordination of boron. It also shows that these square faces prefer to be isolated from each other, much like the isolated pentagons of fullerene molecules. The highest symmetry cluster with such isolated square faces is the snub cube with 24 boron atoms. The associated borane $B_{24}H_{24}^{2-}$ is a minimum with the unique O symmetry (Fig. 7.5a) and is markedly more stable than two other fully deltahedral isomers (Fig. 7.5b, c). The large size of this snub borane cluster allows metal ions to be trapped inside. It is interesting to note that this snub cube geometry is indeed found in the structure of cubic YB_{66} with endohedral boron, though its bonding remains a puzzle due to the rampant partial occupancies present in the structure. Extending similar logic to five-membered open faces resulted in the snub dodecahedron $B_{60}H_{60}^{2-}$. The molecule with its I symmetry is also a minimum (Fig. 7.5d), and is probably the largest possible monocage borane that obeys Wade's rule despite the presence of 12 isolated pentagonal faces much

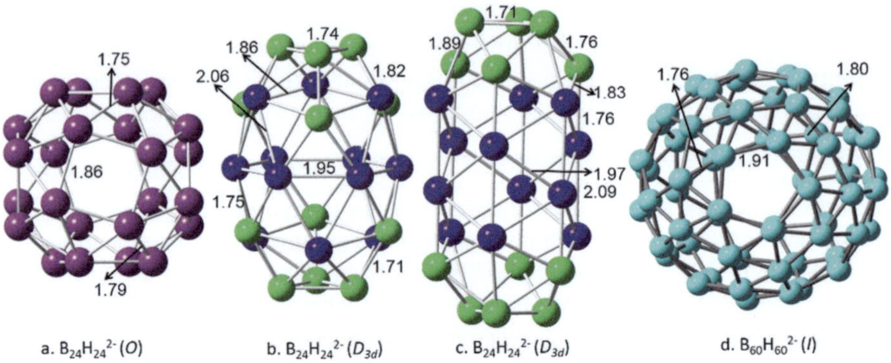

Fig. 7.5 Supraicosahedral boranes in nano dimension. Hydrogens are omitted for clarity. The vertices are colored differently to differentiate the caps and faces

like C_{60}. It is clear that supraicosahedral boranes if characterized will break the deltahedral rule more readily than the electron counting rules.

7.6 Mechanisms for Charge Compensation

One of the major problems associated with large macropolyhedral boranes is the increase in their charge requirements with the size. While linking of polyhedral units through 2c-2e bonding accumulates negative charges, sharing of multiple edges accumulates positive charge. The only topologies that retain constant charge are sharing of a single edge or rhombic linkage between polyhedra. Unfortunately these are not suitable growth mechanisms for many *closo* polyhedra, including the most stable icosahedral borane. The straightforward mechanism for reducing the negative charge is to acquire bridging hydrogen atoms that is ubiquitous in open systems. In addition, there are definite strategies that periodic polyhedral networks adopt in boron rich solids to counter the excess charges which are also found in some discrete polyhedral boranes. Understanding these different mechanisms help in the assessment of electronic requirements and consequently the cluster bonding.

The most ubiquitous phenomenon for compensating the negative charge in boron rich borides is the presence of interstitial boron atoms. Often present as partially occupied sites, these boron atoms cap a deltahedral face or an edge of the polyhedra and donate all their available electrons without stabilizing any additional bonding MO [50]. In molecules, there are examples where the whole polyhedral cage caps an edge of another polyhedron through one of its boron atoms, which is not to be confused with single-vertex sharing [51]. These capping boron atoms can also be present inside the polyhedral cage if there is sufficient space. Calculations done on stuffed polyhedral boranes [52] indeed show that these dopants do not change the electron count of the polyhedra, but merely donate all their valence electrons, reducing the overall negative charge. The most illustrative case is the elemental boron [53]

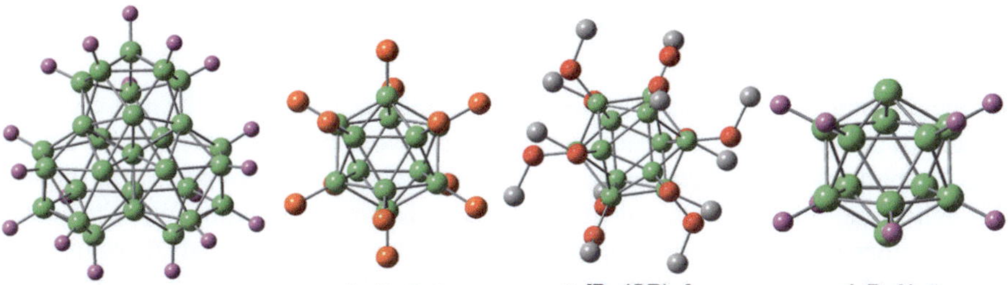

Fig. 7.6 Polyhedral frameworks illustrating the relieving of excess charge by (**a**) endohedral atoms, (**b**) and (**c**) exo-multiple bonding and (**d**) elimination of exo-substituents

itself as its β-rhombohedral polymorph [54] hosts several partially occupied interstitial borons as well as a boron atom in the middle of the condensed polyhedral cluster, B_{27} (Fig. 7.6a) [50, 55]. Similar trapping of boron atoms is found in the structure of YB_{66} in which the snub B_{24} cluster has a boron atom inside the cage, partially occupied and distributed in two symmetrically equivalent sites [56]. This indicates that stuffing may be an effective strategy for synthesizing large polyhedral systems.

Another mechanism that is effectively used in extended structures is partial oxidation. Often, the charge requirements of the polyhedral network are not completely satisfied in many boron rich solids. One ideal representative is boron carbide that shows large variation in carbon concentration [57] in different samples. Theoretical studies on exo-substituted boranes [58, 59] show that elements with higher electronegativity have a strong tendency for partially oxidizing the cluster. This imparts partial exo-polyhedral multiple bonding much like quinones in organic aromatic systems. There are several polyhedral boranes reported with highly electronegative substituents in their partially oxidized forms (Fig. 7.6b, c) indicating this strategy is effective even for zero-dimensional molecules [60, 61]. Presence of these partial oxidations can be easily perceived as they undergo characteristic distortion of the polyhedral cage that depends on the nodal features of the MO that loses electrons.

As a last resort, polyhedral boranes are also capable of eliminating some of their exo-substituents, and remain stable with bare boron atom(s) (Fig. 7.6d). Since the substituents may leave either as a radical, or anion or cation, electron counting may be a difficult task. Theoretical investigation on the effect of removing two hydrogens from the B_{12} polyhedron [62] shows that the skeleton exhibits characteristic distortions that depends on whether the hydrogens are removed as protons, hydrogen atoms or hydride ions. Hence a closer look at the distorted geometry can clearly indicate the electron count associated with the structure. Irrespective of the electron count, the ground state of these even-electron systems are always found to be singlets, albeit with some open-shell character if the frontier levels are too close. Polyhedral boranes having boron atoms without exo-substitutents are reported experimentally [63], with more such cases in heteropolyhedral boranes [64] and are also believed to exist in some boron rich borides such as YB_{66}.

7.7 Conclusion

By the prudent combination of localized and delocalized bonding, the non-classical structures exhibited by the mysterious and mischievous boron in polyhedral boranes and the bulk of boron rich solids can be comprehended. After the initial breakthrough in the understanding of monopolyhedral boranes aided by Wade's cluster bonding formalism, the beginning of the century saw remarkable advances in large polyhedral boranes. The different possible modes of scaling the cluster size beyond 12, quantitative assessment of its electronic requirements, understanding of their structural preferences and various possible mechanisms of charge compensation are explored. Wade's cluster bonding model that uses localized exo-polyhedral bonds and delocalized skeletal bonds can be successfully used as the basis to interpret the nature of bonding in large polyhedral boranes; linking of polyhedra by various mechanisms are understood as perturbation of exo-polyhedral localized bonding, while sharing of vertices, capping, stuffing and hole formation in supraicosahedral boranes are understood as perturbations to the delocalized skeletal bonding. The phenomenon of exo-polyhedral multiple bonding with elements of high electronegativity, however, is viewed as a perturbation involving both skeletal and exo-bonds. With the impressive success in experimental characterization of several large polyhedral systems, it is expected that the quantitative electron-counting rules and qualitative understanding of bonding will help increase the size of these polyhedral clusters further to nano-scale that may reveal several novel applications hitherto unknown. It will also give insights into the structure and bonding of several boron rich solids that remain mysterious, thereby providing a handle to effectively engineer these systems for several technologically important applications.

References

1. Wang Y, Quillian B, Wei P, Wannere CS, Xie Y, King RB, Schaefer HF, Schleyer PR, Robinson GH (2007) A stable neutral diborene containing a BB double bond. J Am Chem Soc 129(41):12412–12413
2. Braunschweig H, Dewhurst RD, Hammond K, Mies J, Radacki K, Vargas A (2012) Ambient-temperature isolation of a compound with a boron-boron triple bond. Science 336(6087):1420–1422
3. Frenking G, Holzmann N (2012) A boron-boron triple bond. Science 336(6087):1394–1395
4. Laszlo P (2000) A diborane story. Angew Chem Int Ed 39(12):2071–2072
5. Raine GP, Schaefer HF (1984) Vibrational frequencies for the classical and nonclassical forms of protonated acetylene-C_2H^{+3}. J Chem Phys 81(9):4034–4037
6. Schleyer PVR, Kos AJ, Pople JA, Balaban AT (1982) Carbenium-carbonium structures, H_2C^+-CH_4^+, for the ethane dication. J Am Chem Soc 104(13):3771–3773
7. Longuet-Higgins HC (1949) Substances hydrogenées avec défaut d'electrons. J Chim Phys 46:268–275
8. Longuet-Higgins HC, de Roberts VM (1954) The electronic structure of the borides MB6. Proc R Soc Lond Ser A 224:336–347

9. Longuet-Higgins HC, de Roberts VM (1955) The electronic structure of an icosahedron of boron atoms. Proc R Soc Lond Ser A 230:110–119
10. Eberhardt WH, Crawford B, Lipscomb WN (1954) The valence structure of the boron hydrides. J Chem Phys 22(6):989–1001
11. Lipscomb WN (1977) The Boranes and their relatives. Science 196(4294):1047–1055
12. Wade K (1971) Electron deficient compounds. Nelson, London
13. Williams RE (1971) Carboranes and boranes; polyhedra and polyhedral fragments. Inorg Chem 10(1):210–214
14. Wade K (1976) Structural and bonding patterns in cluster chemistry. In: Emeléus HJ, Sharpe AG (eds) Advances in inorganic chemistry and radiochemistry, vol 18. Elsevier-Science, pp 1–66
15. Hoffmann R, Lipscomb WN (1962) Theory of polyhedral molecules. I. Physical factorizations of the secular equation. J Chem Phys 36(8):2179–2189
16. Stone AJ (1980) A new approach to bonding in transition metal clusters. Mol Phys 41(6):1339–1354
17. Balakrishnarajan MM, Hoffmann R, Pancharatna PD, Jemmis ED (2003) Magic electron counts and bonding in tubular boranes. Inorg Chem 42(15):4650–4659
18. Mingos DMP (1972) A general theory for cluster and ring compounds of the main group and transition elements. Nat Phys Sci 236(68):99–102
19. Jemmis ED (1982) Overlap control and stability of polyhedral molecules. Closo-carboranes. J Am Chem Soc 104(25):7017–7020
20. Jemmis ED, Schleyer PR (1982) Aromaticity in three dimensions. 4. Influence of orbital compatibility on the geometry and stability of capped annulene rings with six interstitial electrons. J Am Chem Soc 104(18):4781–4788
21. Aihara J (1978) Three-dimensional aromaticity of polyhedral boranes. J Am Chem Soc 100(11):3339–3342
22. Friedman LB, Dobrott RD, Lipscomb WN (1963) Preparation and structure of a New Boron hydride, $B_{20}H_{16}$. J Am Chem Soc 85(21):3505–3506
23. Miller NE, Muetterties EL (1963) A new boron hydride, $B_{20}H_{16}$. J Am Chem Soc 85(21):3506–3506
24. Pitochelli AR, Hawthorne FM (1960) The isolation of the icosahedral $B_{12}H_{12}^{-2}$ Ion. J Am Chem Soc 82(12):3228–3229
25. Londesborough MGS, Hnyk D, Bould J, Serrano-Andrés L, Sauri V, Oliva JM, Kubát P, Polívka T, Lang K (2012) Distinct photophysics of the isomers of $B_{18}H_{22}$ explained. Inorg Chem 51(3):1471–1479
26. Cerdán L, Braborec J, Garcia-Moreno I, Costela A, Londesborough MGS (2015) A borane laser. Nat Commun 6:5958
27. Jemmis ED, Balakrishnarajan MM, Pancharatna PD (2001) Electronic requirements for macropolyhedral boranes. Chem Rev 102(1):93–144
28. Mingos DMP (1984) Polyhedral skeletal electron pair approach. Acc Chem Res 17(9):311–319
29. Balakrishnarajan MM, Jemmis ED (2000) Electronic requirements of polycondensed polyhedral boranes. J Am Chem Soc 122(18):4516–4517
30. Jemmis ED, Balakrishnarajan MM, Pancharatna PD (2001) Missing hydrogens in $B_{19}H_{20}^{-}$? application of electron counting rule for edge-sharing macropolyhedral boranes. Inorg Chem 40(8):1730–1731
31. Oliva JMR, Juanjo H, Drahomír K, John D, Rosenfeld VR (2013) Borane polyhedra as building blocks for unknown but potentially isolatable New molecules – extensions based on computations of the known $B_{18}H_{22}$ isomers. Croat Chem Acta 86(4):485–494
32. Shameema O, Jemmis ED (2008) Orbital compatibility in the condensation of polyhedral boranes. Angew Chem Int Ed 47(30):5561–5564
33. Kiani FA, Hofmann M (2006) Which nido:nido-macropolyhedral boranes are most stable? Inorg Chem 45(17):6996–7003

34. Hnyk D, Holub J, Jelínek T, Macháček J, Londesborough MGS (2010) Revisiting B20H16 by means of a joint computational/experimental NMR approach. Collect Czech Chem C 75(11):1115–1123
35. Jemmis ED, Balakrishnarajan MM, Pancharatna PD (2001) A unifying electron-counting rule for macropolyhedral boranes, metallaboranes, and metallocenes. J Am Chem Soc 123(18):4313–4323
36. Carr N, Mullica DF, Sappenfield EL, Stone FGA (1994) Carborane complexes of nickel and platinum: synthesis and protonation reactions of anionic allyl (carborane) species. Inorg Chem 33(8):1666–1673
37. Maier A, Hofmann M, Pritzkow H, Siebert W (2002) A planar, aromatic bicyclo-tetraborane(4). Angew Chem Int Ed 41(9):1529–1532
38. Präsang C, Hofmann M, Geiseler G, Massa W, Berndt A (2002) Aromatic boranes with planar-tetracoordinate boron atoms and very short B–B distances. Angew Chem Int Ed 41(9):1526–1529
39. Balakrishnarajan MM, Hoffmann R (2004) Electron-deficient bonding in ♦ rhomboid rings. J Am Chem Soc 126(40):13119–13131
40. Albert B (1998) A new "old" sodium boride: linked pentagonal bipyramids and octahedra in Na_3B_{20}. Angew Chem Int Ed 37(8):1117–1118
41. Brown LD, Lipscomb WN (1977) Closo boron hydrides with 13 to 24 boron atoms. Inorg Chem 16(12):2989–2996
42. Schleyer PR, Najafian K, Mebel AM (1998) The large closo-borane dianions, $B_nH_n^{2-}$ (n = 13–17) Are aromatic, why are they unknown? Inorg Chem 37(26):6765–6772
43. Deng L, Chan HS, Xie ZW (2006) Synthesis, structure, and reactivity of 13-vertex carboranes and 14-vertex metallacarboranes. J Am Chem Soc 128(15):5219–5230
44. Deng L, Zhang J, Chan HS, Xie ZW (2006) Synthesis and structure of 14-and 15-vertex ruthenacarboranes. Angew Chem Int Ed 45(26):4309–4313
45. McIntosh RD, Ellis D, Rosair GM, Welch AJ (2006) A 15-vertex heteroborane. Angew Chem Int Ed 45(26):4313–4316
46. Wong EH, Prasad L, Gabe EJ, Gatter MG (1983) Structural characterization of a $B_{11}H_{11}^{2-}$-derivative: molecular structure of $(C_2H_5)_4N^+B_{11}H_{10}S(CH_3)_2$. Inorg Chem 22(7):1143–1146
47. Burke A, Ellis D, Giles BT, Hodson BE, Macgregor SA, Rosair GM, Welch AJ (2003) Beyond the icosahedron: the first 13-vertex carborane. Angew Chem Int Ed 42(2):225–228
48. Grimes RN (2003) Supercarboranes. Angew Chem Int Ed 42(11):1198–1200
49. Pancharatna PD, Marutheeswaran S, Austeria MP, Balakrishnarajan MM (2013) Deltahedra with holes: structural preferences of supraicosahedral boranes. Polyhedron 63:55–59
50. Jemmis ED, Balakrishnarajan MM (2001) Polyhedral boranes and elemental boron: direct structural relations and diverse electronic requirements. J Am Chem Soc 123(18):4324–4330
51. Rathke J, Schaeffer R (1974) Boranes. XXXVIII. Reactions of hexaborane(10) with boron hydride lewis acids. Inorg Chem 13(12):3008–3011
52. Jemmis ED, Balakrishnarajan MM (2000) Ab initio predictions on novel stuffed polyhedral boranes. J Am Chem Soc 122(30):7392–7393
53. Albert B, Hillebrecht H (2009) Boron: elementary challenge for experimenters and theoreticians. Angew Chem Int Ed 48(46):8640–8668
54. Hughes RE, Kennard CHL, Sullenger DB, Weakliem HA, Sands DE, Hoard JL (1963) The structure of -rhombohedral boron. J Am Chem Soc 85(3):361–362
55. Prasad DLVK, Balakrishnarajan MM, Jemmis ED (2005) Electronic structure and bonding of beta-rhombohedral boron using cluster fragment approach. Phys Rev B 72(19):195102
56. Kasper JS, Richards SM (1969) The crystal structure of YB_{66}. Acta Cryst 25(6):237–251
57. Balakrishnarajan MM, Pancharatna PD, Hoffmann R (2007) Structure and bonding in boron carbide: the invincibility of imperfections. New J Chem 31(4):473–485
58. Balakrishnarajan MM, Hoffmann R (2003) Exohedral multiple bonding in polyhedra. 2. Skeletal distortions in ring-stacked boranes. Inorg Chem 43(1):27–32

59. Balakrishnarajan MM, Hoffmann R (2003) Polyhedral boranes with Exo multiple bonds: three-dimensional inorganic analogues of quinones. Angew Chem Int Ed 42(32):3777–3781
60. Finze M, Reiss GJ, Zähres M (2007) $[1\text{-}H_2N-CB_{11}F_{11}]^-$ – synthesis and reactions of a functionalized fluorinated carbadodecaborate anion. Inorg Chem 46(23):9873–9883
61. Farha OK, Julius RL, Lee MW, Huertas RE, Knobler CB, Hawthorne MF (2005) Synthesis of stable dodecaalkoxy derivatives of hypercloso-$B_{12}H_{12}$. J Am Chem Soc 127(51):18243–18251
62. Pancharatna PD, Balakrishnarajan MM, Jemmis ED, Hoffmann R (2012) Polyhedral borane analogues of the benzynes and beyond: bonding in variously charged $B_{12}H_{10}$ isomers. J Am Chem Soc 134(13):5916–5920
63. Lewis JS, Kaczmarczyk A (1966) Polyhedral borane free radicals 1. J Am Chem Soc 88(5):1068–1069
64. Joosten D, Pantenburg I, Wesemann L (2006) Distanna-closo-dodecaborate. Angew Chem Int Ed 45(7):1085–1087

Chapter 8
Applications of Nanocatalysis in Boron Chemistry

Yinghuai Zhu, Amartya Chakrabarti, and Narayan S. Hosmane

Abstract The rapid progress of nanoscience and nanotechnology opened up new vistas of application for nanomaterials. Nanocatalysis is a novel trend in the area of catalysis benefited by the advancements of nanotechnology. Developments in the field of boron chemistry have also been greatly influenced by nanocatalysis. In this chapter, we report recent updates on different boron compounds being either synthesized or utilized with the help of nanocatalysis.

8.1 Introduction

Nanomaterials are one of the modern-day marvels in the world of materials science with at least one of their dimensions on the order of less than 100 nm. The small size of nanomaterials makes their properties vary from those of their bulk counterparts in many respects. The unique improvements to the properties of nanomaterials make them useful for many applications [1]. The significant role of nanomaterials in different scientific and technological fields is being investigated all around the globe, and several major breakthroughs have been achieved over the past few decades. Nonetheless, a variety of nanomaterials have been discovered recently with different morphologies, ranging from nanoparticles to self-assembled complex nanostructures. Discoveries in the field of nanotechnology have rapidly increased mainly due to the scientific advancements in the division of characterization tools, specifically in the area of microscopy. For example, electron microscopes have undoubtedly impacted the development of nanotechnology over the years.

Y. Zhu
Institute of Chemical and Engineering Sciences (ICES),
1 Pesek Road, Jurong Island, Singapore 627833, Singapore

A. Chakrabarti • N.S. Hosmane (✉)
Department of Chemistry and Biochemistry, Northern Illinois University,
1425 W. Lincoln Hwy, Dekalb 60115, IL, USA
e-mail: hosmane@niu.edu

D. Hnyk, M.L. McKee (eds.), *Boron*, Challenges and Advances in Computational Chemistry and Physics 20,
DOI 10.1007/978-3-319-22282-0_8

Advancement of nanotechnology has rendered its application in a variety of fields an extremely promising area in the field of catalysis. Catalysts are defined as substances that enhance the rate of a chemical reaction without going through any chemical alteration. They are an important class of materials in modern-day industries as the availability of a huge amount of inexpensive chemicals and pharmaceutical products largely depends on industrial catalytic processes. Depending on their morphology in the reaction medium, catalysts are generally classified as homogeneous and heterogeneous. Homogeneous catalysts work in the same phase with the reagents, while heterogeneous catalysts reside in a different phase during catalysis. Although homogeneous catalysts are preferred in reactions due to their high activity and selectivity [2], heterogeneous catalysts are also widely used in industries because of their ease of separation from the reaction media, which is considered to be a major advantage for this kind of catalysts [3].

On the other hand, nanomaterials are emerging as attractive choices as catalysts in various chemical processes. A high surface-to-volume ratio of metal nanoparticles with active surface atoms makes them superior catalysts in various reactions [4]. With the decrease of particle size, surface area per unit mass of material increases significantly. As the catalytic reaction takes place on the surface, nanomaterials with higher surface area compared to similar mass of bulk materials are expected to have high efficiency of activity. However, stabilization and separation of the nanoparticles from product mixtures remain big challenges in nanocatalysis.

Nanoparticles based on transition metals are subjects of great interest because of their efficient role as catalysts in several organic syntheses [5]. There are a variety of methods available to synthesize transition metal nanoparticles, including thermal decomposition, wet hydrogen and electrochemical reduction, sonochemical decomposition, UV photolysis, ion implantation, microwave irradiation, biomineralization and different mechanochemical synthetic techniques [5]. Depending on the process of synthesis, the particle size and morphology of the produced nanocatalysts and, consequently, their performance differ. There are extensive reports on nanocatalysts for many organic transformations [6]. This chapter concentrates on the influence and improvements involving nanocatalysis in boron chemistry, such as formation of C-C bonds (Suzuki Miyaura Coupling reaction and homocoupling reactions), formation and breaking of B-H bonds in borohydrides, as well as isotopic exchanges in a variety of molecules containing H-B, B-B and I-B bonds.

8.2 Suzuki Miyaura Cross-coupling Reactions

The Suzuki-Miyaura cross-coupling reaction, popularly known as Suzuki coupling, is one of the most commonly used C-C bond formation reactions in organic synthesis. The reaction involves palladium-catalyzed cross coupling between an aryl halide and an organoboronic acid. The well-characterized mechanism is shown in Scheme 8.1. Although it started with the palladium-catalyzed cross coupling between an organoboronic acid and aryl halides, over time different catalyst systems and methods were investigated, which have widened the applications [7]. The

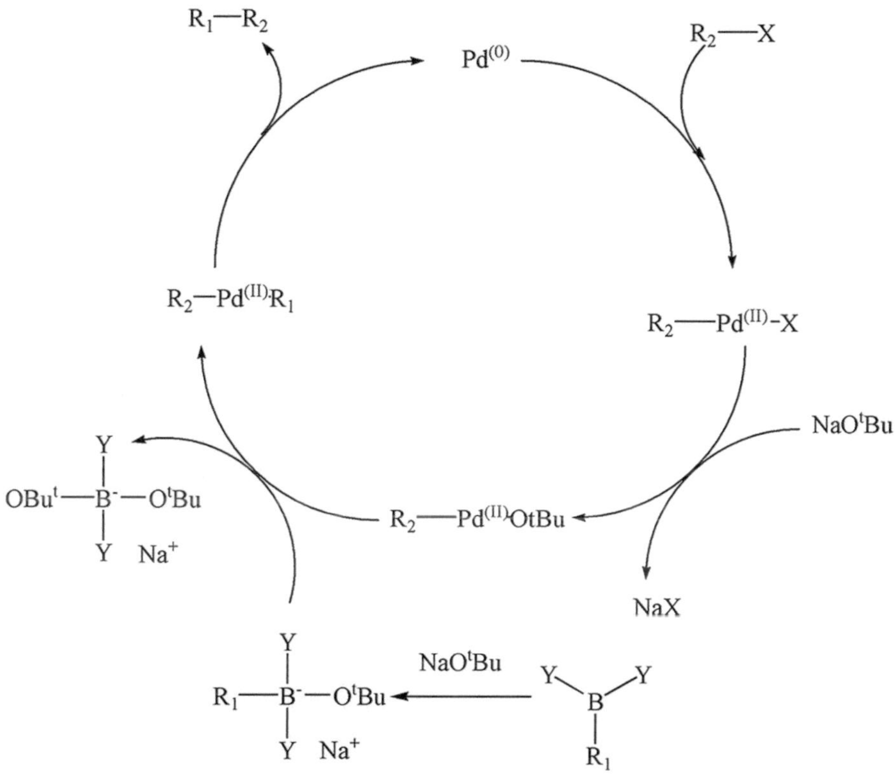

Scheme 8.1 Mechanism of Suzuki coupling

scope of the reaction now covers a variety of compounds, including alkyls, alkenyls and alkynyls.

Among several catalyst systems, palladium-based homogeneous catalysis has been the most investigated and reviewed in the literature [7]. In Suzuki-Miyaura reactions (Scheme 8.1), as the active Pd^0 species are involved in a proposed mechanism, pre-prepared palladium nanoparticles (Pd-NPs) become an excellent choice for the homocatalytic systems. When the Pd-NPs-based catalysts are introduced into homogeneous catalysts as a reservoir of catalytically active species, the resulting catalytic system has proven to be highly active for such reactions [8–14].

Reetz and co-workers have been pioneers in utilizing Pd-NPs for Suzuki reactions with their first reported synthesis involving Pd and Pd/Ni bimetallic clusters [8]. There have been several reports published since then with different morphologies of Pd-based nanomaterials, including spherical, cubic and polyhedral shapes. Many of those catalysts have been prepared and mounted on a variety of supports with different pore size and surface area, such as micro-, meso- and macro-porous carbon-based materials, metal oxides and polymers [8–14]. Although most of the Pd-NPs, either in supported or in colloidal composite morphology, are reported to be spherical or nearly spherical in shape, Pd-NPs of other shapes, such as nanorods

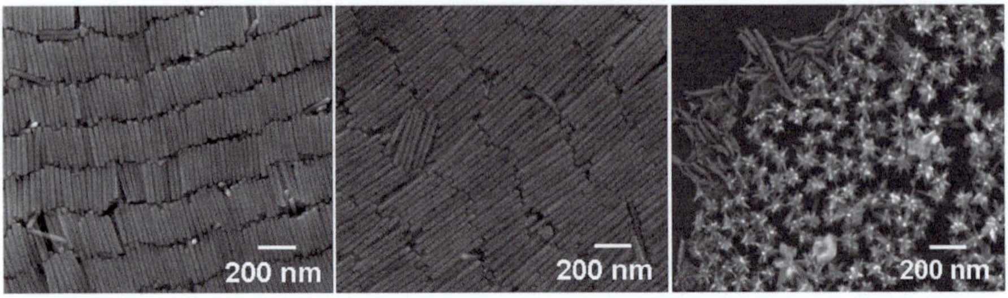

Fig. 8.1 TEM images of Pd-NPs with the shapes of *nanorods* and *branched stars* (Reprinted with permission from Ref. [15]. Copyright (2009) American Chemical Society)

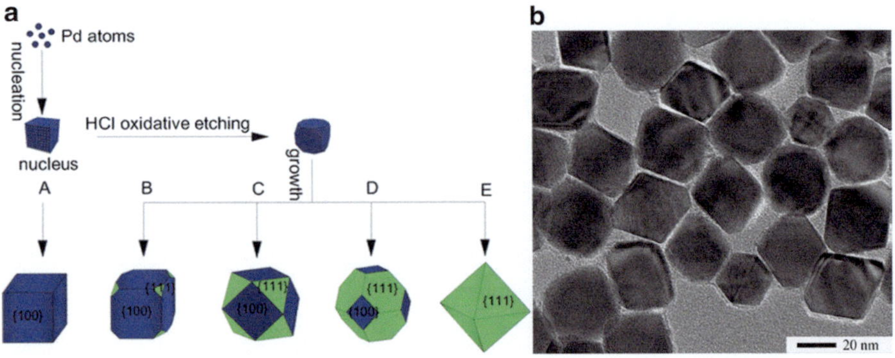

Fig. 8.2 (**a**) The shape evolution of the Pd NPs and (**b**) TEM image of Pd-nanocrystals (Reprinted with permission from Ref. [17]. Copyright (2014) American Chemical Society)

and branched star NPs (Fig. 8.1), have also been synthesized and successfully used in Suzuki-Miyaura coupling reactions [15, 16].

With technological advancements, different synthetic methodologies to prepare Pd-nanostructures with unique morphologies have surfaced. A very recent report 'showed' or 'demonstrated' that the presence of hydrochloric acid has a strong influence over the formation of different shapes of Pd nanocrystals. When a higher concentration of HCl was used, formation of octahedral Pd nanocrystals bounded by [111] facets predominated over the generation of Pd nanocubes bounded by [100] facets (Fig. 8.2) [17]. For the Pd- NPs-catalyzed Suzuki-Miyaura C–C cross-coupling reactions, the cubic Pd nanocrystals were found to be the most effective in terms of catalysis over their cubooctahedral and octahedral counterparts. Under similar reaction conditions and size of the nanoparticles, higher activity was noted with the smaller-sized Pd-NPs [18].

The success and potential of Pd-NPs in Suzuki coupling reactions many times has been restricted by economic factors and environmental concerns, due to the high price of Pd metal and a certain level of toxicity, respectively. Synthesis of Pd-based bi- or multi-metallic nanocatalysts are now being investigated. Bi- and multi-metallo

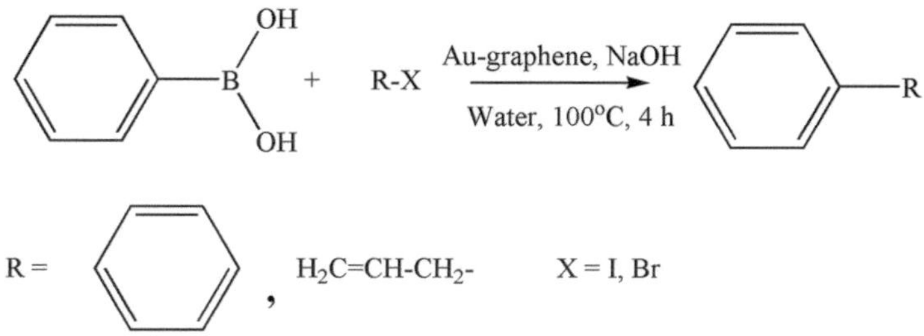

Scheme 8.2 Gold-graphene hybrid catalyzed Suzuki reaction

Pd-NPs were synthesized and examined for their activity in the cross-coupling reactions in colloidal systems [19]. With the monometallic NPs, Pd was found most active followed by Cu and Ru, while Pt exhibited no activity. In the case of bimetallic NPs, Cu/Pd and Ru/Pd were almost equally active, and Pd/Pt followed them with slightly lesser activity. Cu/Ru was found to be less active, and Cu/Pt was not active at all. Among the multi-metallic NPs, the sequence of activity was in the following order: Cu/Pd/Ru >Cu/Pd/Pt >Ru/Pd/Pt >Cu/Ru/Pt/Pd, and it was the combination of copper with palladium that could obtain comparable activity compared to the activity of solely Pd-based catalysts. Under similar reaction conditions, a study revealed that carbon-supported Pd/Cu exhibited higher activity than Pd alone [20]. Carbon-supported Pd/Ag system were less active, followed by Pd-Ni with the least activity in the same study. When ZnO nanopowder was used as support, the activity of the catalysts followed the sequence of Ag/Pd >Cu/Pd >Ni/Pd >Pd for Suzuki reactions, suggesting that the activity of catalysts is controlled not only by the metal center but also the support [21].

Other than Pd, metal nanoparticles including Fe- [22], Ni- [22–25], Rh- [26] and Au- [27, 28] have also been successfully utilized as catalysts in Suzuki coupling reactions. While the recyclability of Ni-based nanoparticles was uncertain because of the possibility of its oxidation during the reaction [25], Rh-based nanoparticles exhibited excellent stability, activity and recyclability for Suzuki coupling reactions [26]. Au-NPs also showed good catalytic performance in colloidal systems and better recyclability [27]. Graphene-supported Au nanoparticles were synthesized and utilized for Suzuki coupling reactions (Scheme 8.2) by Li et al. [29]. The Au-graphene hybrid nanostructures demonstrated high catalytic activity for Suzuki reactions in water under aerobic conditions.

8.3 Oxidative Homo-Coupling Reactions

Oxidative homo-coupling is a well-known methodology to produce symmetrical biaryls. With the availability of a variety of arylboronic acids, this method is becoming more popular lately. A generic mechanism for the coupling reaction is depicted in Scheme 8.3. Although Pd-based homogenous catalysts have been investigated for such reactions, the difficulties associated with catalyst recovery and many ligand-derived side reactions encouraged the rapid development of heterogeneous catalytic systems.

Carbon and protein-supported Pd-NPs have been widely studied as heterogeneous catalysts for the oxidative coupling reactions of arylboronic acids [30–32]. Cravotto and coworkers reported a combination of high-intensity ultrasound and microwave operated Pd/C catalyzed ligand-free homo-coupling as well as cross-coupling reactions, which have been conducted in a flow reactor [30]. Several biaryl compounds were successfully synthesized using the ligand-free catalyst system following the reactions exhibited in Scheme 8.4. The combined application of

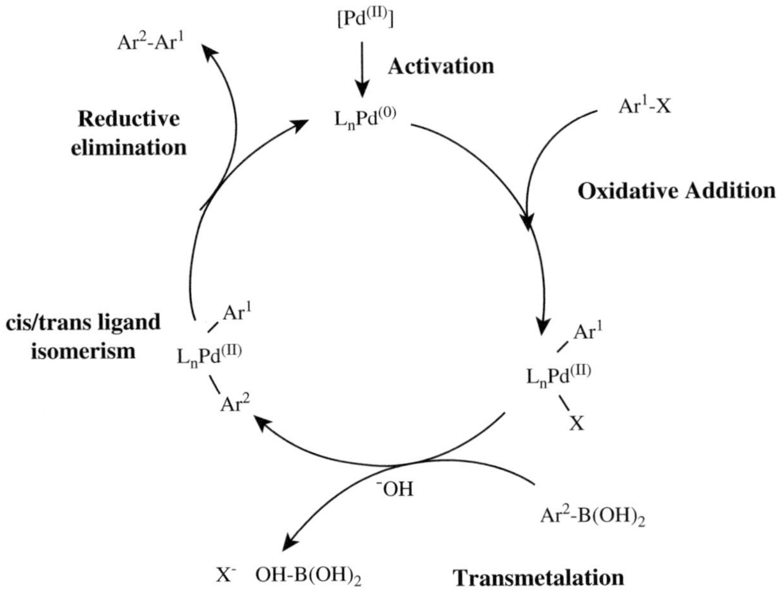

Scheme 8.3 Mechanism for aryl-aryl oxidative homo-coupling

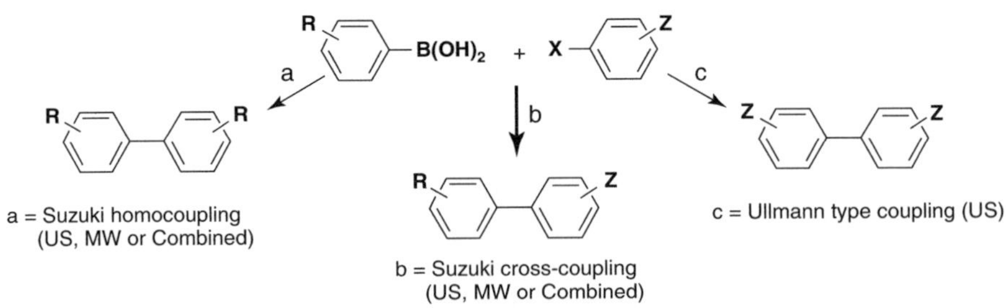

Scheme 8.4 Pd/C-catalyzed aryl–aryl couplings under MW and US, alone or combined (Reprinted with permission from Ref. [30]. Copyright (2005) Elsevier Ltd)

Fig. 8.3 TEM images of PVP-stabilized AuNPs (**a**) Au:PVP (K-15), (**b**) Au:PVP (K-30), and (**c**) Au:PVP (K-90). Three different PVPs were used for the preparation K-15, K-30 and K-90 with average molecular weight of 10, 40 and 630 kDa (Reprinted with permission from Ref. [33]. Copyright (2004) American Chemical Society)

microwave and ultrasound has improved the yield of Suzuki coupling reactions manifold. Another report by Chen et al. showed a very high efficiency and reasonable tolerance of functional groups (except p-MeS) by Pd/C catalysts for homo-coupling reactions in a ligand-free system where a 9:1 v/v ratio of water and 2-propanol was used as solvent [31]. Protein-stabilized Pd-NPs, prepared from a highly thermostable Dps protein from the thermophilic bacterium *Thermosynechoccus elongates*, were reported as highly active catalysts for the homo-coupling reactions of arylboronic acids and potassium aryltrifluoroborates under aerobic phosphine-free conditions in water [32].

Among other metals, apart from Pd, Au has exhibited good catalytic activities in C-C bond formation reactions. Poly(N-vinyl-2-pyrrolidone) or PVP-stabilized Au-NPs were synthesized by Tsunoyama and coworkers with a size distribution between 1.3 and 1.6 nm (Fig. 8.3) [33]. The PVP-stabilized Au-NPs were then tested for their catalytic activities for aerobic homocoupling of phenylboronic acids in aqueous media, and a high yield was reported from the reactions [33]. The molecular oxygen dissolved in water was found to play a crucial role in the catalytic activity through its interaction with the Au-nanoparticles [33]. In a separate report, a high activity of Au-NPs for the homo-coupling reactions of the potassium aryltrifluoroborates (Ar–B, (B = BF_3K)) in water was demonstrated [34]. Moreover, PVP-stabilized bimetallic $Au_{0.5}/Pd_{0.5}$ NPs exhibited good activity for homo-coupling reactions in the presence of aryl chlorides in water at room temperature [35].

Willis et al. investigated the catalytic activities of synthesized Au-NPs (size around 10 nm) and Pd-NPs (size around 15 nm) supported on different nanocrystalline metal oxides, MO_2 (M = Ce, Ti, Zr, Si) [36]. Except for SiO_2, the other oxide-supported Au-NPs produced 100 % phenylboronic acid conversion, and the selectivity towards the biphenyl product was also 100 %. Wang et al. reported formation of Au-NPs of small sizes between 1 and 4 nm on Mg-Al mixed oxides (Au/MAO) (Fig. 8.4) that exhibited excellent catalytic properties in the homo-coupling reaction of phenylboronic acid in 1.5 MPa O_2 at 100 °C [37]. The biphenyl product

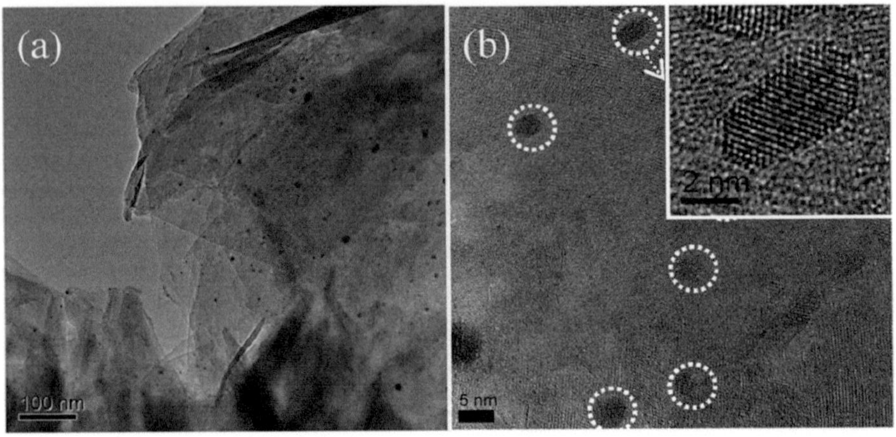

Fig. 8.4 TEM images of (**a**) Au/MAO nanocatalyst, (**b**) enlarged image in the *inset* (Reprinted with permission from Ref. [37]. Copyright (2012) Royal Society of Chemistry)

was obtained in about 90 % yield with 90 % selectivity, and the phenylboronic acid conversion was around 100 % when methanol was used as solvent [37]. On the contrary, Au nanoparticles formed over SiO_2, Al_2O_3 and TiO_2 yielded much lower conversions.

Polymer stabilizers are also effective supports for Au-NPs, which exhibited satisfactory activities for Suzuki homo-coupling reactions. While poly(aniline-*co*-*m*-ferrocenylaniline) [38] and polystyrene cored G3-polyamidoamine dendrimers [39] were proven to be effective polymers, chitosan-stabilized Au-NPs showed remarkable catalyst activity and recyclability under acidic conditions (pH 4.57) [40]. In addition, chitosan-supported bimetallic $Au_{0.81}/Pd_{0.19}$-NPs were found to be more active for the aerobic oxidative homo-coupling reactions under acidic conditions than other chitosan-supported metal NPs including Au-, $Au_{0.91}/Pd_{0.09}$-, $Au_{0.86}/Pd_{0.14}$-, $Au_{0.77}/Pd_{0.23}$- and $Au_{0.72}/Pd_{0.28}$-NPs [41]. The chitosan-supported $Au_{0.81}/Pd_{0.19}$-NPs catalysts have successfully produced biaryl products in quantitative yields from different $ArB(OH)_2$ substrates (Ar=C_6H_5, *p*-OMe-C_6H_4, *p*-Me-C_6H_4, *p*-F-C_6H_4, *p*-Cl-C_6H_4, *p*-CN-C_6H_4) [41]. However, another investigation has shown that the addition of aryl iodides prevents the homo-coupling reactions that are catalyzed by PVP-stabilized Au- and $Au_{0.5}/Pd_{0.5}$-NPs [42]. Chitosan-stabilized Au-NPs have also shown similar behavior in the presence of aryl iodides [42].

8.4 Carbonylative Cross-Coupling

An alternative way of preparing aryl ketones involves carbonylative cross-coupling, in which an reactive Pd(0) complex inserts into the C-X bond of a halide. The following steps include migratory insertion of carbon monoxide into the

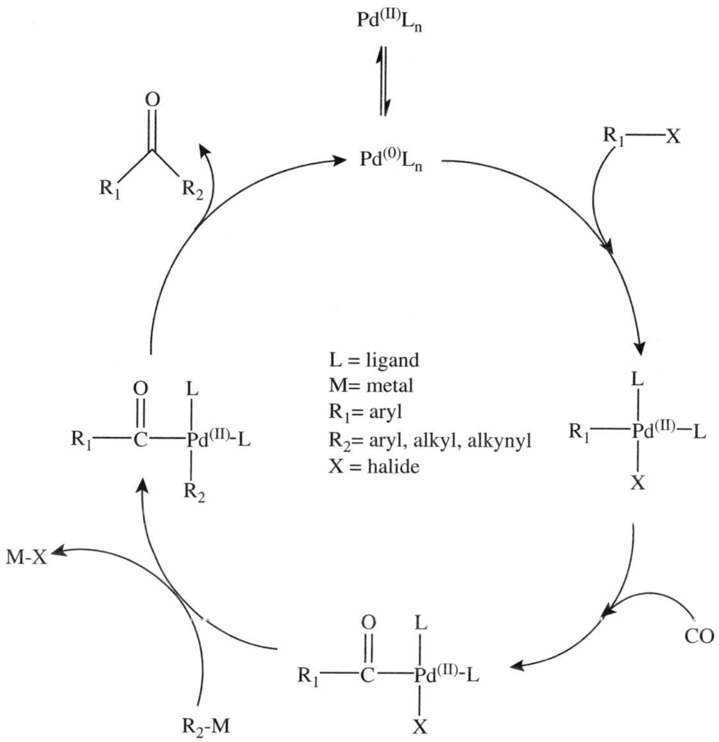

Scheme 8.5 Mechanism of carbonylative cross-coupling

palladium-aryl bond and transmetallation with the nucleophile. Then the process of reductive elimination provides the aryl ketone and regenerates the catalyst (Scheme 8.5). This process is advantageous over other methods like acylation, since it uses less toxic boronic acid as substrates and recyclable catalysts.

Khedkar and coworkers successfully utilized Pd/C heterogeneous recyclable catalyst for carbonylative cross-coupling reactions with a wide variety of aryl and heteroaryl iodides [43]. Very recently, graphene oxide (GO)-supported Pd-NPs were developed with a size averaging around 4.9 nm (Fig. 8.5). This unique catalyst system exhibited good activity for the carbonylative cross-coupling reaction, and was found recyclable at least three times without altering their activity [44].

8.5 Hydroboration Reactions and Dehydrogenation of Boron Compounds

Organoboranes are synthesized by hydroboration of a C=C or C≡C bond; typically the reaction is uncatalyzed. However, some transition metal complexes, including Co-, Au-, and Cu-complexes, were reported to facilitate the reaction at ambient

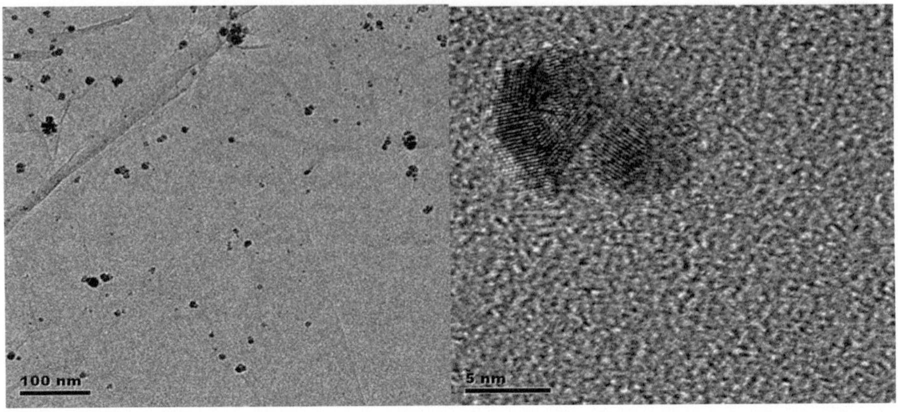

Fig. 8.5 TEM images of GO-supported Pd-NPs (Reprinted with permission from Ref. [44]. Copyright (2013) Royal Society of Chemistry)

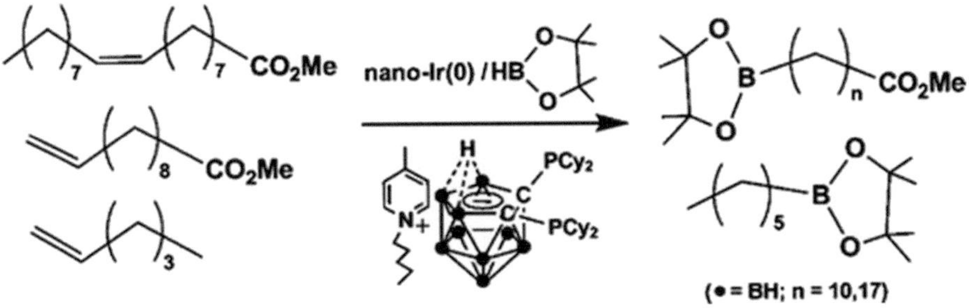

Scheme 8.6 Hydroboration catalyzed by Ir-Nps (Reprinted with permission from Ref. [54]. Copyright (2011) American Chemical Society)

temperature while affecting the chemo-, regio-, and stereoselectivity [45–51]. Selective diboronation of alkenes, e.g., styrene and vinylcyclohexene, was accomplished by utilizing ligand-stabilized Au-NPs that are around 7 nm in dimension [52]. Under similar conditions, Ag-NPs and Cu-NPs exhibited inferior activities. Grirrane et al. reported Pt-NPs/MgO, a mixed composite catalyst, to be efficient for diboronation of alkynes and superior over other comparable supported metal nanoparticles [53]. Zhu and coworkers have successfully demonstrated that Ir-NPs, with dimension around 4 nm and stabilized by a pyridinium salt of a carboranyl-diphosphine ligand, are active and recyclable for hydroboration reactions (Scheme 8.6) [54]. Production of the boronated ester by the hydroboration process is indicative of the involvement of an isomerization step.

In the dehydrogenation of the boron compounds, an H-H bond is formed at the expense of a B-H bond, and the generation of hydrogen gas is extremely important due to the potential application in hydrogen storage. While hydrogen has been identified as a better alternative energy source in comparison to conventional fossil fuels, storage and distribution of hydrogen gas remains a major challenge [55]. Different hydrogen-containing chemicals have attracted attention of the scientific community as a source material for hydrogen production, and of them, certain boron-containing compounds, including metal borohydrides ($NaBH_4$, $LiBH_4$ etc.), ammonia-borane (NH_3BH_3) and hydrazine borane ($N_2H_4BH_3$), have proven to be extremely promising [56–61].

Metal borohydrides react with water to form hydrogen gas; this process can be catalytically favored. Hydrolysis of metal borohydrides is an exothermic process and many times not favored as a hydrogen generator since the formation of MBO_2 (Eq. 8.1) increases the basicity of the medium, lowering the rate of H_2 formation [62, 63]. However, incorporation of a catalyst system has been proven to be helpful, and there are several reports suggesting such improvements [62, 63]. Among the catalysts, water-dispersible ruthenium nanoclusters [64], carbon-supported Pt catalysts [65], a robust Co-W-B hybrid catalyst supported on Ni-foam [66] were reported to generate high rate of hydrogen gas (more than 8000 mL/min/g) via catalytic hydrolysis of $NaBH_4$. On the contrary, $LiBH_4$ is not as promising as $NaBH_4$ in H_2 production through hydrolysis because of its thermodynamic and kinetic limitations. Moreover, both the hydrogenation and dehydrogenation temperatures are too high for $LiBH_4$ to be used for hydrogen storage. Au and coworkers have shown that modifying $LiBH_4$ by doping with 25 % TiO_2 has reduced the dehydrogenation temperature to 100 °C [67]. A nanocomposite of $LiBH_4$ and C_{60} with a 70/30 weight ratio demonstrated H_2 absorption under mild conditions of 150 °C and 10 bar H_2 pressure [68]. Size reduction of hydride particles to the nanometer scale also played a significant role in lowering the dehydrogenation temperature of the materials. This observation can be attributed to high surface to volume ratio and the higher number of surface atoms. While a $LiBH_4$@C microporous-macroporous type material has shown a decreased dehydrogenation temperature of 200 °C [69], a core-shell nanocomposite of nano-$NaBH_4$ (<30 nm) @Ni exhibited hydrogen absorption/desorption capacity at 350 °C and 4 MPa H_2 [70].

$$MBH_4 + 2H_2O \rightarrow 4H_2 \uparrow + MBO_2 (M = Li, Na, K) \quad (8.1)$$

Ammonia-borane or aminoborane is a low molecular weight borane with a high hydrogen capacity (hydrogen gravimetric density of 19.6 wt%, 148 g H_2 L^{-1}), thus a promising hydrogen storage material [71, 72]. It is a very stable compound under inert and dry conditions and suffers dehydrogenation either *via* hydrolysis, methanolysis or thermolysis (Eqs. 8.2, 8.3, and 8.4) [71]. Among several efficient nanocatalysts, Ag@Co/Graphene [73], Pd@Co/Graphene [74, 75], Ru@Co/C [75], Ni@Ru [76], Ru@SiO_2 [77] and Ru/C [78], to name a few, show a very high turnover frequency (TOF) of more than 100 (molH_2·mol catalyst^{-1}·min^{-1}). In comparison to noble metal-based nanocatalysts like Ru-, Pt- or Pd-NPs, non-noble metals failed to exceed performance in terms of activity when they are used. Methanolysis is a process similar to hydrolysis that can be used to break the B-H bond of aminoborane to generate H_2 gas. This process, however, has not been explored significantly, and there are not many nanocatalysts reported to be used. Dai et al. investigated montmorillonite-supported Ru-NPs for methanolysis of aminoborane and found the TOF ($molH_2 \cdot mol\ catalyst^{-1} \cdot min^{-1}$) to be as high as 90.9 [79]. Two other catalyst systems, zeolite-supported Rh-NPs [80] and carbon-supported $Co_{48}Pd_{52}$-NPs [81] have shown lower TOFs of 30 and 27.7, respectively. As shown in Eq. 8.4, it is evident that thermolysis of aminoborane generates H_2 but it requires a very high temperature for the complete conversion. Pd- [82], Rh- [83], and Ru-NPs [84] are found to be active catalysts for dehydrogenation of secondary amine borane, Me_2NHBH_3, with TOFs (h^{-1}) of 75, 60 and 55 at room temperature, respectively.

$$NH_3 \cdot BH_3 + 2H_2O \rightarrow NH_4BO_2 + 3H_2 \uparrow \tag{8.2}$$

$$NH_3 \cdot BH_3 + 4MeOH \rightarrow NH_4B(OMe)_4 + 3H_2 \uparrow \tag{8.3}$$

$$NH_3 \cdot BH_3 \xrightarrow{100\sim200^{o}\,C} 2H_2 \uparrow + [NH^{-}BH] \xrightarrow{>500^{o}\,C} H_2 \uparrow + [NB] \tag{8.4}$$

Chemical modification of aminoborane *via* replacing the NH_3 moiety by hydrazine (N_2H_4) leads to the formation of hydrazine borane ($N_2H_4BH_3$, 15.4 wt%, 149 g H_2 L^{-1}). Hydrolysis of hydrazine borane to produce 5 equivalent of H_2 per mole of compound can be achieved ideally following Eq. 8.5. However, at room temperature only 3 equivalent of H_2 are released, as the dehydrogenation of N_2H_4 is a slower process and also requires elevated temperatures. Different catalyst systems were utilized by several research groups. It has been found that Rh- [85] and Ni-based [86] nanocatalysts provided relatively high TOFs.

$$N_2H_4 \cdot BH_3 + 3H_2O \rightarrow B(OH)_3 + 5H_2 \uparrow + N_2 \uparrow \tag{8.5}$$

8.6 Isotope Exchange Reactions

Boron neutron capture therapy (BNCT) is a binary treatment for cancer. BNCT is based on the nuclear reaction that occurs when ^{10}B isotope is irradiated with low-energy thermal neutrons to yield excited ^{11}B atoms. The unstable ^{11}B atoms then produce high linear energy transfer alpha particles and recoiling lithium-7 nuclei, which eventually destroy the tumor cells [87–89]. Thus, isotope-enriched boron compounds plays a crucial role in medicinal application. Now, different methodologies are available to prepare isotope-enriched boron compounds, and catalytic isotope exchange reactions are more efficient than the conventional multistep synthetic procedures. The multistep synthetic methodologies not only use longer reaction times but also are typically of low yield. On the contrary, catalytic hydrogen-deuterium (H-D) exchange in borane compounds with aluminum, ruthenium and iridium complexes demonstrated good activity [90–95]. Moreover, Pd-, Pt-, Rh- and Ni-based nanoparticles have been thoroughly investigated to catalyze the H-D

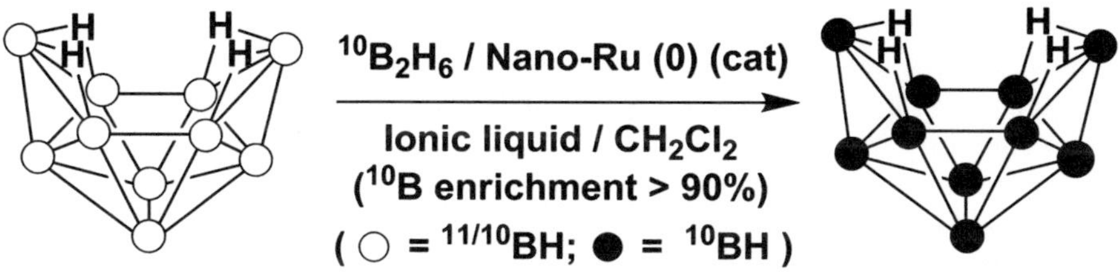

Scheme 8.7 Ru(0)-NPs catalyzed isotope exchange reaction (Reprinted with permission from Ref. [97]. Copyright (2007) American Chemical Society)

isotope exchanges in carbon-based compounds [96]. Zhu et al. reported for the first time a catalytic isotope exchange method to prepare ^{10}B-enriched decaborane (14), one of the most useful precursors of boron hydrides useful in the syntheses of enriched boron-containing compounds [97]. The method involved Ru(0)-NPs dispersed and stabilized in an ionic liquid, trihexyltetradecylphosphonium dodecylbenzenesulfonate, as a catalyst. The isotope exchange reaction between ^{10}B enriched diborane and naturally abundant decaborane produced highly ^{10}B-enriched (~90 %) decaborane (14) (Scheme 8.7). On the other hand, ^{125}I-labeled borane compounds were recognized as potential biomarkers for pharmacokinetic studies in BNCT research [98]. Several ^{125}I-labeled carboranes were reported to be synthesized in high yields by Herrmann's catalyst, a Pd-based complex, *via* isotope exchange in 2-I-p-, 3-I-o-, 9-I-o-, and 9-I-m-1,2-carboranes [99].

8.7 Conclusions and Future Perspectives

Nanocatalysis is an emerging field greatly influenced by the rapid development of nanotechnology. Different areas of boron chemistry benefit by the application of nanocatalysts. Suzuki-Miyaura coupling reactions are one of the areas where the catalytic breakdown of C-B bonds in arylboronic acids is the key step. Although noble metal-based nanoparticles are largely under investigation, several multi metal nanoparticles and non-noble metal nanoparticles are also getting attention. Moreover, catalyst supports and stabilizers are also proven to be vital for the activity of such nanocatalysts. Nanocatalysis is also playing a pivotal role in hydrogen storage by actively taking part in the dehydrogenation process of borohydrides. However, more understanding of the physical phenomena is required for further development of borohydrides-based hydrogen storage materials. Isotope exchange of H-B, B-B and I-B bonds, to produce isotope-enriched boron compounds mediated by different metal-based nanoparticles is also being explored. Application of such compounds in medical science and/or cancer therapy is a major driving force for research in this direction. The high efficiency and recyclability of metal NPs warranties further utilization of these NPs-based catalysts for the isotope exchange reactions in the near future.

We have reviewed several aspects of boron chemistry that are influenced and improved by the application of nanocatalysis. We have occupied our fullest potential to cover this huge area of research and application, but as both nanocatalysis and boron chemistry are developing at a tremendous growth rate, we may fail to cite some of the recent publications. In conclusion, a complete replacement of conventional catalysts by their nanomaterials-based counterpart may not take place in the near future, but a rapid rate of investigative research to ensure such success is highly warranted.

Acknowledgement This work was supported by grants from the Institute of Chemical and Engineering Sciences Ltd. (ICES) Singapore. NSH thanks the support from the National Science Foundation (Grant #CHE-0906179) and the Chinese Academy of Sciences for the visiting professorship for senior international scientists at Ningbo Institute of Materials Technology and Engineering (NIMTE).The timely help of Dr. K. Vijayaraghavan in proofreading the manuscript is gratefully acknowledged.

References

1. Cao G (2004) Nanostructures & nanomaterials: synthesis, properties & applications. Imperial College Press, London
2. (a) Viswanathan B (2009) Catalysis: selected applications, Alpha Science International Ltd., Oxford, UK; (b) Cole-Hamilton D (2003) Homogeneous catalysis – new approaches to catalyst separation, recovery, and recycling. Science 299:1702–1706
3. (a) Rase HF (2000) Handbook of commercial catalysts: Heterogeneous catalysts. CRC Press, New York; (b) Yoon TJ, Lee W, Oh Y-S, Lee J-K (2003) Magnetic nanoparticles as a catalyst vehicle for simple and easy recycling. New J Chem 27:227–229
4. Goesmann H, Feldmann C (2010) Nanoparticulate functional materials. Angew Chem Int Ed 49:1362–1395
5. (a) Ferrando R, Jellinek J, Johnston RL (2008) Nanoalloys: from theory to applications of alloy clusters and nanoparticles. Chem Rev 108:845–910; (b) Cuenya BR (2010) Synthesis and catalytic properties of metal nanoparticles: Size, shape, support, composition, and oxidation state effects. Thin Solid Films 518:3127–3150; (c) Paskevicius M, Webb J, Pitt MP et al (2009) Mechanochemical synthesis of aluminium nanoparticles and their deuterium sorption properties to 2kbar. J Alloys Compd 481:595–599; (d) Huot J, Ravnsbæk DB, Zhang J et al (2013) Mechanochemical synthesis of hydrogen storage materials. Prog Mater Sci 58: 30–75; (e) Yu W, Porosoff MD, Chen JG (2012) Review of Pt-based bimetallic catalysis: From model surfaces to supported catalysts. Chem Rev 112:5780–5817; (f) Jiang H-L, Xu Q (2011) Recent progress in synergistic catalysis over heterometallic nanoparticles. J Mater Chem 21:13705–13725
6. (a) Zhu Y, Lee CN, Kemp RA et al (2008) Latest developments in the catalytic application of nanoscaled neutral group 8–10 metals. Chem Asian J 3:650–662; (b) Yan N, Xiao C, Kou Y (2010) Transition metal nanoparticle catalysis in green solvents. Coord Chem Rev 254:1179–1218
7. Miyaura N, Suzuki A (1995) Palladium-catalyzed cross-coupling reactions of organoboron compounds. Chem Rev 95:2457–2483
8. Reetz MT, Breinbauer R, Wanninger K (1996) Suzuki and fleck reactions catalyzed by preformed palladium clusters and palladium/nickel bimetallic clusters. Tetrahedron Lett 37:4499–4502

9. Reetz MT, Westermann E (2000) Phosphane-free palladium-catalyzed coupling reactions: The decisive role of Pd nanoparticles. Angew Chem Int Ed 39:165–168
10. Yin L, Liebscher J (2007) carbon–carbon coupling reactions catalyzed by heterogeneous palladium catalysts. Chem Rev 107:133–173
11. Fihri A, Bouhrara M, Nekoueishahraki B et al (2011) Nanocatalysts for Suzuki cross-coupling reactions. Chem Soc Rev 40:5181–5203
12. Moreno-Manas M, Pleixats R (2003) Formation of carbon–carbon bonds under catalysis by transition-metal nanoparticles. Acc Chem Res 36:638–643
13. Narayanan R (2010) Recent advances in noble metal nanocatalysts for Suzuki and Heck cross-coupling reactions. Molecules 15:2124–2138
14. Ranu BC, Dey R, Chatterjee T, Ahammed S (2012) Copper nanoparticle-catalyzed carbon-carbon and carbon-heteroatom bond formation with a greener perspective. ChemSusChem 5:22–44
15. Chen Y-H, Hung H-H, Huang MH (2009) Seed-mediated synthesis of palladium nanorods and branched nanocrystals and their use as recyclable Suzuki coupling reaction catalysts. J Am Chem Soc 131:9114–9121
16. Piao Y, Jang Y (2007) Facile aqueous-phase synthesis of uniform palladium nanoparticles of various shapes and sizes. Small 3:255–260
17. Zhang J, Feng C, Deng Y et al (2014) Shape-controlled synthesis of palladium single-crystalline nanoparticles: the effect of HCl oxidative etching and facet dependent catalytic properties. Chem Mater 26:1213–1218
18. Zheng Z, Li H, Liu T, Cao R (2010) Monodisperse noble metal nanoparticles stabilized in SBA-15: synthesis, characterization and application in microwave-assisted Suzuki–Miyaura coupling reaction. J Catal 270:268–274
19. Thathagar MB, Beckers J, Rothenberg G (2002) Copper-catalyzed Suzuki cross-coupling using mixed nanocluster catalysts. J Am Chem Soc 124:11858–11859
20. Kim S-J, Oh S-D, Lee S, Choi S-H (2008) Radiolytic synthesis of Pd-M (M=Ag, Ni, and Cu)/C catalyst and their use in Suzuki-type and Heck-type reaction. J Indus Eng Chem 14:449–456
21. Kim M-R, Choi S-H (2009) One-step synthesis of Pd-M/ZnO (M=Ag, Cu, and Ni) catalysts by γ-irradiation and their use in hydrogenation and Suzuki Reaction. J Nanomater 302919:1–7
22. Jana R, Pathak TP, Sigman MS (2011) Advances in transition metal (Pd, Ni, Fe)-catalyzed cross-coupling reactions using alkyl-organometallics as reaction partners. Chem Rev 111:1417–1492
23. Lipshutz BH, Tasler S, Chrisman W et al (2003) On the nature of the 'heterogeneous' catalyst: Nickel-on-charcoal. J Org Chem 68:1177–1189
24. Park J, Kang E, Son SU et al (2005) Monodisperse nanoparticles of Ni and NiO: synthesis, characterization, self-assembled superlattices, and catalytic applications in the Suzuki coupling reaction. Adv Mater 17:429–434
25. Rosen BM, Quasdorf KW, Wilson DA et al (2011) Nickel-catalyzed cross-couplings involving carbon-oxygen bonds. Chem Rev 111:1346–1416
26. Gniewek A, Trzeciak AM (2013) Rh(0) nanoparticles: Synthesis, structure and catalytic application in Suzuki–Miyaura reaction and hydrogenation of benzene. Top Catal 56:1239–1245
27. Han J, Liu Y, Guo R (2009) Facile synthesis of highly stable gold nanoparticles and their unexpected excellent catalytic activity for Suzuki-Miyaura cross-coupling reaction in water. J Am Chem Soc 131:2060–2061
28. Garcia P, Malacria M, Aubert C et al (2010) Gold-catalyzed cross-couplings: New opportunities for C-C bond formation. Chem Cat Chem 2:493–497
29. Li Y, Fan X, Qi J et al (2010) Gold nanoparticles-graphene hybrids as active catalysts for Suzuki reaction. Mat Res Bull 45:1413–1418
30. Cravotto G, Beggiato M, Penoni A et al (2005) High-intensity ultrasound and microwave, alone or combined, promote Pd/C-catalyzed aryl–aryl couplings. Tetrahedron Lett 46:2267–2271

31. Chen J-S, Krogh-Jespersen K, Khinast JG (2008) Base- and ligand-free heterogeneously catalyzed homocoupling of arylboronic acids. J Mol Catal A285:14–19
32. Prastaro A, Ceci P, Chiancone E et al (2010) Homocoupling of arylboronic acids and potassium aryltrifluoroborates catalyzed by protein-stabilized palladium nanoparticles under air in water. Tetrahedron Lett 51:2550–2552
33. Tsunoyama H, Sakurai H, Ichikuni N et al (2004) Colloidal gold nanoparticles as catalyst for carbon-carbon bond formation: Application to aerobic homocoupling of phenylboronic acid in water. Langmuir 20:11293–11296
34. Sakurai H, Tsunoyama H, Tsukuda T et al (2007) Oxidative homo-coupling of potassium aryltrifluoroborates catalyzed by gold nanocluster under aerobic conditions. J Organomet Chem 692:368–374
35. Dhital RN, Sakurai H (2012) Anomalous efficacy of bimetallic Au/Pd nanoclusters in C–Cl bond activation and formal metathesis-type C–B bond activation at room temperature. Chem Lett 41:630–632
36. Willis NG, Guzman J (2008) Influence of the support during homocoupling of phenylboronic acid catalyzed by supported gold. Appl Catal A 339:68–75
37. Wang L, Zhang W, Su DS et al (2012) Supported Au nanoparticles as efficient catalysts for aerobic homocoupling of phenylboronic acid. Chem Commun 48:5476–5478
38. Chaicharoenwimolkul L, Munmai A, Chairam S et al (2008) Effect of stabilizing ligands bearing ferrocene moieties on the gold nanoparticle-catalyzed reactions of arylboronic acids. Tetrahedron Lett 49:7299–7302
39. Zheng J, Lin S, Zhu X et al (2012) Reductant-directed formation of PS–PAMAM-supported gold nanoparticles for use as highly active and recyclable catalysts for the aerobic oxidation of alcohols and the homocoupling of phenylboronic acids. Chem Commun 48:6235–6237
40. Dhital RN, Murugadoss A, Sakurai H (2012) Dual roles of polyhydroxy matrices in the homocoupling of arylboronic acids catalyzed by gold nanoclusters under acidic conditions. Chem Asian J 7:55–59
41. Sophiphun O, Wittayakun J, Dhital RN et al (2012) Gold/palladium bimetallic alloy nanoclusters stabilized by chitosan as highly efficient and selective catalysts for homocoupling of arylboronic acid. Aust J Chem 65:1238–1243
42. Dhital RN, Kamonsatikul C, Somsook E et al (2013) Aryl iodides as strong inhibitors in gold and gold-based bimetallic quasi-homogeneous catalysis. Chem Commun 49:2542–2544
43. Khedkar MV, Tambade PJ, Qureshi ZS, Bhanage BM (2010) Pd/C: an efficient, heterogeneous and reusable catalyst for phosphane-free carbonylative Suzuki coupling reactions of aryl and heteroaryl iodides. Eur J Org Chem 36:6981–6986
44. Oh AB, Vangala VR, Chen CS et al (2014) Cross-coupling reaction between arylboronic acids and carboranyl iodides catalyzed by graphene oxide (GO)-supported Pd(0) recyclable nanoparticles for the synthesis of carboranylaryl ketones. Dalton Trans 43:5014–5020
45. Smith K, Pelter A (1991) A hydroboration of C=C and C≡C. In: Trost BM, Fleming I (eds) Comprehensive organic synthesis, vol 8. Pergamon, Oxford, pp 703–731
46. Miyaura N (2001) Hydroboration, diboration, silylboration, and stannylboration. In: Togni A, Grutzmacher H (eds) Catalytic hetero-functionalization from hydroboration to hydrozirconation. Wiley-VCH, Weinheim, Germany, p1
47. Bose SK, Fucke K, Liu L et al (2014) Zinc-catalyzed borylation of primary, secondary and tertiary alkyl halides with alkoxy diboron reagents at room temperature. Angew Chem Int Ed 53:1799–1803
48. Zhang L, Zuo Z, Leng X, Huang Z (2014) A cobalt-catalyzed alkene hydroboration with pinacolborane. Angew Chem Int Ed 53:2696–2700
49. Takaya J, Iwasawa N (2012) Catalytic, direct synthesis of bis(boronate) compounds. ACS Catal 2:1993–2006
50. Luo Y, Roy ID, Madec AGE, Lam HW (2014) Enantioselective synthesis of allylboronates and allylic alcohols by copper-catalyzed 1,6-boration. Angew Chem Int Ed 53:4186–4190
51. Greenhalgh MD, Thomas SP (2013) Chemo-, regio-, and stereoselective iron-catalysed hydroboration of alkenes and alkynes. Chem Commun 49:11230–11232

52. Ramírez J, Sanaú M, Fernández E (2008) Gold(0) nanoparticles for selective catalytic diboration. Angew Chem Int Ed 47:5194–5197
53. Grirrane A, Corma A, Garcia H (2011) Stereoselective single (copper) or double (platinum) boronation of alkynes catalyzed by magnesia-supported copper oxide or platinum nanoparticles. Chem Eur J 17:2467–2478
54. Zhu Y, Jang SHA, Tham YH et al (2012) An efficient and recyclable catalytic system comprising nano-iridium(0) and a pyridinium salt of nido-carboranyldiphosphine for the synthesis of one-dimensional boronate esters via hydroboration reaction. Organometallics 31:2589–2596
55. Armaroli N, Vincenzo B (2011) The Hydrogen Issue. ChemSusChem 4:21–36
56. Li C, Peng P, Zhou DW, Wan L (2011) Research progress in $LiBH_4$ for hydrogen storage: A review. Int J Hydrogen Energy 36:14512–14526
57. Goudon JP, Bernard F, Renouard J, Yvart P (2010) Experimental investigation on lithium borohydride hydrolysis. Int J Hydrogen Energy 35:11071–11076
58. Sahin Ö, Dolaş H, Özdemir M (2007) The effect of various factors on the hydrogen generation by hydrolysis reaction of potassium borohydride. Int J Hydrogen Energy 32:2330–2336
59. Xu D, Wang H, Guo Q, Ji S (2011) Catalytic behavior of carbon supported Ni–B, Co–B and Co–Ni–B in hydrogen generation by hydrolysis of KBH_4. Fuel Process Tech 92:1606–1610
60. Akdim O, Demirci UB, Miele P (2011) Deactivation and reactivation of cobalt in hydrolysis of sodium borohydride. Int J Hydrogen Energy 36:13669–13675
61. Liu BH, Li ZP (2009) A review: Hydrogen generation from borohydride hydrolysis reaction. J Power Sour 187:527–534
62. Schlesinger HI, Brown HC, Finholt AE et al (1953) Sodium borohydride, its hydrolysis and its use as a reducing agent and in the generation of hydrogen. J Am Chem Soc 75:215–219
63. Merrero-Alfonso E, Gray JR, Davis TA, Matthews MA (2007) Hydrolysis of sodium borohydride with steam. Int J Hydrogen Energy 32:4717–4722
64. Ozkar S, Zahmakiran M (2005) Hydrogen generation from hydrolysis of sodium borohydride using Ru(0) nanoclusters as catalyst. J Alloys Compd 404–406:728–731
65. Bai Y, Wu C, Wu F, Yi B (2006) Carbon-supported platinum catalysts for on-site hydrogen generation from $NaBH_4$ solution. Mater Lett 60:2236–2239
66. Dai HB, Liang Y, Wang P et al (2008) High-performance cobalt–tungsten–boron catalyst supported on Ni foam for hydrogen generation from alkaline sodium borohydride solution. Int J Hydrogen Energy 33:4405–4412
67. Au M, Spencer W, Jurgensen A, Zeigler C (2008) Hydrogen storage properties of modified lithium borohydrides. J Alloys Compd 462:303–309
68. Ward PA, Teprovich JA Jr, Peters B et al (2013) Reversible hydrogen storage in a $LiBH_4$–C60 Nanocomposite. J Phys Chem C 117:22569–22575
69. Brun N, Janot R, Sanchez C et al (2010) Preparation of $LiBH_4$@carbon micro–macrocellular foams: tuning hydrogen release through varying microporosity. Energy Env Sci 3:824–830
70. Christian ML, Aguey-Zinsou K-F (2012) core-shell strategy leading to high reversible hydrogen storage capacity for $NaBH_4$. ACS Nano 6:7739–7751
71. Moussa G, Moury R, Demirci UB et al (2013) Boron based hydrides for chemical hydrogen storage. Int J Energy Res 37:825–842
72. Petit J-F, Moussa G, Demirci UB et al (2014) Hydrazine borane-induced destabilization of ammonia borane, and vice versa. J Hazard Mater 278:158–162
73. Yang L, Luo W, Cheng G (2013) Graphene-supported Ag-based core–shell nanoparticles for hydrogen generation in hydrolysis of ammonia borane and methylamine borane. ACS Appl Mater Interfaces 5:8231–8240
74. Wang J, Qin YL, Liu X, Zhang XB (2012) In situ synthesis of magnetically recyclable graphene-supported Pd@Co core–shell nanoparticles as efficient catalysts for hydrolytic dehydrogenation of ammonia borane. J Mater Chem 22:12468–12470
75. Cao N, Su J, Hong X et al (2014) In situ facile synthesis of Ru-based core-shell nanoparticles supported on carbon black and their high catalytic activity in the dehydrogenation of amine-boranes. Chem An Asian J 9:562–571

76. Chen GZ, Desinan S, Nechache R et al (2011) Bifunctional catalytic/magnetic Ni@Ru core-shell nanoparticles. Chem Commun 47:6308–6310
77. Yao Q, Shi W, Feng G et al (2014) Ultrafine Ru nanoparticles embedded in SiO_2 nanospheres: highly efficient catalysts for hydrolytic dehydrogenation of ammonia borane. J Power Sour 257:293–299
78. Liang HY, Chen GZ, Desinan S et al (2012) In situ facile synthesis of ruthenium nanocluster catalyst supported on carbon black for hydrogen generation from the hydrolysis of ammonia-borane. Int J Hydrogen Energy 37:17921–17927
79. Dai HB, Kang XD, Wang P (2010) Ruthenium nanoparticles immobilized in montmorillonite used as catalyst for methanolysis of ammonia borane. Int J Hydrogen Energy 35:10317–10323
80. Caliskan S, Zahmakıran M, Özkar S (2010) Zeolite confined rhodium(0) nanoclusters as highly active, reusable, and long-lived catalyst in the methanolysis of ammonia-borane. Appl Catal B 93:387–394
81. Sun D, Mazumder V, Metin Ö, Sun S (2012) Methanolysis of ammonia borane by CoPd nanoparticles. ACS Cat 2:1290–1295
82. Gulcan M, Zahmakiran M, Özkar S (2014) Palladium(0) nanoparticles supported on metal organic framework as highly active and reusable nanocatalyst in dehydrogenation of dimethylamine-borane. Appl Catal B Env 147:394–401
83. Zahmakiran M, Özkar S (2009) Dimethylammonium hexanoate stabilized rhodium(0) nanoclusters identified as true heterogeneous catalysts with the highest observed activity in the dehydrogenation of dimethylamine–borane. Inorg Chem 48:8955–8964
84. Zahmakiran M, Tristany M, Philippot K et al (2010) Aminopropyltriethoxysilane stabilized ruthenium(0) nanoclusters as an isolable and reusable heterogeneous catalyst for the dehydrogenation of dimethylamine–borane. Chem Commun 46:2938–2940
85. (a) Karahan S, Zahmakran M, Özkar S (2011) Catalytic hydrolysis of hydrazine borane for chemical hydrogen storage: Highly efficient and fast hydrogen generation system at room temperature. Int J Hydrogen Energy 36:4958–4966; (b) Clik D, Karahan S, Zahmakran M, Özkar S (2012) Hydrogen generation from the hydrolysis of hydrazine-borane catalyzed by rhodium(0) nanoparticles supported on hydroxyapatite. Int J Hydrogen Energy 37:5143–5151; (c) Zhang DC, Aranishi K, Singh AK et al (2012) The synergistic effect of Rh–Ni catalysts on the highly-efficient dehydrogenation of aqueous hydrazine borane for chemical hydrogen storage. Chem Commun 48:11945–11947
86. (a) Sencanli S, Karahan S, Özkar S (2013) Poly(4-styrenesulfonic acid-co-maleic acid) stabilized nickel(0) nanoparticles: Highly active and cost effective catalyst in hydrogen generation from the hydrolysis of hydrazine borane. Int J Hydrogen Energy 38:14693–14703; (b) Hannauer J, Akdim O, Demirci UB et al (2011) High-extent dehydrogenation of hydrazine borane $N_2H_4BH_3$ by hydrolysis of BH_3 and decomposition of N_2H_4. Energy Environ Sci 4:3355–3358; (c) Li C, Dou Y, Liu J et al (2013) Synthesis of supported Ni@(RhNi-alloy) nanocomposites as an efficient catalyst towards hydrogen generation from $N_2H_4BH_3$. Chem Commun 49:9992–9994
87. Nilsson GN, Kerr WJ (2010) The development and use of novel iridium complexes as catalysts for *ortho*-directed hydrogen isotope exchange reactions. J Labelled Compd Radiopharm 53:662–667
88. Soloway AH, Tjarks W, Barnum BA et al (1998) The chemistry of neutron capture therapy. Chem Rev 98:1515–1562
89. Hosmane NS, Maguire JA, Zhu Y (2011) Boron and Gadolinium Neutron Capture Therapy for Cancer Treatment. World Scientific Pub. Co. Inc., New Jersey.
90. Hoel EL, Hawthorne MF (1974) Transition metal catalyzed exchange of deuterium gas with terminal boron-hydrogen bonds in carboranes, metallocarboranes, and other boron compounds. J Am Chem Soc 96:4676–4677
91. Hoel EL, Talebinsab-Savari M, Hawthorne MF (1977) Deuterium exchange at terminal boron-hydrogen bonds catalyzed by certain transition metal complexes. A qualitative study of selectivity and mechanism. J Am Chem Soc 99:4356–4367

92. Gaines DF, Heppert JA, Kunz JC (1985) Hydrogen isotope exchange between boranes and deuterated aromatic hydrocarbons: evidence for reversible hydroboration of benzene. Inorg Chem 24:621–624
93. Dopke JA, Gaines DF (1999) Deuteration of decaborane(14) via exchange with deuterated aromatic solvents. Inorg Chem 38:4896–4897
94. D'Ulivo A, Dedina J, Mester Z et al (2011) Mechanisms of chemical generation of volatile hydrides for trace element determination (IUPAC Technical Report). Pure Appl Chem 83:1283–1340
95. Nelson DJ, Egbert JD, Nolan SP (2013) Deuteration of boranes: catalysed *versus* non-catalysed processes. Dalton Trans 42:4105–4109
96. Atzrodt J, Derdau V, Fey T, Zimmermann J (2007) The renaissance of H/D exchange. Angew Chem Int Ed 46:7744–7765 and references therein.
97. Zhu Y, Widjaja E, Shirley LPS et al (2007) Ruthenium(0) nanoparticle-catalyzed isotope exchange between ^{10}B and ^{11}B nuclei in decaborane(14). J Am Chem Soc 129:6507–6512
98. Tolmachev V, Sjőberg S (2002) Polyhedral boron compounds as potential linkers for attachment of radiohalogens to targeting proteins and peptides. A review. Collect Czech Chem Commun 67:913–935
99. Winberg KJ, Barberà G, Eriksson L et al (2003) High yield [$^{125}I^-$]-labeling of iodinated carboranes by palladium catalyzed isotopic exchange. J Organomet Chem 680:188–192

Chapter 9
Noncovalent Interactions of Heteroboranes

Robert Sedlak, Jindřich Fanfrlík, Adam Pecina, Drahomír Hnyk, Pavel Hobza, and Martin Lepšík

Abstract This chapter deals with noncovalent interactions between heteroboranes and their various organic or biomolecular partners. At first, the physical essence of noncovalent interactions in general is discussed. Focusing then on boron clusters, their contacts are discussed based on the unusual electron distribution within the boron hydride cages and especially around the heteroatoms (i.e. non-boron atoms incorporated in the cluster framework) or the substituents replacing the terminal hydrogens. The bare (i.e. not linked to hydrogen) heteroatoms within the cage bear prevailingly a partial positive charge. This results in an opposite direction of the compound dipole moments (as proved experimentally), contrary to what would be expected from the electronegativity concept.

The anisotropic distribution of the electron density around the heteroatoms gives rise to the so-called σ-holes, regions of positive electrostatic potential (ESP). This can be viewed as a driving force for noncovalent interactions with, e.g. organic aromatics or Lewis bases. Examples of σ-hole bonding of heteroboranes incorporated in the cluster cages are chalcogen or pnictogen bonding. Heteroatoms as *exo-substituents* can also be centers of σ-hole bonding, in this case halogen bonding. An important characteristics of the σ-hole bonding is that it can be tuned, e.g. by other *exo*-substituents or by the point of attachment to the cage – whether the *exo-halogens* are bonded to boron or carbon atoms within a particular cluster.

R. Sedlak • J. Fanfrlík • A. Pecina • M. Lepšík (✉)
Institute of Organic Chemistry and Biochemistry (IOCB), Academy of Sciences of the Czech Republic, v.v.i. and Gilead Sciences and IOCB Research Center, Flemingovo nám. 2, Prague 6 16610, Czech Republic
e-mail: lepsik@uochb.cas.cz

D. Hnyk
Institute of Inorganic Chemistry of the Academy of Sciences of the Czech Republic, v.v.i., Husinec-Řež 250 68, Czech Republic

P. Hobza
Institute of Organic Chemistry and Biochemistry (IOCB), Academy of Sciences of the Czech Republic, v.v.i. and Gilead Sciences and IOCB Research Center, Flemingovo nám. 2, Prague 6 16610, Czech Republic

Department of Physical Chemistry, Regional Center of Advanced Technologies and Materials, Palacký University, Olomouc 77146, Czech Republic

D. Hnyk, M.L. McKee (eds.), *Boron*, Challenges and Advances in Computational Chemistry and Physics 20,
DOI 10.1007/978-3-319-22282-0_9

Apart from the σ-hole bonding, the hydride character of the terminal hydrogens of the heteroboranes is responsible for forming unique dihydrogen H ··· H bonds, which provides the essence of heteroborane noncovalent interactions with various organic, inorganic and molecules of biological interest. The influence of various molecular shapes on the strength of these nonclassical contacts is discussed. Likewise, the choices of optimal mathematical models and computational protocols to study the crucial energy terms contributing to these interactions are reviewed.

9.1 Introduction

The variability of boron cluster compounds with a broad range of unique properties has been asking for practical applications [1]. Three major areas of use on an industrial scale have been radionuclide waste extraction [2], boron neutron capture therapy [3] and potential use as rocket fuels, discontinued in the early 1960 [1]. In the new millennium, novel applications have emerged in materials science [4, 5] and medicinal chemistry [6–9 and references therein].

In materials sciences, the nonclassical noncovalent interactions underlying the crystal packing of heteroboranes has recently started to be understood. Thus, B–H ··· H–C and B–H ··· H–S dihydrogen bonds were assigned as the stabilizing factor for self-assembly processes [10, 11]. Similarly, B–S ··· π interactions were found to have a dominant role in stabilizing the crystal of 12-Ph-1-$SB_{11}H_{10}$ [12].

Traditional medicinal chemistry uses organic compounds to bind to a biomolecular target and modulate its function. After many successful decades of producing drugs the number of new drug entities on the market have been declining. New approaches and molecular entities are thus urgently sought for. It was noticed that parent dicarbaboranes (*closo*-$C_2B_{10}H_{12}$ having *ortho*- 1,2-; *meta*- 1,7- and *para*- 1,12- isomers [13]) could be viewed as three-dimensional substitutes for planar organic moieties [6, 7, 14]. Examples can be found for all three isomers; asborin, o-carborane analogue of aspirin [15], m-carborane sulfamides as inhibitors of human carbonic anhydrase [16] or *p*-carborane analogues of nonsecosteroidal vitamin D receptor ligands [17]. The replacement of cyclic aliphatic or two-dimensional aromatic moieties with three-dimensional aromatics [18] alters the properties of the compounds, such as transmission of electronic substituent effects [19] leading to distinct noncovalent interactions with the target [20] or even change of specificity [21]. The introduction of boron clusters into medicinal chemistry thus greatly expands the possibilities of the current drug design.

9.2 Physical Basis of Noncovalent Interactions

To understand the behavior of heteroboranes in the above mentioned applications, their noncovalent interactions including their physical basis need to be understood. Noncovalent interactions appear due to interactions of permanent, induced and time-dependent instantaneous electric multipoles between the partner molecules in a "reversible" manner, i.e. no covalent bonds are formed. They are thus relatively weak (interaction energies of up to a few tens of kcal·mol^{-1}) but can be numerous. Within the framework of the intermolecular perturbation theory (IPT), the interaction energy is defined as the sum of several components arising from specific physical phenomena: electrostatics (Coulomb), induction (polarization), dispersion and exchange (Pauli) repulsion. These components will be described in the following paragraphs.

9.2.1 *Electrostatics*

Electrostatic interaction represents the energy of the Coulombic interaction between the charge distributions of the isolated monomers. A convenient way of depicting the electron distribution within a molecule is by plotting the electrostatic potential on the surface of the molecule. The molecular surface is usually defined as 0.001 a.u. (e.Bohr^{-3}) isodensity surface [22]. The electrostatic potential in space can be explicitly evaluated using Eq. 9.1

$$V(r) = \sum_{A=1}^{N} \frac{Z_A}{(R_A - r)} - \int \frac{\rho(r')dr'}{(r' - r)} \tag{9.1}$$

where V(r) is the potential created at any point *r* by the nuclei and the electrons of the molecule, Z_A is charge of nucleus A located at R_A and ρ(r) is the molecule's electron density.

The electrostatic potential is a real physical property, an observable, and can be determined both experimentally (by diffraction techniques) and computationally (by quantum chemical computations) [23, 24]. Its sign at any point of space depends upon whether the effect of the nuclei (positive) or the electrons (negative) predominates. Nowadays, the mapping of the electrostatic potential on the isodensity surface is widely used as a powerful tool in rational drug design.

Electrostatic potentials around heteroboranes are peculiar in several aspects. First, due to the electropositivity of boron, the boron-bound hydrogens are partially negatively charged. This allows dihydrogen bonding (see below) with the slightly positively charged carbon-bound hydrogens in organic compounds, biomolecules as well as carbons in carboranes [10, 25, 26]. Second, electron-rich heteroatoms from main groups 15 and 16 (pnictogens and chalcogens, respectively) will bear a partial positive charge and display anisotropy in the spatial electron arrangement. The areas of positive electrostatic potential are called σ-holes (see below) [27]. In analogy to the halogen-containing molecules which are able to form halogen bonds (X-bonds) in biomolecular systems [28], neutral chalcogen- and pnictogen-containing heteroboranes are capable of forming chalcogen or pnictogen bonds (see below) [12, 29].

Finally, it should also be noticed that a large number of borane molecules are anions. In such cases, the electrostatic interactions with their surroundings and specially with counterions have a major effect on the properties of charged boranes. These ion…ion interactions underlie their surfactant behavior [30] and may be also in part responsible for their aggregation [31, 32]. Also in biomolecular systems, specific interactions with ions can play a role, as exemplified by the effect of sodium counterions on the rotational profile of cobalt bis(dicarbollide) in complex with HIV protease [33].

9.2.2 Dispersion

The origin of (London) dispersion energy can be interpreted as the energy among instantaneously induced electrostatic multipole moments of partner molecules. Dispersion energy represents dominant source of interaction mainly among neutral and non-polar molecules, as these molecules do not possess permanent multipole moments and therefore interact only through induced multipole moments. The dispersion energy decreases as R^{-6}, where R is the interparticle center of mass distance. Dispersion energy, as well as the induction, is always an attractive component of the total interaction energy and generally exclusive source of stabilization in $\pi \cdots \pi$, aliphatic C–H ⋯ π and aliphatic ⋯ aliphatic complexes [34]. Further, dispersion is an important stabilizing term in dihydrogen bonding [26] or σ-hole bonding [12, 29]. This non-specific interaction becomes stronger with increasing polarizability of the partner molecules. Therefore neutral heteroboranes with large three-dimensional aromatic cages are perfect candidates of strong dispersion attraction.

9.2.3 Induction (Polarization)

Induction energy can be understood as the interaction among permanent and induced electric multipole moments of interacting molecules. This term is usually not a major player in the stabilization of noncovalent complexes. In some cases, however, it can be quite significant as in the $B_{12}H_{12}^{2-}$ ⋯ benzene complex (61 % of the total stabilization) [26]. The reason is that the permanent monopole of the borane induces electron redistribution in the polarizable benzene. This effect was apparent in ^{1}H NMR chemical shifts as well as in the QM calculated electron densities [26].

9.2.4 *Charge-Transfer*

In some cases, charge is transferred between the partner molecules. The associated charge transfer (CT) energy represents, within the IPT, a short range ("overlap") part of the induction (polarization) energy. The origin of the CT energy can be easily interpreted through the following scheme: $D^0 \cdots A^0 \rightarrow D^{\delta+}\text{–}A^{\delta-}$, which shows the transfer of an electron from the neutral donor molecule (D^0) to the neutral acceptor molecule (A^0) in the donor-acceptor (DA) complex. The DA complexes are CT complexes between a molecule with a low ionization potential (donor) and a molecule with a high electron affinity (acceptor). The electron transfer causes that partner molecules carry opposite partial charges (D positive and A negative) and consequently the attractive electrostatic interaction appears. The magnitude of the transferred charge and the strength of the interaction depends on the ionization potential of D, electron affinity of A as well as on the intermolecular distance and mutual orientation of the interacting molecules. CT complexes are not rare and play important roles in material [35, 36] and biological sciences [37, 38]. The character and strength of the CT complexes can be studied by computational methods.

The concept of the charge transfer is easy to understand intuitively, however hard to define rigorously. This is an intrinsic property of any CT analyses or CT decomposition scheme which separate the wave function (or an electron density) of the complex into parts (i.e. interacting monomers/fragments). As the complex is an inseparable entity, from the quantum mechanical point of view, any of these methods relies on an arbitrary partitioning scheme.

The CT concept within noncovalent interactions was developed by Mulliken more than 60 years ago [39]. Since then, many different methods for the quantification of the CT and CT energy have been developed, e.g. Natural Bond Orbitals (NBO) [40], Morokuma scheme [41, 42] or Misquita decomposition [43]. The symmetry adapted perturbation theory (SAPT) of Jeziorski et al. can also be used for estimation of the charge-transfer energy [44]. As there is no exact solution for the charge separation in noncovalent complexes, the goal is to find physically meaningful partitioning scheme. One of the latest promising approaches has been developed by Řezáč and de la Lande [45].

9.3 A Plethora of Noncovalent Interactions

Characteristics of noncovalent interactions are predetermined by the electron structure of the interacting molecules. The classical concepts known from organic chemistry (such as covalent bonding, partial charges, aromaticity) take on a different meaning in the world of boron hydride chemistry. The low electronegativity of boron atoms results in the hydridic character of their bound *exo*-skeletal hydrogens [25]. Additionally, heteroboranes can form a wide variety of structures (*closo*-, *nido*-, etc.) with varying total charges. Their electronic structure is determined by the three-center two-electron bonding within each B–B–B triangle (including those where the boron atom(s) are replaced by heteroatoms). Any BH group can be formally replaced by an isoelectronic group with a different contribution to the total charge, such as CH^+, NH^{2+} or S^{2+}. Metal atoms in metallaboranes become acceptors of the delocalized surplus electrons and thus their formal positive charge is reduced close to neutrality [46].

9.3.1 *Classical Hydrogen Bonds*

The best known example of noncovalent interaction in chemistry and biology is a classical hydrogen bond (H-bond). It can be schematically depicted as X–H ··· Y, where X–H represents the hydrogen bond donor and Y the hydrogen bond acceptor (Y). X is more electronegative atom than H and Y is typically an electronegative atom such as halogen, O, N or S possessing excess of electrons.

The forces responsible for stabilization of H-bonds mainly include those of electrostatic origin. However, charge-transfer (CT) as well as dispersion forces also participate in the stabilization.

Classical H-bond is an example of strong, directional and specific noncovalent interaction, which plays an important role in structure determination of polymers and biomacromolecules. We mention it here as a "standard" interaction against which all other types of noncovalent interactions will be compared.

Heteroboranes have only limited possibilities to form classical H-bond. Boron clusters are inherently electron deficient and they are poor H-bond acceptors. The S, Se or P vertices have usually partial positive charge [29], even though the S, Se or P are electronegative elements.

It is also hard to find a boron cluster that would be a good H-bond donor. The classical H-bond can hypothetically be formed by *closo*-1-$NB_{11}H_{12}$ and μ-6,9-O-5,10-*arachno*-$C_2B_8H_{11}$ but has never been observed.

The strength of H-bonds varies significantly from a few kcal·mol^{-1} (e.g. N–H ··· O, O–H ··· O 5 – 7 kcal·mol^{-1}) in the gas phase[47] up to several tens of kcal·mol^{-1} (e.g. O–H ··· O^-, F–H ··· F^- 23–39 kcal·mol^{-1}) [34]. The important factors which strongly influence the strength of the X–H ··· Y hydrogen bond are:

1. The strength of the X–H···Y bond correlates positively with the increase of the proton donation ability of the atom X, i.e. electronegativity (e.g. proton donation ability of carbon atom depends on its hybridization state: C(sp^3) <C(sp^2) <C(sp)) [48].
2. The arrangement of the X–H ··· Y atoms should ideally be linear (180°), i.e. the more linear it is, the stronger the H-bond. The formation of a hydrogen bond is usually connected with several characteristic features. The length of the X–H bond usually increases on hydrogen bond formation leading to a red shift in the infrared X–H stretching frequency and an increase in the infrared absorption cross-section for the X–H stretching vibration. The greater the lengthening of the

X–H bond in X– H ··· Y complex, the stronger the H-bond [49]. There are, however, certain hydrogen bonds in which the X–H bond length decreases, the so-called blue-shifting or improper H-bond [50].

9.3.2 Weak Hydrogen Bonds

In cases where one or two of the interaction partners is weak, as for example, when the H-bond donor is C–H and/or the H-bond acceptor is the π electron density of aromatic molecules, we talk about weak or non-conventional H-bonds [34]. The C–H ··· O, N, Cl [51] or N,O,C–H ··· π [52] interactions are reported in literature. The C–H ··· O, N, F interactions are important, especially in the absence of classical H-bonds, despite the fact they are considerably weaker than conventional (classical) H-bonds. This, at first glance surprising fact, can be understood as a consequence of a high abundance of this interaction in (bio)macromolecules. The importance of the C–H ··· O interaction in crystal engineering [53, 54] or biological systems [55, 56] is indisputable.

The N,O,C–H ··· π interactions represent another subset of weak H-bonds so called π H-bonds. The hydrogen atoms point toward the electron rich region and the relatively short distance (smaller than sum of van der Waals radii of H and C(sp^2)) is typically observed. This type of H-bond, similarly to the previously discussed interactions (C–H ··· O, N, F), is also weaker than conventional H-bonds. Nevertheless, these also play crucial role in formation of molecular assemblies [57], stability of the peptides [58], proteins [59] and conformation of small organic molecules [60].

Comprehensive structural study done by Steiner and Koellner [61] revealed that about 1 out of 11 aromatic amino acids act as π acceptors for H bonding with the N, O, S–H donors. This relatively high frequency, although small compared to classical hydrogen bonds, supports the postulated role which these interactions play in stabilizing the secondary structure of proteins [62]. The frequently observed contraction of the N, O, C–H bond is opposite to what is observed for classical H-bonds. This contraction leads to a shift of the bond stretching movement to higher frequencies in the infra-red spectrum, so called blue-shifting H-bonds [50].

The stabilization energy of weak H-bonds is of a few kcal·mol^{-1} (e.g. C–H ··· π has 1–1.5 kcal·mol^{-1}) [34] and has a reduced electrostatic and an increased dispersion contribution with respect to conventional H-bonds. This is the reason why weak H-bonds occur in different conformations.

Weak H-bonds of icosahedral carboranes have been well known from solid state structures [63]. The first weak H-bond of carborane was reported in 1977 for 1-$Me_2NC(S)$-1,2-$C_2B_{10}H_{11}$. Here, an intramolecular C–H ··· S H-bond with the distance of 2.53 Å was formed [64]. The first intermolecular carborane H-bonds were found for the dimethylsulfoxide adduct of decachloro-ortho-carborane in 1986 [63]. A C–H ··· π interaction was found with a carborane molecule located above an aromatic ring [65–67]. These interactions were studied computationally in model complexes of monocarbaborane and benzene. The major stabilizing component was shown to be dispersion, followed by electrostatics [26].

9.3.3 Dihydrogen Bonds

Dihydrogen bonds are noncovalent interactions known especially from organometallic chemistry where proton donors Y–H (Y = F, O, N, C) of a (bio)organic molecule interact with a σ-bonding electron pair of M–H bonds of main-group or transition metal (M = B, Al, Ga, Li, Be, Xe, Ir, Mo, Mn, Os, Re, Ru, W) hydrides (Fig. 9.1) [69]. Here, we list several properties of dihydrogen bonds:

I. the H ··· H distance varies between 1.7 and 2.3 Å; it is significantly shorter than the sum of the van der Waals radii of two hydrogen atoms (2.4 Å) [70];
II. the Y–H_Y ··· H_B–B structure prefers strongly bent type of the conformation: the H_Y ··· H_B–B angles is usually in the range of 90–150°; the Y–H_Y ··· H_B angle amounts to 150–180° [69];
III. the origin of the stabilization comes from both electrostatic and dispersion;
IV. the stabilization energy increases proportionally with proton donor's acidity [69]; it ranges from 1 to 7 kcal·mol^{-1}.

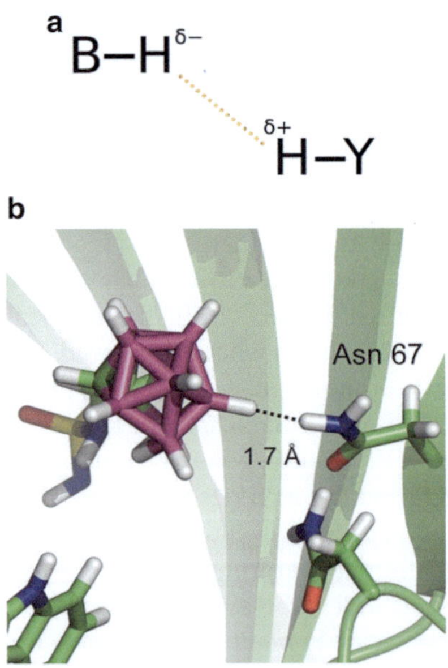

Fig. 9.1 Dihydrogen bonding of heteroboranes. (**a**) A Schematic picture; partial negative charge on boron-bound hydrogen, partial positive charge on hydrogen of the partner (Y = C, O, N, S). (**b**). Dihydrogen bond in the crystal structure of human carbonic anhydrase (hCAII) complexed with a carborane-based inhibitor 7-methylenesulfamide-(7,8-*nido*-dicarbaundecaborate). Color coding: Inhibitor boron atoms are in magenta, hCAII in green. Hydrogen atoms are white. Figure was prepared by PyMol, version 1.5 [68] (Adapted from Ref. [20])

Many studies have shown the importance of H ··· H interaction in crystal packing processes [71] or supramolecular chemistry [72]. Furthermore, boron hydrides combine the ability in solid state to self-assemble into extended dihydrogen-bonded networks together with specific reactivity in solutions [73–75]. This makes them potentially powerful tools for rational assembling of new crystalline covalent materials [76]. Moreover, nowadays the dihydrogen bonding is studied in connection with biomolecular chemistry and medicinal chemistry.

Heteroboranes were found to form B–H ··· H–X dihydrogen bonds with water [77], biomolecules [20, 25, 46, 78] or in materials [11]. Crystal structures of two novel carborane-sulfamide inhibitors in the complex with human carbonic anhydrase II (hCAII) have been published[16] and studied using QM/MM computations [20]. The inhibitors differed in the carborane part, one contained *closo*, while the other a *nido* cage. Although the studied inhibitors bind mainly via the sulfamide moiety to the zinc ion, the nature of binding of the carborane part of the inhibitors differed significantly. The *closo*-carborane cage, which is neutral, was bound mainly via dispersion interactions and formed only very weak dihydrogen bonds (the H ··· H distance greater than 2.2 Å) only with nonpolar C–H groups. The *nido* cage, which is negatively charged, interacted with the protein mainly via electrostatic interactions. It formed short and strong dihydrogen bonds (the H ··· H distances as short as 1.7 Å) with the polar hydrogen of NH_2 groups.

9.3.4 σ-Hole Bonding

The most common representative of this class in organic chemistry are halogen bonds of C–X ··· Y type, where X is a halogen atom (Cl, Br, I) and Y is an electron donor (O, N, S or Ph). In this case, X atoms are more electronegative than C atom. The X atom is thus negatively charged and the interaction with electron donor is counterintuitive. It is enabled by the region of positive (less negative) electrostatic potential (ESP), which is called a σ-hole [27]. The σ-hole is located along the extension of the covalent C–X in the outer region of the X atom (Fig. 9.2a). The short contacts between N/O and X atoms in inorganic crystals have been known from 1960s [79]. However, the interpretation of binding via the quantum chemical concept of σ-hole was introduced quite recently, in 2007 [27]. The σ-hole is characterized by its size and magnitude [80]. The magnitude is defined as the most positive value of the ESP, on the isodensity surface (usually 0.001 e·Bohr^{-3}). The size of the σ-hole is the spatial extent of the σ-hole.

The size and magnitude of the σ-hole and consequently the strength of X-bonds, can be increased by introducing heavier halogen and/or electron withdrawing group(s) in the vicinity of the X atom. This has already been shown not only for small model systems [81] but also for biomolecular complexes. Specifically, the modulation of the X-bond has been used to tune the activity of aldose reductase [82] and cathepsin inhibitors [83]. Binding via σ-holes is applicable not only to X atoms, but also for main group 15 and 16 elements. The respective σ-hole interactions are

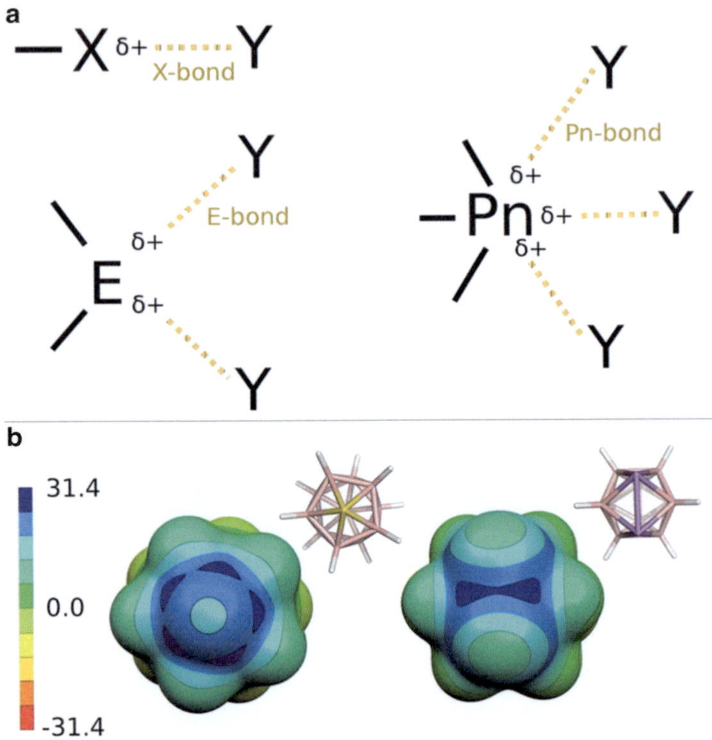

Fig. 9.2 (**a**). Schematic figure of halogen (*X*), chalcogen (*E*), and pnictogen (*Pn*) bonds. Y stands for an electron donor. (**b**) Electrostatic potentials (ESPs) on a 0.001 a.u. isodensity surface, computed at the HF/cc-pVDZ level. The molecular surfaces of $SB_{11}H_{11}$ (*left*) and $P_2B_{10}H_{10}$ (*right*) are shown. ESP values in kcal·mol^{-1}

called chalcogen and pnictogen bonds (Fig. 9.2a). In heteroborane chemistry, heavier halogens can be bound either to carbon or boron atoms. Recently, it has been demonstrated that the σ-hole on the Br atom is positive only if Br is bound to the C vertex [29]. In case the Br atom is bound to B vertex, however, the σ-hole of the Br atom is less negative than the belt surrounding it. Nevertheless, its magnitude is negative (Fig. 9.2b).

Novel types of σ-holes were found in heteroboranes incorporating main group 15 chalcogen (E) and 16 pnictogen (Pn) elements. Recently, the E-bond has been found in the X-ray structure of phenyl-substituted thiaborane [12]. The quantum chemical analysis revealed that the E-bond in this structure was considerably stronger than in its organic counterparts as well as in known X-bonds. It was due to the highly positive σ-holes on the overall positively charged E atom. A theoretical study of heteroboranes showed that E and Pn atoms incorporated into the borane clusters systematically carry partial positive charge and consequently highly positive σ-holes [29]. In contrast to X atoms, E and Pn atoms posses several σ-holes on their sides (Fig. 9.2b). This has an important consequence for the preferential geometrical arrangement, as the E- and Pn-bonds are bent contrary to the X-bonds which are linear (Fig. 9.2a).

9.4 Methods of Study

Experimental The methods of study can be direct or indirect. The former group comprise structural methods, such as X-ray, NMR or gas-phase electron diffraction (GED). The indirect methods prove the existence of noncovalent interactions, by measuring a frequency shifts in IR spectroscopy of dihydrogen bonded complexes [84].

Theoretical A principal tool to study noncovalent interactions are various quantum mechanical (QM) calculations [85]. The interaction energy represents the guidance to determine the strength of the noncovalent interaction. Thus, the accurate evaluation of the interaction energy is of crucial importance. Precise quantification of the correlation part of the interaction energy represents the most challenging part of the calculation from the computational point of view. It is well known that the different types of noncovalent interactions require application of different computational approaches. For example, the quantitative description of X-bonds, weak H-bonds or $\pi \cdots \pi$ interactions is very demanding, as these interactions are mostly dominated by dispersion, which is a purely correlation effect, hence not covered at the Hartree-Fock (HF) level of theory. Therefore, the use of more elaborate *ab initio* methods (post-HF methods) is required. The Møller-Plesset method (MP2) is the simplest method able to describe long-range correlation effects from the right reason. However, it is well known that MP2 overestimates the stacking interaction [86]. Further, the dispersion is weak effect and at van der Waals distances is significantly compensated by Pauli repulsion, what can lead to error amplification. Moreover, for small and middle size basis set superposition error can be as big as dispersion. Therefore, use of more elaborate ab initio methods in combination with sufficiently large basis set is needed in order to obtain directly the correct description of this type of interaction. The situation is not so complicated for H-bonded complexes, as it is known that MP2 provides accurate description of hydrogen-bonding. In the next sections, different methods, which are used for the accurate description of noncovalent interactions, will be briefly described. Before that let us make small comments on the convergence of the energy with respect to the size of the basis set. It is well known that this convergence is relatively slow and errors resulting from the use of incomplete basis set are typically the most significant ones. Therefore, different extrapolation schemes toward the complete basis set (CBS) limit are frequently utilized [87, 88].

9.4.1 Coupled Cluster Theory

Coupled Cluster theory with Single, Double and perturbative Triple excitations (CCSD(T)) is the most commonly used method for reference computations due to its outstanding accuracy/performance ratio [89]. Coupled Cluster (CC) methods are size-extensive and their convergence toward the full configuration interaction limit is faster compared to the methods with the same asymptotic scaling with respect to system size. Moreover, CC methods can be systematically improved upon inclusion of higher excitation operators. The scaling of the CCSD method is N^6, where N reflects the system size. The inclusion of perturbative triple excitations into the computation increases the scaling to N^7. Nowadays, systems of size approximately 30 atoms can be routinely treated; however, systems with more than about 50 atoms are still significantly impractical. So far, the largest published systems for which regular CCSD(T) computations were performed have approximately 70 atoms: the coronene dimer [90] and the guanine-cytosine step from DNA [91]. The largest error in CCSD(T) computations is mostly attributed to the incompleteness of the basis sets; there are thus many approaches which deal with extrapolation of CCSD(T) from a finite basis set towards the complete basis set (CBS) limit [87, 88, 92]. The overall accuracy of the CCSD(T)/CBS values goes beyond the "chemical accuracy" (1 kcal·mol^{-1}) but usually hardly reaches the "subchemical accuracy" (0.1 kcal·mol^{-1}) [93].

9.4.2 Symmetry-Adapted Perturbation Theory (SAPT)

Contrary to CCSD(T) or DFT (see below) methods, which are variational and their interaction energy is defined as the difference between the energy of complex and subsystems, the interaction energy within SAPT [44] is defined as the sum of the physically meaningful terms (cf. Eq. 9.2)

$$E_{int} = E^1_{El} + E^1_{Ex} + E^2_{I} + E^2_{Ex-I} + E^2_{D} + E^2_{Ex-D} + \delta HF \tag{9.2}$$

where E^1_{El} is the first-order electrostatic (Coulomb) term, E^1_{Ex} is the first-order exchange-repulsion term, $E^2_{Ex\text{-}I}$ and $E^2_{Ex\text{-}D}$ are respectively the second-order exchange induction and the dispersion terms, E^2_I and E^2_D are respectively the second-order induction and the dispersion terms and δHF is the Hartree-Fock higher order correction term. The Coulomb, induction, dispersion and exchange interaction energy components, which arise from low orders of perturbation theory, provide insight into origin of the interaction and help interpreting the interaction based on the monomer properties.

In order to partially overcome one of the essential drawbacks of the many-body SAPT, which is the extremely high computational cost (SAPT scales as N^7 with the system size), the Density Functional Theory SAPT (DFT-SAPT) method has been developed [94]. The monomer correlation energy is treated within the DFT-SAPT method through a less expensive DFT method, contrary to regular many-body SAPT. The scaling of the DFT-SAPT method is N^6 with respect to system size [95]. However, this enhanced scaling properties of the DFT-SAPT compared to the many-body SAPT needs to be improved further, in order to develop a method which would be computationally feasible for an investigation of extended complexes (dozens of atoms). This can be achieved through density fitting (also called resolution of the identity) procedure. The resulting method is abbreviated as DF-DFT-SAPT and it scales as N^5 [95].

The DFT-SAPT method works very well, as long as a balanced exchange-correlation functional with an asymptotic correction is employed, e.g. PBE0AC [96, 97] and LPBE0AC [98]. Several studies have confirmed that the DFT-SAPT method provides accurate estimates of interaction energies for various types of noncovalent interactions [99–102]. However, one should bear in mind that these functionals in combination with the aug-cc-pVDZ basis set provide accurate first-order terms as well as induction and respective exchange-repulsion terms [103, 104], while dispersion is underestimated approximately by 10–20 % [105].

Further progress in speeding up the DFT-SAPT computation was introduced by Hesselmann, who employed the empirical form for the dispersion terms of the interaction energy in a similar fashion as Grimme did for DFT [106]. The empirical correction was designed to mimic the effects of both the E^2_D as well as the $E^2_{Ex\text{-}D}$ terms (cf. Eq. 9.2). The parameters of empirical correction were fitted toward the CCSD(T)/CBS interaction energies of complexes from the S22 data set [92]. By introducing the empirical correction, the method is feasible for complexes of over hundred atoms, as the evaluation of 2nd order dispersion terms, which have scaling of $\mathbf{N}^5$, is circumvented.

9.4.3 Density Functional Theory Augmented with Empirical Dispersion (DFT-D)

It is well known that standard density functionals (LDA, GGA, meta-GGA, hybrids) are not able to account for long-range, van der Waals (dispersion) interactions, because of their local (semi-local) dependence on the electron density. There are several approaches able to tackle this drawback of DFT methods, e.g.: development of truly non local functionals [107], reparametrization of current functionals [108], double hybrid functionals [109] or long-range corrected functionals [110]. However, the most widely used approach for better incorporation of the dispersion into DFT, is its simple addition to the plain DFT result [111, 112]. The dispersion energy term can be, within this approach, expressed by the following empirical atom-atom pair-wise functional form

$$E_{disp}[\rho] = -\sum_{i\leq j}\sum_{n=6,8,\ldots} C_n^{ij}(\rho)R_{ij}^{-n} f_{damp}^{n}(R_{ij},\rho) \tag{9.3}$$

where i and j are atom labels, $C_n^{ij}(\rho)$ are, in general, density-dependent dispersion coefficients of particular atomic pairs (ij), R_{ij} is the interatomic separation and $f_{damp}(R_{ij}, \rho)$ is, in general, a density-dependent damping function. Nowadays, a variety of methods, which utilize the above mentioned form for dispersion energy (cf. Eq. 9.3), are commonly used. Forms of the empirical dispersion correction utilized by most methods differ mostly in following three aspects: (i) the number of terms included in eqn. 6.3 (i.e. $n=6,(8,10)$), (ii) the type of damping function, and (iii) the degree of approximations used to derive the $C_n^{ij}(\rho)$ coefficients.

It was shown that several versions of the DFT-D method, e.g. DFT-D3 [111], provide binding energies with respect to reference values for small- medium- and large-size complexes at van der Waals distances, with an error roughly around 10 % [113, 114]. The regular DFT computation employing nonhybrid functionals scales as N^3 with respect to the size of the system, an order of magnitude faster than the regular HF method. Another advantage of DFT based methods, contrary to *ab initio* methods, is their faster convergence toward the CBS limit. These features and the above mentioned robustness of the DFT-D methods makes them often methods of choice, especially when the reasonably accurate interaction energies are required and post-MP2 methods are not feasible/applicable.

9.4.4 Quantum Theory of Atoms in Molecules (QTAIM)

The "Quantum Theory of Atoms In Molecules" [115] is a method based on the splitting of the electron density around atoms using determined and classified critical points, following a separation of subsystems by "zero-flux" surfaces and integration of electron density within each region. This topological analysis provides quantitative/qualitative information about molecular structure. For example complete description of chemical bonds such as bond orders and bond bending characteristics of 8,9,10,12-tetra-fluoro-o-carborane [116] and H ··· H short contacts of metallabis(dicarbollide) systems [10].

9.4.5 Partial Atomic Charges

The atomic charge is not a physical observable and, thus, there is not a unique way of assigning a charge to the atom in a given molecule. The intrinsic drawback of the atomic charge concept is the neglect of the anisotropy of the electron density around atoms in molecules, i.e. it automatically assumes spherically-symmetrical charge distribution around a given atom. Consequently, such phenomena as halogen bonding can not be explained by the atomic charge approach since covalently bonded halogens carry mostly negative charge. Despite all these facts the concept of atomic charges is commonly used among chemists mainly due to its ease of interpretation.

There are many approaches which can be used for determining the atomic charges: (i) methods based on the partitioning of the wave function [39], Natural Population Analysis (NPA) [40], (ii) methods based on the partitioning of the electron density – Bader [22], Hirshfeld [117], (iii) methods based on the reproduction of the electrostatic potential (RESP) [118]. The above-mentioned population analyses differ in the computational cost, physical relevance, and basis set dependence; however none of them is unique and robust enough to provide atomic charges consistent with experimental measurements for all spectrum of applications. For example the RESP method was shown to correctly describe the hydridic character of the boron-bound hydrogens [8, 25]. In contrast, the NPA approach provides positively charged hydrogens [8], incompatible with the experimental observations of dihydrogen bonding [84].

9.5 Conclusions and Outlook

In this book chapter, we have reviewed the current knowledge of the rapidly growing area of noncovalent interactions of heteroboranes. The expanding wealth of both, the heteroborane compounds and their nonclassical noncovalent interactions follows directly in innovative and original research lines within materials sciences and drug design.

Possible discoveries of other types of nonclassical bonding including heteroboranes will strongly correlate with the synthetic availability of new types of these unique clusters; let us have in mind that not all main-group elements have so far been incorporated in boron cages. Since all deltahedral heteroboranes are man-made and have never been found in Nature, one could expect that there are no enzymes able to modify or destroy such materials. This fact provides a tantalizing possibility of getting a new way of thinking about boron clusters in various applications, e.g. drug design. The recent investigation of amino acids with a carborane core [119] could serve as an example of such endeavour.

References

1. Plesek J (1992) Potential applications of the boron cluster compounds. Chem Rev 92:269–278
2. Gruner B, Rais J, Selucky P, Lucanıkova M (2011) Recent progress in extraction agents based on cobalt bis(dicarbollides) for partitioning of radionuclides from high level nuclear waste. In: Hosmane NS (ed) Boron science-new technologies and applications. CRC Press, Boca Raton/New York, pp 463–490
3. Hawthorne MF (1993) The role of chemistry in the development of boron neutron-capture therapy of cancer. Angew Chem Int Ed 32:950–984
4. Grimes RN (2002) Putting small metallacarboranes to work: a review. Collect Czech Chem C 67:728–750

5. Grimes RN (2000) Metallacarboranes in the new millennium. Coord Chem Rev 200:773–811
6. Issa F, Kassiou M, Rendina LM (2011) Boron in drug discovery: carboranes as unique pharmacophores in biologically active compounds. Chem Rev 111:5701–5722
7. Scholz M, Hey-Hawkins E (2011) Carbaboranes as pharmacophores: properties, synthesis, and application strategies. Chem Rev 111:7035–7062
8. Farras P, Juarez-Perez EJ, Lepsik M, Luque R, Nunez R, Teixidor F (2012) Metallacarboranes and their interactions: theoretical insights and their applicability. Chem Soc Rev 41:3445–3463
9. Sivaev IB, Bregadze VV (2009) Polyhedral boranes for medical applications: current status and perspectives. Eur J Inorg Chem 2009:1433–1450
10. Juárez-Pérez EJ, Núñez R, Viñas C, Sillanpää R, Teixidor F (2010) The role of C–H···H–B interactions in establishing rotamer configurations in metallabis(dicarbollide) systems. Eur J Inorg Chem 2010:2385–2392
11. Planas JG, Vinas C, Teixidor F, Comas-Vives A, Ujaque G, Lledos A, Light ME, Hursthouse MB (2005) Self-assembly of mercaptane-metallacarborane complexes by an unconventional cooperative effect: a C-H···S-H, ···H-B hydrogen/dihydrogen bond interaction. J Am Chem Soc 127:15976–15982
12. Fanfrlik J, Prada A, Padelkova Z, Pecina A, Machacek J, Lepsik M, Holub J, Ruzicka A, Hnyk D, Hobza P (2014) The dominant role of chalcogen bonding in the crystal packing of 2D/3D aromatics. Angew Chem Int Edit 53:10139–10142
13. Bregadze VI (1992) Dicarba-*closo*-dodecaboranes $C_2B_{10}H_{12}$ and their derivatives. Chem Rev 92:209–223
14. Lesnikowski ZJ (2007) Boron units as pharmacophores – new applications and opportunities of boron cluster chemistry. Collect Czech Chem C 72:1646–1658
15. Scholz M, Bensdorf K, Gust R, Hey-Hawkins E (2009) Asborin: the carbaborane analogue of aspirin. ChemMedChem 4:746–748
16. Brynda J, Mader P, Šícha V, Fábry M, Poncová K, Bakardiev M, Grüner B, Cígler P, Řezáčová P (2013) Carborane-based carbonic anhydrase inhibitors. Angew Chem Int Ed 52:13760–13763
17. Fujii S, Masuno H, Taoda Y *et al* (2011) Boron cluster-based development of potent nonsecosteroidal vitamin D receptor ligands: direct observation of hydrophobic interaction between protein surface and carborane. J Am Chem Soc 133:20933–20941
18. King RB (2001) Three-dimensional aromaticity in polyhedral boranes and related molecules. Chem Rev 101:1119–1152
19. Campanelli AR, Domenicano A, Hnyk D (2015) Transmission of electronic substituent effects across the 1,12-dicarba- *closo* -dodecaborane cage: a computational study based on structural variation, atomic charges, and ^{13}C NMR chemical shifts. J Phys Chem A 119:205–214
20. Pecina A, Lepšík M, Řezáč J, Brynda J, Mader P, Řezáčová P, Hobza P, Fanfrlík J (2013) QM/MM calculations reveal the different nature of the interaction of two carborane-based sulfamide inhibitors of human carbonic anhydrase II. J Phys Chem A 119:1388–1395
21. Scholz M, Steinhagen M, Heiker JT, Beck-Sickinger AG, Hey-Hawkins E (2011) Asborin inhibits aldo/keto reductase 1A1. ChemMedChem 6:89–93
22. Bader RFW, Carroll MT, Cheeseman JR, Chang C (1987) Properties of atoms in molecules: atomic volumes. J Am Chem Soc 109:7968–7979
23. Stewart RF (1979) On the mapping of electrostatic properties from bragg diffraction data. Chem Phys Lett 65:335–342
24. Politzer P, Truhlar DG (1981) Chemical applications of atomic and molecular electrostatic potentials: reactivity, structure, scattering, and energetics of organic, inorganic, and biological systems. Springer, New York
25. Fanfrlik J, Lepsik M, Horinek D, Havlas Z, Hobza P (2006) Interaction of carboranes with biomolecules: formation of dihydrogen bonds. Chem Phys Chem 7:1100–1105

26. Sedlák R, Fanfrlík J, Hnyk D, Hobza P, Lepšík M (2010) Interactions of boranes and carboranes with aromatic systems: CCSD(T) complete basis set calculations and DFT-SAPT analysis of energy components. J Phys Chem A 114:11304–11311
27. Clark T, Hennemann M, Murray JS, Politzer P (2007) Halogen bonding: the sigma-hole. J Mol Model 13:291–296
28. Auffinger P, Hays FA, Westhof E, Ho PS (2004) Halogen bonds in biological molecules. Proc Natl Acad Sci U S A 101:16789–16794
29. Pecina A, Lepsik M, Hnyk D, Hobza P, Fanfrlik J (2015) Chalcogen and pnicogen bonds in complexes of neutral icosahedral and bicapped square-antiprismatic heteroboranes. J Phys Chem A 119:1388–1395
30. Chevrot G, Schurhammer R, Wipff G (2006) Surfactant behavior of "ellipsoidal" dicarbollide anions: a molecular dynamics study. J Phys Chem B 110:9488–9498
31. Matejicek P, Cigler P, Prochazka K, Kral V (2006) Molecular assembly of metallacarboranes in water: light scattering and microscopy study. Langmuir 22:575–581
32. Gassin P-M, Girard L, Martin-Gassin G, Brusselle D, Jonchère A, Diat O, Viñas C, Teixidor F, Bauduin P (2015) Surface activity and molecular organization of metallacarboranes at the air–water interface revealed by nonlinear optics. Langmuir 31:2297–2303
33. Fanfrlík J, Brynda J, Řezáč J, Hobza P, Lepšík M (2008) Interpretation of protein/ligand crystal structure using QM/MM calculations: case of HIV-1 protease/metallacarborane complex. J Phys Chem B 112:15094–15102
34. Desiraju GR, Steiner T (2001) The weak hydrogen bond: In structural chemistry and biology. Oxford University Press/International Union of Crystallography, Oxford
35. Hoppe H, Sariciftci NS (2004) Organic solar cells: an overview. J Mater Res 19:1924–1945
36. Schmidt-Mende L, Fechtenkötter A, Müllen K, Moons E, Friend RH, MacKenzie JD (2001) Self-organized discotic liquid crystals for high-efficiency organic photovoltaics. Science 293:1119–1122
37. Krapf S, Koslowski T, Steinbrecher T (2010) The thermodynamics of charge transfer in DNA photolyase: using thermodynamic integration calculations to analyse the kinetics of electron transfer reactions. Phys Chem Chem Phys 12:9516–9525
38. Warshel A (2002) Molecular dynamics simulations of biological reactions. Acc Chem Res 35:385–395
39. Mulliken RS (1955) Electronic population analysis on LCAO–MO molecular wave functions I. J Chem Phys 23:1833–1840
40. Reed AE, Curtiss LA, Weinhold F (1988) Intermolecularinteractions from a natural bond orbital, donor-acceptor viewpoint. Chem Rev 88:899–926
41. Kitaura K, Morokuma K (1976) A new energy decomposition scheme for molecular interactions within the Hartree-Fockapproximation. Int J Quantum Chem 10:325–340
42. Mo Y, Gao J, Peyerimhoff SD (2000) Energy decomposition analysis of intermolecular interactions using a block-localized wave function approach. J Chem Phys 112:5530–5538
43. Stone AJ, Misquitta AJ (2009) Charge-transfer in symmetry-adapted perturbation theory. Chem Phys Lett 473:201–205
44. Jeziorski B, Moszynski R, Szalewicz K (1994) Perturbation theory approach to intermolecular potential energy surfaces of van der waals complexes. Chem Rev 94:1887–1930
45. Řezáč J, de la Lande A (2015) Robust, basis-set independent method for the evaluation of charge-transfer energy in noncovalent complexes. J Chem Theory Comput 11:528–537
46. Fanfrlik J, Hnyk D, Lepsik M, Hobza P (2007) Interaction of heteroboranes with biomolecules – part 2. The effect of various metal vertices and *exo*-substitutions. Phys Chem Chem Phys 9:2085–2093
47. Dixon DA, Dobbs KD, Valentini JJ (1994) Amide-water and amide-amide hydrogen bond strengths. J Phys Chem 98:13435–13439
48. Gonnade RG, Shashidhar MS, Bhadbhade MM (2012) "Halogen bonding" interactions in molecular crystals: from early recognition to recent developments. J Indian I Sci 87: 149–165

49. Arunan E, Desiraju GR, Klein RA *et al* (2011) Definition of the hydrogen bond (IUPAC recommendations 2011). Pure Appl Chem 83:1637–1641
50. Hobza P, Havlas Z (2000) Blue-shifting hydrogen bonds. Chem Rev 100:4253–4264
51. Taylor R, Kennard O (1982) Crystallographic evidence for the existence of CH···O, CH···N and CH···Cl hydrogen bonds. J Am Chem Soc 104:5063–5070
52. Meyer EA, Castellano RK, Diederich F (2003) Interactions with aromatic rings in chemical and biological recognition. Angew Chem Int Ed 42:1210–1250
53. Jeffrey GA (2003) Hydrogen-bonding: an update. Crystallogr Rev 9:135–176
54. Steiner T (2003) C–H···O hydrogen bonding in crystals. Crystallogr Rev 9:177–228
55. Bella J, Berman HM (1996) Crystallographic evidence for Cα–H···O=C hydrogen bonds in a collagen triple helix. J Mol Biol 264:734–742
56. Berger I, Egli M, Rich A (1996) Inter-strand CH...O hydrogen bonds stabilizing four-stranded intercalated molecules: stereoelectronic effects of O4' in cytosine-rich DNA. Proc Natl Acad Sci 93:12116–12121
57. Nishio M (2004) CH/π hydrogen bonds in crystals. CrystEngComm 6:130–158
58. Tatko CD, Waters ML (2004) Comparison of C-H...pi and hydrophobic interactions in a beta-hairpin peptide: impact on stability and specificity. J Am Chem Soc 126:2028–2034
59. Brandl M, Weiss MS, Jabs A, Sühnel J, Hilgenfeld R (2001) C-H...pi-interactions in proteins. J Mol Biol 307:357–377
60. Yamakawa M, Yamada I, Noyori R (2001) CH/π attraction: the origin of enantioselectivity in transfer hydrogenation of aromatic carbonyl compounds catalyzed by chiral η6-arene-ruthenium(II) complexes. Angew Chem Int Ed 40:2818–2821
61. Steiner T, Koellner G (2001) Hydrogen bonds with pi-acceptors in proteins: frequencies and role in stabilizing local 3D structures. J Mol Biol 305:535–557
62. Tóth G, Watts CR, Murphy RF, Lovas S (2001) Significance of aromatic-backbone amide interactions in protein structure. Proteins 43:373–381
63. Fox MA, Hughes AK (2004) Cage C-H···X interactions in solid-state structures of icosahedral carboranes. Coord Chem Rev 248:457–476
64. Beall H (1977) The structure of 1-N, N-dimethylthiocarbamoyl-1,2-dicarba-closo-dodecaborane (12) a propos the bonding of carbonyl groups to the carborane cage. Inorg Nucl Chem Lett 13:111–114
65. Blanch RJ, Williams M, Fallon GD, Gardiner MG, Kaddour R, Raston CL (1997) Supramolecular complexation of 1,2-dicarbadodecaborane(12). Angew Chem Int Ed Engl 36:504–506
66. Raston CL, Cave GW (2004) Nanocage encapsulation of two ortho-carborane molecules. Chem-Eur J 10:279–282
67. Hamilton EJ, Kultyshev RG, Du B, Meyers EA, Liu SM, Hadad CM, Shore SG (2006) A stacking interaction between a bridging hydrogen atom and aromatic pi density in the n-B18H22-benzene system. Chem-Eur J 12:2571–2578
68. DeLano WL (2002) The PyMOL molecular graphics system. Oxford University Press/International Union of Crystallography, Oxford
69. Custelcean R, Jackson JE (2001) Dihydrogen bonding: structures, energetics, and dynamics. Chem Rev 101:1963–1980
70. Richardson TB, deGala S, Crabtree RH, Siegbahn PEM (1995) Unconventional hydrogen bonds: intermolecular B-H···H-N interactions. J Am Chem Soc 117:12875–12876
71. Campbell JP, Hwang J-W, Young VG, Von Dreele RB, Cramer CJ, Gladfelter WL (1998) Crystal engineering using the unconventional hydrogen bond. Synthesis, structure, and theoretical investigation of cyclotrigallazane. J Am Chem Soc 120:521–531
72. Gusev DG, Lough AJ, Morris RH (1998) New polyhydride anions and proton-hydride hydrogen bonding in their ion pairs. X-ray crystal structure determinations of Q[mer-Os(H)3(CO)(PiPr3)2], Q=[K(18-crown-6)] and Q=[K(1-aza-18-crown-6)]. J Am Chem Soc 120:13138–13147

73. Lee J, Peris E, Rheingold A, Crabtree R (1994) An unusual type of H...H interaction – Ir-H... h-O and Ir-H...h-N hydrogen-bonding and its involvement in sigma-bond metathesis. J Am Chem Soc 116:11014–11019
74. Ayllón JA, Gervaux C, Sabo-Etienne S, Chaudret B (1997) First NMR observation of the intermolecular dynamic proton transfer equilibrium between a hydride and coordinated dihydrogen: (dppm)2HRuH $\cdots$ H – OR = [(dppm)2HRu(H2)] + (OR)-. Organometallics 16:2000–2002
75. Epstein LM, Shubina ES (1998) Proton transfer and hydrogen bonding with transition metal atoms and hydride hydrogen by IR and NMR studies. Berich Bunsen Gesell 102:359–363
76. Custelcean R, Jackson JE (1998) Topochemical control of covalent bond formation by dihydrogen bonding. J Am Chem Soc 120:12935–12941
77. Karki K, Gabel D, Roccatano D (2012) Structure and dynamics of dodecaborate clusters in water. Inorg Chem 51:4894–4896
78. Mader P, Pecina A, Cigler P, Lepsik M, Sicha V, Hobza P, Gruener B, Fanfrlik J, Brynda J, Rezacova P (2014) Carborane-based carbonic anhydrase inhibitors: insight into CAII/CAIX specificity from a high-resolution crystal structure, modeling, and quantum chemical calculations. Biomed Res Int 2014:389869
79. Bent HA (1968) Structural chemistry of donor-acceptor interactions. Chem Rev 68:587–648
80. Kolář M, Hostaš J, Hobza P (2014) The strength and directionality of a halogen bond are codetermined by the magnitude and size of the σ-hole. Phys Chem Chem Phys 16:9987–9996
81. Riley KE, Murray JS, Fanfrlík J, Řezáč J, Solá RJ, Concha MC, Ramos FM, Politzer P (2011) Halogen bond tunability I: the effects of aromatic fluorine substitution on the strengths of halogen-bonding interactions involving chlorine, bromine, and iodine. J Mol Model 17: 3309–3318
82. Fanfrlík J, Kolář M, Kamlar M *et al* (2013) Modulation of aldose reductase inhibition by halogen bond tuning. ACS Chem Biol 8:2484–2492
83. Hardegger LA, Kuhn B, Spinnler B *et al* (2011) Systematic investigation of halogen bonding in protein–ligand interactions. Angew Chem Int Ed 50:314–318
84. Epstein LM, Shubina ES, Bakhmutova EV, Saitkulova LN, Bakhmutov VI, Chistyakov AL, Stankevich IV (1998) Unusual hydrogen bonds with a hydride atom in boron hydrides acting as proton acceptor. Spectroscopic and theoretical studies. Inorg Chem 37:3013–3017
85. Riley KE, Pitoňák M, Jurečka P, Hobza P (2010) Stabilization and structure calculations for noncovalent interactions in extended molecular systems based on wave function and density functional theories. Chem Rev 110:5023–5063
86. Sinnokrot MO, Valeev EF, Sherrill CD (2002) Estimates of the Ab initio limit for $\pi - \pi$ interactions: the benzene dimer. J Am Chem Soc 124:10887–10893
87. Halkier A, Helgaker T, Jørgensen P, Klopper W, Koch H, Olsen J, Wilson AK (1998) Basis-set convergence in correlated calculations on Ne, N2, and H2O. Chem Phys Lett 286:243–252
88. Lee EC, Kim D, Jurecka P, Tarakeshwar P, Hobza P, Kim KS (2007) Understanding of assembly phenomena by aromatic-aromatic interactions: benzene dimer and the substituted systems. J Phys Chem A 111:3446–3457
89. Bartlett RJ, Musiał M (2007) Coupled-cluster theory in quantum chemistry. Rev Mod Phys 79:291–352
90. Janowski T, Ford AR, Pulay P (2010) Accurate correlated calculation of the intermolecular potential surface in the coronene dimer. Mol Phys 108:249–257
91. Pitonak M, Neogrady P, Hobza P (2010) Three- and four-body nonadditivities in nucleic acid tetramers: a CCSD(T) study. Phys Chem Chem Phys 12:1369–1378
92. Jurečka P, Šponer J, Černý J, Hobza P (2006) Benchmark database of accurate (MP2 and CCSD(T) complete basis Set limit) interaction energies of small model complexes, DNA base pairs, and amino acid pairs. Phys Chem Chem Phys 8:1985–1993
93. Rode M, Sadlej J, Moszynski R, Wormer PES, van der Avoird A (1999) The importance of high-order correlation effects for the CO–CO interaction potential. Chem Phys Lett 314:326–332

94. Hesselmann A, Jansen G (2003) The helium dimer potential from a combined density functional theory and symmetry-adapted perturbation theory approach using an exact exchange-correlation potential. Phys Chem Chem Phys 5:5010–5014
95. Hesselmann A, Jansen G, Schütz M (2005) Density functional theory symmetry adapted intermolecular perturbation theory with density fitting: a new efficient method to study intermolecular interaction energies. J Chem Phys 122:014103
96. Adamo C, Barone V (1999) Toward reliable density functional methods without adjustable parameters: the PBE0 model. J Chem Phys 110:6158–6170
97. Grüning M, Gritsenko OV, van Gisbergen SJA, Baerends EJ (2001) Shape corrections to exchange-correlation potentials by gradient-regulated seamless connection of model potentials for inner and outer region. J Chem Phys 114:652–660
98. Sala FD, Görling A (2001) Efficient localized Hartree-Fock methods as effective exact-exchange Kohn-Sham methods for molecules. J Chem Phys 115:5718–5732
99. Biedermannova L, Riley KE, Berka K, Hobza P, Vondrasek J (2008) Another role of proline: stabilization interactions in proteins and protein complexes concerning proline and tryptophane. Phys Chem Chem Phys 10:6350–6359
100. Kolar M, Berka K, Jurecka P, Hobza P (2010) On the reliability of the AMBER force field and its empirical dispersion contribution for the description of noncovalent complexes. Chem Phys Chem 11:2399–2408
101. Riley KE, Pitoňák M, Černý J, Hobza P (2010) On the structure and geometry of biomolecular binding motifs (hydrogen-bonding, stacking, $X-H\cdots\pi$): WFT and DFT calculations. J Chem Theory Comput 6:66–80
102. Adhikari U, Scheiner S (2012) Sensitivity of pnicogen, chalcogen, halogen and H-bonds to angular distortions. Chem Phys Lett 532:31–35
103. Hesselmann A, Jansen G (2002) First-order intermolecular interaction energies from Kohn-Sham orbitals. Chem Phys Lett 357:464–470
104. Hesselmann A, Jansen G (2002) Intermolecular induction and exchange-induction energies from coupled-perturbed Kohn-Sham density functional theory. Chem Phys Lett 362: 319–325
105. Hesselmann A, Jansen G (2003) Intermolecular dispersion energies from time-dependent density functional theory. Chem Phys Lett 367:778–784
106. Hesselmann A (2011) Comparison of intermolecular interaction energies from SAPT and DFT including empirical dispersion contributions. J Phys Chem A 115:11321–11330
107. Vydrov OA, Van Voorhis T (2009) Nonlocal van der waals density functional made simple. Phys Rev Lett 103:063004
108. Zhao Y, Schultz NE, Truhlar DG (2006) Design of density functionals by combining the method of constraint satisfaction with parametrization for thermochemistry, thermochemical kinetics, and noncovalent interactions. J Chem Theory Comput 2:364–382
109. Schwabe T, Grimme S (2007) Double-hybrid density functionals with long-range dispersion corrections: higher accuracy and extended applicability. Phys Chem Chem Phys 9:3397–3406
110. Yanai T, Tew DP, Handy NC (2004) A new hybrid exchange–correlation functional using the coulomb-attenuating method (CAM-B3LYP). Chem Phys Lett 393:51–57
111. Grimme S, Antony J, Ehrlich S, Krieg H (2010) A consistent and accurate ab initio parametrization of density functional dispersion correction (DFT-D) for the 94 elements H-Pu. J Chem Phys 132:154104
112. Jurečka P, Černý J, Hobza P, Salahub D (2007) Density functional theory augmented with an empirical dispersion term. Interaction energies and geometries of 80 noncovalent complexes compared with Ab initio quantum mechanics calculations. J Comput Chem 28:555–569
113. Goerigk L, Kruse H, Grimme S (2011) Benchmarking density functional methods against the S66 and S66x8 datasets for Non-covalent interactions. Chem Phys Chem 12:3421–3433
114. Sedlak R, Janowski T, Pitoňák M, Řezáč J, Pulay P, Hobza P (2013) Accuracy of quantum chemical methods for large noncovalent complexes. J Chem Theory Comput 9:3364–3374

115. Bader RFW (1990) Atoms in molecules: a quantum theory. Oxford University Press, Oxford
116. Lyssenko KA, Antipin MY, Lebedev VN (1998) Topological analysis of the electron density distribution in the crystal of 8,9,10,12-tetrafluoro-o-carborane on the basis of the high-resolution X-ray diffraction data at 120 K. Inorg Chem 37:5834–5843
117. Hirshfeld FL (1977) Bonded-atom fragments for describing molecular charge densities. Theor Chim Acta 44:129–138
118. Bayly CI, Cieplak P, Cornell WD, Kollman PA (1993) A well-behaved electrostatic potential based method using charge restrains for deriving atomic charges – the RESP model. J Phys Chem 97:10269–10280
119. Dávalos JZ, González J, Ramos R, Hnyk D, Holub J, Santaballa JA, Canle-L M, Oliva JM (2014) Acidities of *closo*-1-COOH-1,7-$C_2B_{10}H_{11}$ and amino acids based on icosahedral carboranes. J Phys Chem A 118:2788–2793

MIX
Papier aus verantwortungsvollen Quellen
Paper from responsible sources
FSC® C105338

If you have any concerns about our products,
you can contact us on
ProductSafety@springernature.com

In case Publisher is established outside the EU,
the EU authorized representative is:
Springer Nature Customer Service Center GmbH
Europaplatz 3, 69115 Heidelberg, Germany

Printed by Libri Plureos GmbH
in Hamburg, Germany